T0210666

Foundations in Signal Processing, Communications and Networking

Volume 10

Series editors

Wolfgang Utschick, Ingolstadt, Germany
Holger Boche, München, Germany
Rudolf Mathar, Aachen, Germany

For further volumes:
http://www.springer.com/series/7603

Rudolf Ahlswede

Storing and Transmitting Data

Rudolf Ahlswede's
Lectures on Information Theory 1

Edited by

Alexander Ahlswede
Ingo Althöfer
Christian Deppe
Ulrich Tamm

 Springer

Author
Rudolf Ahlswede (1938–2010)
Universität Bielefeld
  Fakultät für Mathematik
Bielefeld
Germany

Editors
Alexander Ahlswede
Bielefeld
Germany

Ingo Althöfer
Friedrich-Schiller University
Jena
Germany

Christian Deppe
University of Bielefeld
Bielefeld
Germany

Ulrich Tamm
Bielefeld University of Applied Sciences
Bielefeld
Germany

ISSN 1863-8538   ISSN 1863-8546   (electronic)
ISBN 978-3-319-35238-1   ISBN 978-3-319-05479-7   (eBook)
DOI 10.1007/978-3-319-05479-7
Springer Cham Heidelberg New York Dordrecht London

Mathematics Subject Classification (2010): 94xx, 94Axx

Springer is part of Springer Science+Business Media (www.springer.com)

Preface

Classical information processing consists of the main tasks **gaining** knowledge, **storage**, **transmission**, and **hiding** data.[1]

The first named task is the prime goal of Statistics and for the two next Shannon presented an impressive mathematical theory, called Information Theory, which he based on probabilistic models.

Based on this theory are concepts of codes—lossless and lossy—with small error probabilities in spite of noise in the transmission, which is modeled by channels.

Another way to deal with noise is based on a combinatorial concept of error correcting codes, pioneered by Hamming. This leads to another way to look at Information Theory, which instead of being looked at by its tasks can be also classified by its mathematical structures and methods: primarily **probabilistic** versus **combinatorial**.

Finally, Shannon also laid foundations of a theory concerning hiding data, called Cryptology. Its task is in a sense dual to transmission and we therefore prefer to view it as a subfield of Information Theory.

Viewed by mathematical structures there is again already in Shannon's work a **probabilistic** and a combinatorial or **complexity**-theoretical model.

The lectures are suitable for graduate students in Mathematics, and also in Theoretical Computer Science, Physics, and Electrical Engineering after some preparations in basic Mathematics.

The lectures can be selected for courses or supplements of courses in many ways.

Rudolf Ahlswede

---

[1] This is the original preface written by Rudolf Ahlswede for the first 1,000 pages of his lectures. This volume consists of the first third of these pages.

Contents

1 Introduction. 1

Part I   Storing Data

2 Data Compression 9

3 The Entropy as a Measure of Uncertainty. 39

4 Universal Coding 79

Part II   Transmitting Data

5 Coding Theorems and Converses for the DMC 113

6 Towards Capacity Functions 169

7 Error Bounds. 215

Part III   Appendix

8 Inequalities 265

Rudolf Ahlswede 1938–2010. 291

Comments by Holger Boche. 295

Index 297

Author Index 301

Words and Introduction of the Editors

Rudolf Ahlswede was one of the world-wide accepted experts on Information Theory. Many main developments in this area are due to him. Especially, he made big progress in Multi-User Theory. Furthermore, with Identification Theory and Network Coding he introduced new research directions. Rudolf Ahlswede died in December 2010. The biography "Rudolf Ahlswede 1938–2010" is provided at the end of this book, describing the main parts of his scientific life. It was previously published in the book "Information Theory, Combinatorics, and Search Theory; In Memory of Rudolf Ahlswede" which provides a comprehensive picture of the vision Rudolf Ahlswede had and includes obituaries and several stories and anecdotes to Rudolf Ahlswede's life.

In 1975, Rudolf Ahlswede was appointed professor at the Department of Mathematics at the University of Bielefeld where he regularly gave lectures on Information Theory until he became emeritus professor in 2003. In these almost 30 years, he collected his lecture notes—first as note cards, slides for overhead projectors and even a typed script in German language (with Gunter Dueck). In the 1990s, this material was transferred to TEX files and further lecture notes were steadily added. Over the years several plans (and even concrete offers by the leading publishing companies) for book publications out of this material were considered. However, he decided to delay these plans in order to further work on a "General Theory of Information Transfer" according to which he finally wanted to shape the book.

Starting from 2008 he worked very intensively on this plan. He had already structured and ordered about 4,000 pages of lecture notes and provided a lot of corrections and remarks before his sudden and unexpected death. We (Ingo Althöfer, Ulrich Tamm, Christian Deppe) were the last three research assistants at his chair in Bielefeld. Together with his son (Alexander "Sascha" Ahlswede) we think that we should publish these lecture notes. To this aim we divide them into 10 volumes of approximately 400 pages per book. The titles of the volumes are "1. Storing and Transmitting Data," "2. Transmitting and Gaining Data," "3. Hiding Data: Selected Topics," "4. Combinatorial Methods and Models," "5. Probabilistic Methods," "6. New Directions for Probabilistic Models," "7. Memories and Communication," "8. Communication and Complexity," "9. Quantum Information I," and "10. Quantum Information II." Since for each book revisions will be necessary, it will take some time until all volumes will appear.

Changes of the manuscript are carried out according to the revisors' suggestions or in the case of obvious misprints, errors in the enumeration of formulae, etc. Otherwise we leave Rudolf Ahlswede's original text unchanged. We are aware that after these revisions the text will and cannot be in the shape that Rudolf Ahlswede would have given to it. However, (a) we know that the publication was his strong wish and discussions with the publisher had already reached a final stage and (b) over the years there was a strong interest and steady requests by many information theorists in these lecture notes.

This first volume "Storing and Transmitting Data" starts with an introduction into Shannon Theory. It basically includes the material of his regular lecture "Information Theory I" and may hence also serve as a textbook for undergraduate and graduate students. Characteristic for Rudolf Ahlswede's lectures was that he often provided more than one proof for a theorem (especially, the important ones). This helps to better understand the topic.

Rudolf Ahlswede was able to give normal talks and courses. But from time to time he taught without etiquette and in very clear words. The resulting sections of his courses and scripts were and are highlights. Chapter 6 in this volume is of such a calibre. Even a reader without understanding for the technical intricacies of Information Theory can benefit a lot from going through the chapter with its lively comments and uncensored insider views from the world of science and research.

For each volume, we asked a well-known expert to provide some comments. For this book, Holger Boche wrote the comments which can be found in the supplement. In 2009, Rudolf Ahlswede started an intensive cooperation with Holger Boche, who was a professor in Berlin at that time. This cooperation with Holger Boche and his group finally led to very good results in Quantum Information Theory, which already have been published. Holger also invited Rudolf Ahlswede to give lectures for graduate and Ph.D. students at the Technical University Berlin. For these lectures, he already used the material from his lecture notes.

Our thanks go to Regine Hollmann, Carsten Petersen, and Christian Wischmann for helping us typing, typesetting, and proofreading.

Alexander Ahlswede
Ingo Althöfer
Christian Deppe
Ulrich Tamm

Chapter 1
Introduction

In the last decades great progress has been made in the technologies of information and communication (e.g. smaller and faster microchips, storage media like CD's and usb sticks, glass fibre cables, TV satellites, etc.).

The mathematical discipline dealing with *communication* and *information* is called Information Theory. Information Theory originated from Electrical Engineering, as the above examples suggest. Research was initiated by Claude E. Shannon in his pioneering paper "A Mathematical Theory of Communication" in 1948.

One object of Information Theory is the *transmission of data*. So we have the typical situation that a *sender* and a *receiver* are communicating via a communication system (mail, telephone, Internet, etc.). The semantical aspect of the messages sent is excluded. A further important aspect, is the security of the data transmitted. This is the object of Cryptology.

A *communication system* can be described by the following diagram:

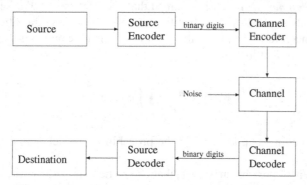

The source output—think for example of a Chinese text—has to be prepared for the transmission system, which of course does not know the Chinese letters. So during the source encoding procedure the text is translated into a sequence of binary digits.

The channel encoder now translates this binary sequence into a sequence of letters over the appropriate input alphabet $\mathcal{X}$. If the channel is for example a telephone line,

A. Ahlswede et al. (eds.), *Storing and Transmitting Data*,
Foundations in Signal Processing, Communications and Networking 10,
DOI: 10.1007/978-3-319-05479-7_1, © Springer International Publishing Switzerland 2014

then the messages are transmitted in form of electrical pulses and the alphabet size could depend on the possible pulse durations.

The channel usually is noisy, i.e. the letters can be disturbed by the channel.

The channel decoder receives a sequence of letters over an output alphabet $\mathcal{Y}$. Observe that input and output alphabet are not necessarily the same.

This sequence is translated into binary digits, which finally are retranslated into a Chinese text.

Observe that source encoder and channel encoder have contrasting tasks.

The source encoder wants to produce a short sequence of bits. So his concern is *data compression*.

The channel encoder wants to assure that the message transmitted can be correctly decoded at the receiving end in spite of the disturbances caused by the channel. So his concern is to build in *redundancy* into the sequences sent.

In the diagram presented above only one sender and one receiver are communicating. There are also several *multi-user* channel models, for example the broadcast channel (one sender, many receivers) and the multiple access channel (many senders, one receiver).

There are two basic approaches to describe a *channel*.

1.1 The Probabilistic Model

Let the input $x^n = (x_1, x_2, \ldots, x_n) \in \mathcal{X}^n$ be a sequence of n letters from an input alphabet $\mathcal{X}$. Analogously, the output $y^n = (y_1, y_2, \ldots, y_n) \in \mathcal{Y}^n$, where $\mathcal{Y}$ denotes the output alphabet, is defined.

The most commonly used channel in this model of communication is the *Discrete Memoryless Channel* (DMC). It is assumed that if the letter $x \in \mathcal{X}$ is sent, then the channel will output $y \in \mathcal{Y}$ with probability $w(y|x)$, where these transition probabilities are known a priori. The probability $W^n(y^n|x^n)$ that the output is y^n on input x^n is given by the product formula

$$W^n(y^n|x^n) = \prod_{t=1}^{n} w(y_t|x_t)$$

Channel encoding and decoding in the probabilistic model can be described as follows.

We are given M possible messages, that can be transmitted over the channel.

To each message $i \in \{1, \ldots, M\}$ we assign a *codeword* $u_i \in \mathcal{X}^n$ and a set $D_i \subset \mathcal{Y}^n$ of words, which are decoded as message i. The D_i's are disjoint ($D_i \cap D_j = \varnothing$ for $i \neq j$). The set of pairs $\{(u_i, D_i), i = 1, \ldots, M\}$ is denoted as (n, M)-Code.

An (n, M)-Code is called (n, M, λ)-*Code*, $\lambda \in (0, 1)$, if additionally $W^n(D_i| u_i) \geq 1 - \lambda$ for all $i = 1, \ldots, M$. Hence, the smaller λ the better is the code.

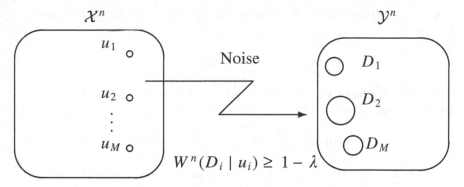

The question now is:

Which triples (n, M, λ) are possible?

Observe that there are two conflicting conditions on the quality of a code. We want to make M large in order to be able to transmit many messages, while keeping the probability of error λ small. So two standardizations are possible:

(1) We require, that a certain number $M(n)$ of messages can be transmitted over the channel. It is reasonable to set them exponential in $n \left(M(n) = e^{R \cdot n} \right)$, since in the noiseless case $(w(y|x) = 0$ for $y \neq x) \, |\mathcal{X}|^n$ messages are possible. R is denoted as the *rate* of the code. Let $\lambda(R, n)$ be the smallest error probability for (n, e^{Rn}) codes and define the largest error exponent as $E(R) = \lim_{n \to \infty} \frac{1}{n} \log \lambda(R, n)$.

(2) We require that $\lambda \in (0, 1)$ is fixed. Let $M(n, \lambda)$ denote the maximum number of messages that can be transmitted over the channel for given word length n and probability of error λ.

In his Fundamental Theorem Shannon proved that $M(n, \lambda)$ grows exponentially in n. More exactly, he proved that $\lim \inf_{n \to \infty} \frac{\log M(n, \lambda)}{n}$ exists and does not depend on $\lambda \in (0, 1)$. Shannon defined this limit as the *capacity* of the channel.

1.2 The Non-probabilistic Model

The idea here is that usually only few letters $(\leq d$, say) of a codeword are disturbed by the channel. If this is indeed the case, the output will be correctly decoded. For simplicity now assume that both, input and output alphabet, are binary, hence $\mathcal{X} = \mathcal{Y} = \{0, 1\}$. The *Hamming distance* $d_H(x^n, y^n)$ of $x^n = (x_1, \dots, x_n)$ and $y^n = (y_1, \dots, y_n)$ is the number of components in which $x_t \neq y_t$. More formally,

$$d_H(x^n, y^n) \triangleq \sum_{t=1}^{n} h(x_t, y_t), \quad \text{where } h : \{0, 1\} \times \{0, 1\} \to \{0, 1\}$$

$$\text{is given by } h(x, y) = \begin{cases} 0, & x = y \\ 1, & x \neq y \end{cases}.$$

An $[n, M, d]$-*Code* is a set of codewords $\{u_1, \ldots, u_M\} \subset \{0, 1\}^n$ with the property

$$d_H(u_i, u_j) \geq 2d + 1 \quad \text{for all} \quad i \neq j.$$

So if only few letters ($\leq d$) of the codeword u_i are disturbed during the transmission, the output is contained in

$$S_d^n(u_i) \triangleq \{z^n \in \{0, 1\}^n : d_H(u_i, z^n) \leq d\},$$

the *sphere* in $\{0, 1\}^n$ with radius d and center u_i. Since $d_H(u_i, u_j) \geq 2d + 1$, $S_d^n(u_i)$ and $S_d^n(u_j)$ do not overlap, and the output is correctly decoded.

So the construction of a good code is a sphere packing problem in the n-dimensional (Hamming) space.

Information is an important notion in various parts of Mathematics and Computer Science. So there are relations between Information Theory and Statistics, Game Theory, Image Processing, Speech, Radar, etc.

We will take a look at Search Theory, Memories, and Communication Complexity.

1.2.1 Search Theory

We are given a set $\{1, \ldots, N\}$ of N elements, one of which is defect. We are allowed to ask questions of the type "Is the defect element in the subset $A \subset \{1, \ldots, N\}$?". Our concern is to find the element with a minimum number of questions. The best strategy in our example is to subsequently divide the sets into halves. So our first question is "Is the defect element in $\{1, \ldots, \lfloor \frac{N}{2} \rfloor\}$?", etc. Search problems of this kind are equivalent to problems of Data Compression (Source Coding). More exactly, optimal strategies for a search problem yield optimal codes for a corresponding problem of Data Compression, and vice versa. Even if we introduce noise, we can find an equivalent problem in Coding Theory.

Searching is an important topic in Theoretical Computer Science. So one volume of Donald Knuth's famous monograph "The Art of Computer Programming" is completely devoted to "Searching and Sorting". We also draw attention to the book [1] from 1979.

1.2.2 Memories

We are given a storage device (a tape, a hard disk, a CD Rom, etc.), a person W who can write on this device, and a person R who can read from it. The data is usually stored as binary sequences.

Some of these storage devices have the property that information, once stored, cannot be deleted anymore. Think for example of a punched card. To each position on the card we assign a 1 or a 0 depending on its state (punched or not punched).

Now it might be possible to use the card several times. Since it is not possible to repair the punched positions, a transition from 1 to 0 in the corresponding binary sequence is forbidden.

The question now is: How many bits can be stored? There are four models, depending on the information W and R have about the last state of the device— W^+R^+, W^+R^-, W^-R^+, W^-R^-—where R^+ (R^-) means that R knows (does not know) what was written on the device before W began to write. Accordingly W^+ and W^- are defined.

It turns out that even in the case W^-R^-, where W and R are both uniformed, it is possible to store more bits than we can store, if we can write on this device only once.

Of course, usually the storage media are rewritable. However, instead of completely writing the new sequence of bits updating the information stored is often more efficient. So if for example we want to store the string 0001, and 0011 was stored before, only one bit has to be changed.

Since a rewriting (transition from 1 to 0 and from 0 to 1) always costs time and energy, whereas the other two transitions ($0 \rightarrow 0$ and $1 \rightarrow 1$) cost very little, our concern is to minimize the costs of updating information.

Another model of updating information stored on a rewritable medium was motivated by laser technology. Here the printing of the same letter can be done fast, but changing the directions of the magnets necessary for a variation of the letters is rather slow. So here for a long time period of updating it is only possible to write either 1's or 0's, but not the combination of both of them.

1.2.3 Communication Complexity

Our last example demonstrates that there are other connections between Information Theory and Computer Science.

In parallel computers the processors are connected via a communication network. Since local computation usually is very fast, the interprocessor communication often turns out to be the bottleneck of the computation. So the number of bits exchanged during the execution of a programme is an important measure. It is denoted as communication complexity. Since complexity is an important notion in Theoretical Computer Science—e.g. the amount of time and space a computational task requires are important complexity measures—and since communication is the object of Information Theory, both disciplines overlap here. Let us now formally describe the simplest model of such a computer network, which has been intensively studied in the last years.

We are given two processors, say $P_\mathcal{X}$ and $P_\mathcal{Y}$, both of which have unlimited computational capacities. $P_\mathcal{X}$ knows an element $x \in \mathcal{X}$, $P_\mathcal{Y}$ knows an element

$y \in \mathcal{Y}$. Their common task now is to compute the function value $f(x, y)$, where $f : \mathcal{X} \times \mathcal{Y} \to \mathcal{Z}$ and $\mathcal{X}, \mathcal{Y}$, and $\mathcal{Z}$ are finite sets. The number of bits they have to exchange in the worst case (taken over all inputs $(x, y) \in \mathcal{X} \times \mathcal{Y}$) is called (two-way) *communication complexity* of the function f, further denoted as $C(f, 1 \leftrightarrow 2)$.

Of course, there is always a trivial strategy for the two processors to solve their task. One processor can send all the bits of its input to the second processor, which is then able to compute $f(x, y)$ and in turn transmits the function value.

Formally, a *strategy* σ, say, is defined as a sequence of functions $f_1 : \mathcal{X} \to \{0, 1\}^* \triangleq \bigcup_{n=1}^{\infty} \{0, 1\}^n$, $g_1 : \mathcal{Y} \times \{0, 1\}^* \to \{0, 1\}^*$, $f_2 : \mathcal{X} \times \{0, 1\}^* \to \{0, 1\}^*$, $g_2 : \mathcal{Y} \times \{0, 1\}^* \to \{0, 1\}^*, \ldots$. So the processors transmit alternately binary sequences $f_1(x), g_1(y, f_1), f_2(x, f_1 g_1), g_2(y, f_1 g_1 f_2), \ldots$, in knowledge of the preceding sequences and their input values x and y, respectively. Communication stops, when both processors know the result. A strategy σ is said to be *successful*, if the function value $f(x, y)$ will be computed for all possible inputs $(x, y) \in \mathcal{X} \times \mathcal{Y}$.

So the communication complexity is defined as

$$C(f, 1 \leftrightarrow 2) \triangleq \min_{\sigma \text{ successful}} \max_{(x,y)\in\mathcal{X}\times\mathcal{Y}} L\big(\sigma, (x, y)\big),$$

where $L\big(\sigma, (x, y)\big)$ denotes the number of bits the strategy σ requires on input (x, y).

It can be shown that in most cases the trivial strategy is almost optimal. But if we allow *randomization*, i.e. the processors may flip a coin to decide on the next message, it is sometimes possible to save a lot of bits of transmission.

Reference

1. R. Ahlswede, I. Wegener, Suchprobleme, Russian Edition with Appendix by M. Maljutov, 1981; English Edition with Supplement of recent Literature, in *Wiley-Interscience Series in Discrete Mathematics and Optimization, 1987*, ed. by R.L. Graham, J.K. Leenstra, R.E. Tarjan (Teubner Verlag, Stuttgart, 1979)

Part I
Storing Data

Chapter 2
Data Compression

2.1 Lecture on Noiseless Source Coding

2.1.1 Discrete Sources

The source model discussed throughout this chapter is the *Discrete Source* (DS). Such a source is a pair $(\mathcal{X}, P)$, where $\mathcal{X} \triangleq \{1, \ldots, a\}$, say, is a finite alphabet and $P \triangleq (P(1), \ldots, P(a))$ is a probability distribution on $\mathcal{X}$. A discrete source can also be described by a random variable $X : \mathcal{X} \to \mathcal{X}$, where $\text{Prob}(X = x) = P(x)$ for all $x \in \mathcal{X}$.

In Lectures 2.2 and 2.3 sources of a special structure, called discrete memoryless sources (DMS), are analyzed. There a word $(x_1, x_2, \ldots, x_n) \in \mathcal{X}^n$ is the realization of the random variable $X^n \triangleq X_1 \ldots X_n$, where the X_i's are identically distributed ($X_i = X$ for all i) and independent of each other. So the probability $P^n(x_1, x_2, \ldots, x_n) = P(x_1) \cdot P(x_2) \cdots \cdots P(x_n)$ is the product of the probabilities of the single letters.

Observe that this source model is often not realistic. If we consider the English language for example and choose for $\mathcal{X}$ the latin alphabet with an additional symbol for Space and punctuation marks, the probability distribution can be obtained from the following frequency table (taken from D. Welsh [12]).

Frequency of letters in 1000 characters of English							
A	64	H	42	N	56	U	31
B	14	I	63	O	56	V	10
C	27	J	3	P	17	W	10
D	35	K	6	Q	4	X	3
E	100	L	35	R	49	Y	18
F	20	M	20	S	56	Z	2
G	14	T	71				
SPACE/PUNCTUATION MARK 166							

A. Ahlswede et al. (eds.), *Storing and Transmitting Data*,
Foundations in Signal Processing, Communications and Networking 10,
DOI: 10.1007/978-3-319-05479-7_2, © Springer International Publishing Switzerland 2014

Figures were taken from the copy-fitting tables used by professional printers based on counts involving hundreds of thousands of words in a very wide range of English language printed matter.

So $P(A) = 0.064, P(B) = 0.014$, etc. But in English texts the combination "th", for example, occurs more often than "ht". This could not be the case, if an English text was produced by a DMS, since then $P(th) = P(t) \cdot P(h) = P(ht)$. Several source models have been examined, in which those dependencies are considered.

Estimations for the letter probabilities in natural languages are obtained by statistical methods. For example, the above frequency table was obtained by counting the letters in English texts. The same method of course also yields results for pairs, triples, etc. of letters. Shannon proposed another method to estimate these probabilities. This method makes use of the knowledge a person has about his native language. If for example the statistician wants to know the probabilities of pairs of letters, he presents the first letter to some test persons and lets them guess the second letter. The relative frequencies obtained this way are quite good estimations.

In the discussion of the communication model it was pointed out that the source encoder wants to compress the original data into a short sequence of binary digits, hereby using a binary code, i.e. a function $c : \mathcal{X} \to \{0, 1\}^* = \bigcup_{n=0}^{\infty} \{0, 1\}^n$. To each element $x \in \mathcal{X}$ a code word $c(x)$ is assigned.

The three most basic approaches to data compression are

(1) Noiseless Source Coding (Sect. 2.1.4)

All possible messages, which the source can produce, are encoded. The aim of the source encoder is to minimize the average length of the codewords.

(2) Noisy Source Coding for the DMS (Lecture 2.2)

The Source Encoder only encodes a set of messages $A \subset \mathcal{X}^n$, which has probability $P^n(A) \geq 1 - \epsilon, \epsilon \in (0, 1)$. Such a set is denoted as (n, ϵ)-compression of the source $(\mathcal{X}^n, P^n)$.

(3) Rate Distortion Theory (Lecture 2.3)

The data only has to be reproduced with a certain degree of accuracy. Think for example of a photograph stored in a database as a matrix of binary digits (black, white). The objects on the photograph can probably still be recognized, if we allow that 0,1% of the bits are disturbed.

It turns out that in all models the best possible data compression can be described in terms of the *entropy* $H(P)$ of the probability distribution P. The entropy is given by the formula

$$H(P) = - \sum_{x \in \mathcal{X}} P(x) \cdot \log P(x).$$

We will also use the notation $H(X)$ according to the interpretation of the source as a random variable. If not further specified all logarithms are to the base 2. Accordingly, the inverse function $\exp(x)$ usually denotes 2^x. Natural logarithm and exponential function are denoted as $\ell n(x)$ and e^x, respectively.

2.1.2 Uniquely Decipherable Codes, Prefix Codes

Throughout this paragraph the codes are not restricted to be binary. So we allow an arbitrary finite code alphabet, denoted $\mathcal{Y}$. $\mathcal{Y}^* \triangleq \bigcup_{n=0}^{\infty} \mathcal{Y}^n$ is the set of all finite sequences (words) over $\mathcal{Y}$. (Convention: the word of length 0 is called the empty word)

Definition 1 A **code** (of variable length) is a function $c : \mathcal{X} \to \mathcal{Y}^*$, $\mathcal{X} = \{1, \ldots, a\}$. So $\{c(1), c(2), \ldots, c(a)\}$ is the set of **codewords**, where for $x = 1, \ldots, a$ $c(x) = (c_1(x), c_2(x), \ldots, c_{L(x)}(x))$ and $L(x)$ is denoted as the **length** of the codeword $c(x)$.

In the following example some binary codes ($\mathcal{Y} \triangleq \{0, 1\}$) for the latin alphabet are presented. (SP = Space/punctuation mark)

1. $c_1 : a \to 1, b \to 10, c \to 100, d \to 1000, \ldots, z \to \underbrace{10 \ldots 0}_{26}$,

   $SP \to \underbrace{10 \ldots 0}_{27}$.

2. $c_2 : a \to 00000, b \to 00001, c \to 00010, \ldots, z \to 11001, SP \to 11010$. So $c_2(x)$ is the binary representation of the position of the letter x in the alphabetic order.

3. $c_3 : a \to 0, b \to 00, c \to 1, \ldots$ (the further codewords are not important for the following discussion)

The last code presented has an undesirable property. Observe that the sequence 00 could either be decoded as b or as aa. Hence the messages encoded using this code are not uniquely decipherable.

Definition 2 A code c is **uniquely decipherable** (UDC), if every word in $\mathcal{Y}^*$ is representable by at most one sequence of codewords. More formally, it is required that the function $C : \mathcal{X}^* \to \mathcal{Y}^*$ defined by $C(x_1, x_2, \ldots, x_t) = c(x_1)c(x_2) \ldots c(x_t)$ ($t \in \mathbb{N}$ arbitrary, $(x_1, x_2, \ldots, x_t) \in \mathcal{X}^*$) is injective.

The code c_1 is uniquely decipherable, since the number of 0's between two 1's determines the next letter in a message encoded using c_1. Code c_2 is uniquely decipherable, since every letter is encoded with exactly five bits. Hence the first five bits of a sequence of binary digits are decoded as the first letter of the original text, the bits 6 to 10 as the second letter, etc.

Definition 3 A code c is a **prefix code**, if for any two codewords $c(x)$ and $c(y)$, $x \neq y$, with $L(x) \leq L(y)$ holds $(c_1(x), c_2(x), \ldots, c_{L(x)}(x)) \neq (c_1(y), c_2(y), \ldots, c_{L(x)}(y))$. So in at least one of the first $L(x)$ components $c(x)$ and $c(y)$ differ.

Messages encoded using a prefix code are uniquely decipherable. The decoder proceeds by reading the next letter until a codeword $c(x)$ is formed. Since $c(x)$ cannot be the beginning of another codeword, it must correspond to the letter $x \in \mathcal{X}$. Now the decoder continues until another codeword is formed. The process may be repeated until the end of the message. So after having found the codeword $c(x)$ the decoder instantaneously knows that $x \in \mathcal{X}$ is the next letter of the message. Because of this property a prefix code is also denoted as *instantaneous* code.

Observe that the code c_1 is not instantaneous. Every codeword is the beginning of the following codewords. Another example for a uniquely decipherable code, which is not instantaneous, is $c : \{a, b\} \rightarrow \{0, 1\}^*$ with $c(a) = 0$ and $c(b) = \underbrace{0 \ldots 0 1}_{n}$. If

for example we received the sequence $\underbrace{0 \ldots 0 1}_{n+1}$ we would have to wait until the end of the sequence to find out that the corresponding message starts with a.

As pointed out in the previous paragraph the criterion for data compression in Noiseless Coding Theory is to minimize the average length of the codewords. So if we are given a source $(\mathcal{X}, P)$, where $\mathcal{X} = \{1, \ldots, a\}$ and $P = \big(P(1), P(2), \ldots, P(a)\big)$ is a probability distribution on $\mathcal{X}$, the *average length* $\overline{L}(c)$ is defined by

$$\overline{L}(c) \triangleq \sum_{x \in \mathcal{X}} P(x) \cdot L(x).$$

So, if in English texts all letters (incl. space/punctuation mark) would occur with the same frequency, then the code c_1 had average length $\frac{1}{27}(1 + 2 + \cdots + 27) = \frac{1}{27} \cdot \frac{27 \cdot 28}{2} = 14$. Hence the code c_2, with average length 5 would be more appropriate in this case. From the frequency table of the previous paragraph we know that the occurrence of the letters in English texts cannot be modeled by the uniform distribution. In this case it is possible to find a better code by assigning short codewords to letters with high probability as demonstrated by the following prefix code c with average length $\overline{L}(c) = 3 \cdot 0.266 + 4 \cdot 0.415 + 5 \cdot 0.190 + 6 \cdot 0.101 + 7 \cdot 0.016 + 8 \cdot 0.012 = 4.222$.

$a \rightarrow 0110$,	$b \rightarrow 010111$,	$c \rightarrow 10001$,	$d \rightarrow 01001$,
$e \rightarrow 110$,	$f \rightarrow 11111$,	$g \rightarrow 111110$,	$h \rightarrow 00100$,
$i \rightarrow 0111$,	$j \rightarrow 11110110$,	$k \rightarrow 1111010$,	$l \rightarrow 01010$,
$m \rightarrow 001010$,	$n \rightarrow 1010$,	$o \rightarrow 1001$,	$p \rightarrow 010011$,
$q \rightarrow 01011010$,	$r \rightarrow 1110$,	$s \rightarrow 1011$,	$t \rightarrow 0011$,
$u \rightarrow 10000$,	$v \rightarrow 0101100$,	$w \rightarrow 001011$,	$x \rightarrow 01011011$,
$y \rightarrow 010010$,	$z \rightarrow 11110111$,	$SP \rightarrow 000$.	

We can still do better, if we do not encode single letters, but blocks of n letters for some $n \in \mathbb{N}$. In this case we replace the source $(\mathcal{X}, P)$ by $(\mathcal{X}^n, P^n)$ for some $n \in \mathbb{N}$. Remember that $P^n(x_1, x_2, \ldots, x_n) = P(x_1) \cdot P(x_2) \cdots \cdots P(x_n)$ for a word $(x_1, x_2, \ldots, x_n) \in \mathcal{X}^n$, since the source is memoryless. If for example we are given an alphabet with two letters, say $\mathcal{X} = \{a, b\}$ and $P(a) = 0.9$, $P(b) = 0.1$, then the code c defined by $c(a) = 0$, $c(b) = 1$ has average length $\overline{L}(c) = 0.9 \cdot 1 + 0.1 \cdot 1 = 1$. Obviously we cannot find a better code. The combinations of two letters now have the following probabilities:

$$P(aa) = 0.81, \quad P(ab) = 0.09, \quad P(ba) = 0.09, \quad P(bb) = 0.01.$$

The prefix code c^2 defined by

$$c^2(aa) = 0, \quad c^2(ab) = 10, \quad c^2(ba) = 110, \quad c^2(bb) = 111$$

has average length $\overline{L}(c^2) = 1 \cdot 0.81 + 2 \cdot 0.09 + 3 \cdot 0.09 + 3 \cdot 0.01 = 1.29$. So $\frac{1}{2}\overline{L}(c^2) = 0.645$ could be interpreted as the average length the code c^2 requires per letter of the alphabet $\mathcal{X}$. It can be verified that 1.29 is the minimal average length a code defined on pairs of letters can assume. If we encode blocks of n letters we are interested in the behaviour of

$$L(n, P) \triangleq \min_{c^n \text{ UDC}} \frac{1}{n} \sum_{(x_1, \ldots, x_n) \in \mathcal{X}^n} P^n(x_1, \ldots, x_n) L(x_1, \ldots, x_n) = \min_{c^n \text{ UDC}} \overline{L}(c).$$

It follows from the Noiseless Coding Theorem, which is stated in Sect. 2.1.4, that $\lim_{n \to \infty} L(n, P) = H(P)$ the entropy of the source $(\mathcal{X}, P)$.

In our example for the English language we have $H(P) \approx 4.19$. So the code presented above, where only single letters are encoded, is already nearly optimal in respect of $L(n, P)$. Further compression is possible, if we consider the dependencies between the letters.

2.1.3 Kraft's Inequality

We will now introduce a necessary and sufficient condition for the existence of a prefix code for $\mathcal{X} = \{1, \ldots, a\}$ with prescribed word lengths $L(1), \ldots, L(a)$.

Theorem 1 (Kraft's inequality [8]) *Let $\mathcal{X} = \{1, \ldots, a\}$ and let $\mathcal{Y}$ be the code alphabet with $|\mathcal{Y}| = b$. A prefix code $c : \mathcal{X} \to \mathcal{Y}^*$ with word lengths $L(1), \ldots, L(a)$ exists, if and only if*

$$\sum_{x \in \mathcal{X}} b^{-L(x)} \leq 1.$$

Proof We will only prove Kraft's inequality for the binary code alphabet $\mathcal{Y} = \{0, 1\}$. The proof for arbitrary $\mathcal{Y}$ follows exactly the same lines.

The central idea is to interpret the codewords as nodes of a rooted binary tree with depth $T \triangleq \max_{x \in \mathcal{X}} \{L(x)\}$. The tree is required to be complete (every path from the root to a leaf has length T) and regular (every inner node has degree 3). The following example for $T = 3$ may serve as an illustration.

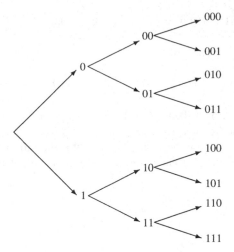

So the nodes with distance n from the root are labeled with the words $x^n \in \{0, 1\}^n$. The upper successor of $x_1 x_2 \ldots x_n$ is labeled $x_1 x_2 \ldots x_n 0$, its lower successor is labeled $x_1 x_2 \ldots x_n 1$.

The *shadow* of a node labeled by $x_1 x_2 \ldots x_n$ is the set of all the leaves which are labeled by a word (of length T) beginning with $x_1 x_2 \ldots x_n$. In other words, the shadow of $x_1 \ldots x_n$ consists of the leaves labeled by a sequence with prefix $x_1 \ldots x_n$. In our example $\{000, 001, 010, 011\}$ is the shadow of the node labeled by 0.

Now assume that we are given a prefix code with word lengths $L(1), \ldots, L(a)$. Every codeword corresponds to a node in the binary tree of depth T. Observe that the shadows of any two codewords are disjoint. If this was not the case, we could find a word $x_1 x_2 \ldots x_T$, which has as prefix two codewords of length s and t, say (w.l.o.g. $s < t$). But these codewords are $x_1 x_2 \ldots x_t$ and $x_1 x_2 \ldots x_s$, which obviously is a prefix of the first one.

The shadow of the codeword $c(x)$ has size $2^{T-L(x)}$. There are 2^T words of length T. For the sum of the shadow sizes follows $\sum_{x \in \mathcal{X}} 2^{T-L(x)} \leq 2^T$, since none of these words can be a member of two shadows. Division by 2^T yields the desired inequality $\sum_{x \in \mathcal{X}} 2^{-L(x)} \leq 1$.

Conversely, suppose we are given positive integers $L(1), \ldots, L(a)$. We further assume that $L(1) \leq L(2) \leq \cdots \leq L(a)$. As first codeword $c_1 = \underbrace{00 \ldots 0}_{L(1)}$ is chosen.

Since $\sum_{x \in \mathcal{X}} 2^{T-L(x)} \leq 2^T$, we have $2^{T-L(1)} < 2^T$ (otherwise only one letter has to be encoded). Hence there are left some nodes on the Tth level, which are not in the shadow of $c(1)$. We pick the first of these remaining nodes and go back $T - L(2)$ steps in direction to the root. Since $L(2) \geq L(1)$ we will find a node labeled by a sequence of $L(2)$ bits, which is not a prefix of $c(1)$. So we can choose this sequence as $c(2)$. Now again, either $a = 2$, and we are ready, or by the hypothesis $2^{T-L(1)} + 2^{T-L(2)} < 2^T$ and we can find a node on the Tth level, which is not contained in the shadows of $c(1)$ and $c(2)$. We find the next codeword as shown above. The process can be continued until all codewords are assigned. $\qquad\square$

Remark In the proof of Kraft's inequality it is essential to order the lengths $L(1) \leq \cdots \leq L(a)$. For example, the construction of a prefix code for given lengths $2, 1, 2$ was not possible with the procedure presented in the proof, if we would not order the lengths.

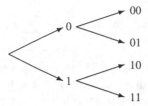

00 would then be chosen as the first codeword $c(1)$. The next node not contained in the shadow of 00 is the one labeled by 01. But according to our procedure we would have to choose its predecessor 0 as $c(2)$. But 0 is a prefix of $c(1) = 00$.

Kraft's inequality gives a necessary and sufficient condition for the existence of a prefix code with codewords of lengths $L(1), \ldots, L(a)$. In the following theorem of [9] it is shown that this condition is also necessary for the existence of a uniquely decipherable code.

Theorem 2 (Kraft's inequality for uniquely decipherable codes [9]) *A uniquely decipherable code with prescribed word lengths $L(1), \ldots, L(a)$ exists, if and only if*

$$\sum_{x \in \mathcal{X}} b^{-L(x)} \leq 1$$

Proof Since every prefix code is uniquely decipherable the sufficiency part of the proof is immediate. Now observe that $\sum_{x \in \mathcal{X}} b^{-L(x)} = \sum_{j=1}^{T} w_j b^{-j}$, where w_j is the number of codewords with length j in the uniquely decipherable code and T again denotes the maximal word length. The sth power of this term can be expanded as

$$\left(\sum_{j=1}^{T} w_j b^{-j} \right)^s = \sum_{k=s}^{T \cdot s} N_k b^{-k}.$$

Here $N_k = \sum_{i_1 + \cdots + i_s = k} w_{i_1} \ldots w_{i_s}$ is the total number of messages whose coded representation is of length k. Since the code is uniquely decipherable, to every sequence of k letters corresponds at most one possible message. Hence $N_k \leq b^k$ and $\sum_{k=s}^{T \cdot s} N_k b^{-k} \leq \sum_{k=s}^{T \cdot s} 1 = T \cdot s - s + 1 \leq T \cdot s$. Taking sth root this yields $\sum_{j=1}^{T} w_j b^{-j} \leq (T \cdot s)^{\frac{1}{s}}$.

Since this inequality holds for any s and $\lim_{s \to \infty} (T \cdot s)^{\frac{1}{s}} = 1$, we have the desired result

$$\sum_{j=1}^{T} w_j b^{-j} = \sum_{x \in \mathcal{X}} b^{-L(x)} \leq 1. \qquad \Box$$

2.1.4 The Noiseless Coding Theorem

Before stating the Noiseless Coding Theorem it is necessary to introduce some further notions. Remember that for a source $(\mathcal{X}, P)$ the entropy was defined as

$$H : \mathcal{P}(\mathcal{X}) \to \mathbb{R}_+ \quad \text{by} \quad H(P) = -\sum_{x \in \mathcal{X}} P(x) \cdot \log P(x),$$

where the logarithm was to the base 2. Throughout this paragraph all logarithms are to the base $b = |\mathcal{Y}|$. Observe that because of $\log_2 x = \log_b x \cdot \log_2 b$ the entropy (to the base 2) is obtained by multiplication with the constant factor $\log_2 b$.

Lemma 1

(i) *The entropy H is concave in P. So if $P_1, P_2, \ldots, P_r$ are probability distributions and $a_1, \ldots, a_r$ are non-negative numbers with $\sum_{i=1}^{r} a_i = 1$, for the entropy of the distribution $\overline{P}$, defined by $\overline{P}(x) \triangleq \sum_{i=1}^{r} a_i P_i(x)$, we have $H(\overline{P}) \geq \sum_{i=1}^{r} a_i H(P_i)$.*
(ii) $0 \leq H(P) \leq \log |\mathcal{X}|$.

Proof (i) Observe that the function $f(t) = -t \cdot \log t$ is concave on the interval $[0, 1]$ (convention $f(0) = 0$), since $f''(t) = -\frac{1}{t} \cdot \frac{1}{\ln b} \leq 0$. $H(P) = \sum_{x \in \mathcal{X}} (-P(x) \cdot \log P(x))$ now is concave, because it is a sum of concave functions.

(ii) Since $f(t) \geq 0$ for all $t \in [0, 1]$, $H(P)$ is a sum of non-negative terms and hence $H(P) \geq 0$. Now let $\mathcal{X} = \{1, \ldots, a\}$, hence $|\mathcal{X}| = a$, and let π be a permutation on $\mathcal{X}$. This permutation defines a probability distribution P_π on $\mathcal{X}$ by $P_\pi(x) \triangleq P(\pi x)$ for all $x \in \mathcal{X}$.

Of course,

$$H(P_\pi) = -\sum_{x \in \mathcal{X}} P(\pi x) \cdot \log P(\pi x) = -\sum_{x \in \mathcal{X}} P(x) \cdot \log P(x) = H(P).$$

The probability distribution P^* defined by $P^*(x) \triangleq \frac{1}{a!} \sum_\pi P_\pi(x)$ is the uniform distribution since for all $x \in \mathcal{X}$ holds $\frac{1}{a!} \sum_\pi P_\pi(x) = \frac{1}{a}$. The uniform distribution has entropy $H(P^*) = \sum_{x \in \mathcal{X}} \frac{1}{a} \log a = \log a$. Hence

$$\log a = H(P^*) = H\Big(\frac{1}{a!} \sum_\pi P_\pi\Big) \geq \frac{1}{a!} \sum_\pi H(P_\pi),$$

since H is concave. As pointed out $H(P_\pi) = H(P)$ for every permutation π. Since there are exactly $a!$ permutations on $\mathcal{X} = \{1, \ldots, a\}$ the right-hand side is $H(P)$ and we have the desired result. $\square$

Definition 4 For two probability distributions P and Q on $\mathcal{X}$ the *relative entropy* $D(P||Q)$ is defined by

$$D(P||Q) \triangleq \sum_{x \in \mathcal{X}} P(x) \log \frac{P(x)}{Q(x)}.$$

It is an important notion in Information Theory. Although it is not a metric, it is a good measure for the distance of two probability distributions. Observe that $D(P||P) = 0$, since

$$D(P||Q) = -H(P) - \sum_{x \in \mathcal{X}} P(x) \log Q(x).$$

In the proof of the Noiseless Coding Theorem we will make use of the fact that $D(P||Q) \geq 0$ for all P and Q, which follows from the log-sum inequality (see Chap. 8).

Now let us return to Noiseless Coding. We are given a source $(\mathcal{X}, P)$ and we want to find a uniquely decipherable code $c : \mathcal{X} \rightarrow \mathcal{Y}^*$ with minimum average word length $\bar{L}(c) = \sum_{x \in \mathcal{X}} P(x)L(x)$, where $L(x)$ denotes the length of the codeword $c(x), x \in \mathcal{X}$. So

$$L_{\min}(P) \triangleq \min_{c \ UDC} \bar{L}(c)$$

is our measure for data compression.

Theorem 3 (Noiseless Coding Theorem) *For a source $(\mathcal{X}, P)$, $\mathcal{X} = \{1, \dots, a\}$ it is always possible to find a uniquely decipherable code $c : \mathcal{X} \rightarrow \mathcal{Y}^*$ with average length*

$$H(P) \leq L_{\min}(P) < H(P) + 1.$$

Proof We will only prove the theorem for $\mathcal{Y} = \{0, 1\}$. For an arbitrary code alphabet $\mathcal{Y}$ the proof follows the same lines. Let $L(1), \dots, L(a)$ denote the codeword lengths of an optimal uniquely decipherable code. Now we define a probability distribution Q on $\mathcal{X} = \{1, \dots, a\}$ by $Q(x) = \frac{2^{-L(x)}}{\rho}$ for $x \in \mathcal{X}$, where

$$\rho \triangleq \sum_{x=1}^{a} 2^{-L(x)}.$$

By Kraft's inequality $\rho \leq 1$. Remember that always the relative entropy $D(P||Q) \geq 0$. So for any probability distribution P

$$D(P||Q) = -H(P) - \sum_{x \in \mathcal{X}} P(x) \cdot \log\left(2^{-L(x)} \cdot \rho^{-1}\right) \geq 0.$$

From this it follows that

$$H(P) \leq - \sum_{x \in \mathcal{X}} P(x) \cdot \log\left(2^{-L(x)} \cdot \rho^{-1}\right)$$

$$= \sum_{x \in \mathcal{X}} P(x) \cdot L(x) - \sum_{x \in \mathcal{X}} P(x) \cdot \log \rho^{-1}$$

$$= L_{\min}(P) + \log \rho.$$

Since $\rho \leq 1$, $\log \rho \leq 0$ and hence $L_{\min}(P) \geq H(P)$.

In order to prove the RHS of the Noiseless Coding Theorem for $x = 1, \ldots, a$ we define $L'(x) \triangleq \lceil -\log P(x) \rceil$. Observe that $-\log P(x) \leq L'(x) < -\log P(x) + 1$ and hence $P(x) \geq 2^{-L'(x)}$.

So $1 = \sum_{x \in \mathcal{X}} P(x) \geq \sum_{x \in \mathcal{X}} 2^{-L'(x)}$ and from Kraft's inequality we know that there exists a uniquely decipherable code with word lengths $L'(1), \ldots, L'(a)$. This code has average length

$$\sum_{x \in \mathcal{X}} P(x) \cdot L'(x) < \sum_{x \in \mathcal{X}} P(x)(-\log P(x) + 1) = H(P) + 1. \qquad \square$$

An analysis of the proof shows that we did not only demonstrate the existence of a code with $\overline{L}(c) < H(P) + 1$. Moreover, we know quite well how such a "good" code must look like, since in the proof of Kraft's inequality a procedure for the construction of a code with prescribed lengths $L(x)$ of the codewords $c(x)$, $x \in \mathcal{X}$, was introduced. However this procedure is not very efficient and the codes obtained this way do not necessarily have the optimal average length $L_{\min}(P)$. In Sect. 2.1.5 we shall present the Huffman Coding Algorithm, which enables us to construct an optimal prefix code for a given probability distribution.

A further direction of research is a methodical one. The central idea in the proof of Kraft's inequality for prefix codes was to embed the codewords into a tree. The tree induces an order relation.

Finally, one might wonder why the entropy arises as a measure for data compression. Entropy was originally introduced in Physics as a measure of disorder by Clausius and Boltzmann (Lecture 3.6). Further properties and applications of the entropy will be discussed in the Chap. 3.

2.1.5 The Huffman Coding Algorithm

In the proof of the Noiseless Coding Theorem it was explicitly shown how to construct a prefix code c to a given probability distribution $P = (P(1), \ldots, P(a))$. The idea was to assign to each $x \in \{1, \ldots, a\}$ a codeword of length $L(x) = \lceil -\log P(x) \rceil$ by choosing an appropriate vertex in the tree introduced. However, this procedure does not always yield an optimal code. If for example we are given the probability distribution $(\frac{1}{3}, \frac{1}{3}, \frac{1}{3})$, we would encode $1 \to 00, 2 \to 01, 3 \to 10$ and thus achieve an average codeword length 2. But the code with $1 \to 00, 2 \to 01, 3 \to 1$ has only average length $\frac{5}{3}$.

A more efficient procedure was introduced by Shannon and Fano [5, 11].

The Shannon-Fano encoding procedure will be illustrated by the following example [10]:

Messages	Probabilities	Encoded messages	Length
1	0.2500	0 0	2
2	0.2500	0 1	2
3	0.1250	1 0 0	3
4	0.1250	1 0 1	3
5	0.0625	1 1 0 0	4
6	0.0625	1 1 0 1	4
7	0.0625	1 1 1 0	4
8	0.0625	1 1 1 1	4
Average length			2.75

After listing the messages in order of nonincreasing probabilities the message set is partitioned into the two most equiprobable subsets X_0 and X_1. A 0 is assigned to each message contained in the first subset X_0 and a 1 to each message in X_1. We repeat this procedure for the respective subsets of X_0 and X_1, hence, X_0 will be partitioned into two subsets X_{00} and X_{01}, etc.. Then the code word for a message in X_{00} begins with 00 and that corresponding to a message in X_{01} starts with 01. This algorithm is iterated until the final subsets consist of only one message.

However, the Shannon-Fano procedure neither yields an optimal code in general, as the probability distribution $(0.25, 0.2, 0.11, 0.11, 0.11, 0.11, 0.11)$ demonstrates. Here the Shannon-Fano procedure yields the code $1 \to 00, 2 \to 01, 3 \to 100, 4 \to 101, 5 \to 110, 6 \to 1110, 7 \to 1111$ with average length 2.77, whereas the prefix code $1 \to 11, 2 \to 100, 3 \to 101, 4 \to 000, 5 \to 001, 6 \to 010, 7 \to 011$ has average length 2.75.

The last code was obtained by applying the Huffman Coding Algorithm [7], which always yields a prefix code with minimal average length. It is based on the following theorem.

Theorem 4 *We are given a source $(\mathcal{X}, P)$, where $\mathcal{X} = \{1, \ldots, a\}$ and the probabilities are ordered non-increasingly $P(1) \geq P(2) \geq \cdots \geq P(a)$. A new probability distribution is defined by*

$$P' = \big(P(1), \ldots, P(a-2), P(a-1) + P(a)\big).$$

Let $c' = \big(c'(1), c'(2), \ldots, c'(a-1)\big)$ be an optimal prefix code for P'. Now we define a code c for the distribution P by

$$c(x) \triangleq c'(x) \ for \ x = 1, \ldots, a - 2,$$

$$c(a - 1) \triangleq c'(a - 1)0,$$

$$c(a) \triangleq c'(a - 1)1.$$

Then c is an optimal prefix code for P and
$L_{\min}(P) - L_{\min}(P') = p(a - 1) + p(a).$

This theorem suggests the following recursive algorithm, which we will illustrate with an example taken from [2].

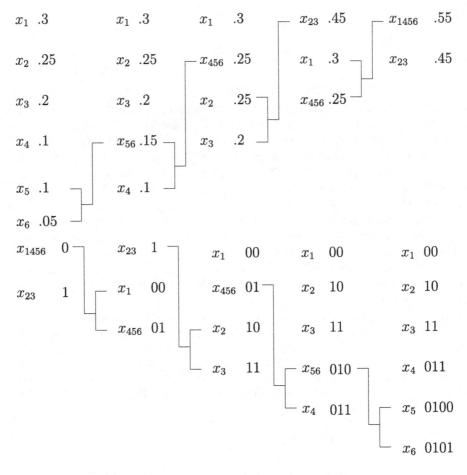

So the source is successively reduced by one element. In each reduction step we add up the two smallest probabilities and insert $p(a') + p(a' - 1)$ in the increasingly ordered sequence $P(1) \geq \cdots \geq P(a' - 2)$, thus obtaining a new probability distribution P' with $P'(1) \geq \cdots \geq P'(a' - 1)$. Finally we arrive at a source with two elements ordered according to their probabilities. The first element is assigned a 0, the second element a 1. Now we again "blow up" the source until the original source is restored. In each step $c(a' - 1)$ and $c(a')$ are obtained by appending 0 or 1, respectively, to the codeword corresponding to $p(a') + p(a' - 1)$.

We still have to prove the above theorem, which is based on the following lemma

Lemma 2 *For a probability distribution P on $X = \{1, \ldots, a\}$ with $P(1) \geq P(2) \geq \cdots \geq P(a)$ there exists an optimal prefix code c with*

(i) $L(1) \leq L(2) \leq \cdots \leq L(a)$
(ii) $L(a - 1) = L(a)$
(iii) $c(a - 1)$ and $c(a)$ differ exactly in the last position.

Proof

(i) Assume that there are $x, y \in X$ with $P(x) \geq P(y)$ and $L(x) > L(y)$. Then the code c' obtained by interchanging the codewords $c(x)$ and $c(y)$ has average length $\overline{L}(c') \leq \overline{L}(c)$, since

$$\overline{L}(c') - \overline{L}(c) = P(x) \cdot L(y) + P(y) \cdot L(x) - P(x) \cdot L(x) - P(y) \cdot L(y)$$
$$= (P(x) - P(y)) \cdot (L(y) - L(x)) \leq 0$$

(ii) Assume we are given a code c' with $L(1) \leq \cdots \leq L(a - 1) < L(a)$. Because of the prefix property we may drop the last $L(a) - L(a - 1)$ bits of $c'(a)$ and thus obtain a new code c with $L(a) = L(a - 1)$.
(iii) If no two codewords of maximal length agree in all places but the last, then we may drop the last digit of all such codewords to obtain a better code. □

Proof of Theorem 4 From the definition of c and c' we have

$$L_{\min}(P) \leq \overline{L}(c) = \overline{L}(c') + p(a - 1) + p(a).$$

Now let c'' be an optimal prefix code with the properties (ii) and (iii) from the preceding lemma. We define a prefix code c''' for

$$P' = (P(1), \ldots, P(a - 2), P(a - 1) + P(a))$$

by $c'''(x) = c''(x)$ for $x = 1, \ldots, a - 2$ and $c'''(a - 1)$ is obtained by dropping the last bit of $c''(a - 1)$ or $c''(a)$.

Now

$$L_{\min}(P) = \overline{L}(c'') = \overline{L}(c''') + P(a - 1) + P(a)$$

$$\geq L_{\min}(P') + P(a - 1) + P(a)$$

and hence $L_{\min}(P) - L_{\min}(P') = P(a - 1) + P(a)$, since $\overline{L}(c') = L_{\min}(P')$. □

2.2 Lecture on the Coding Theorem for the Discrete Memoryless Source

In Noiseless Coding all possible messages are encoded error–free. Short codewords are assigned to messages with high probability. Therefore the codewords have variable length. In this lecture a different model is considered. The messages are encoded with fixed block length and errors are allowed. In order to compress the data only words of a set $A \subset \mathcal{X}^n$ with $P^n(A) \geq 1 - \epsilon, 0 < \epsilon < 1$ are encoded.

Definition 5 $A \subset \mathcal{X}^n$ is an (n, ϵ)-**compression** of the source $(\mathcal{X}^n, P^n)$, if $P^n(A) \geq 1 - \epsilon, 0 < \epsilon < 1$.

The measure for data compression now is

$$K(n, \epsilon) \triangleq \min\{|A| : A \text{ is } (n, \epsilon)\text{-compression}\}.$$

The behaviour of $K(n, \epsilon)$ is described in the following **Source Coding Theorem for the Discrete Memoryless Source**.

Theorem 5 (Shannon 1948 [11]) *Given the source* $(\mathcal{X}^n, P^n)_{n=1}^{\infty}$, *then*

$$\lim_{n \to \infty} \frac{1}{n} \log K(n, \epsilon) = H(P)$$

for all $0 < \epsilon < 1$.

Before we prove the theorem let us first discuss the result. The words $x^n \in A$ can be correctly reproduced, whereas errors occur in the reproduction of a word $x^n \in \mathcal{X}^n \setminus A$. Think for example of a storage device, where we store each word $x^n \in \mathcal{X}^n$ using a fixed block length code c with $c(x^n) \neq c(y^n)$ for all $x^n \neq y^n \in A$. An additional codeword $b \neq c(x^n)$ for all $x^n \in A$ is introduced onto which all $x^n \in \mathcal{X}^n \setminus A$ are mapped. So these words cannot be correctly reproduced. But the philosophy is that such errors occur only very rarely. In fact, it is shown in the Source Coding Theorem for the DMS that we can choose ϵ arbitrarily close to 0 and still have the compression rate $H(P)$. So the limit $H(P)$ does not depend on the choice of ϵ, which only has influence on the speed of the convergence.

Some examples may further illustrate the theorem.

1. P is the uniform distribution. So $H(P) = \log |\mathcal{X}|$ and compression is not possible ($K(n, \epsilon)$ asymptotically has the same size as $\mathcal{X}^n$). This may also become clear by a simple calculation. $(1 - \epsilon) \cdot |\mathcal{X}|^n$ words have to be encoded, so

$$\lim_{n \to \infty} \frac{1}{n} \log K(n, \epsilon) = \lim_{n \to \infty} \frac{1}{n} \log(1 - \epsilon) \cdot |\mathcal{X}|^n$$

$$= \lim_{n \to \infty} \frac{1}{n} \cdot \log(1 - \epsilon) + \lim_{n \to \infty} \log |\mathcal{X}|$$

$$= \log |\mathcal{X}|.$$

2. the concentration on one point $x_0 \in \mathcal{X}$. So $P(x_0) = 1$ and $P(x) = 0$ for all $x \in \mathcal{X} \setminus \{x_0\}$. Here $H(P) = 0$ and indeed one bit suffices to encode the messages in this case $(x_0, x_0, \ldots, x_0) \to 1, x^n \to 0$ for all

$$x^n \in \mathcal{X}^n \setminus (x_0, x_0, \ldots, x_0).$$

3. $P = (\frac{1}{2}, \frac{1}{4}, \frac{1}{4})$. So $H(P) = \frac{3}{2}$ and this is obviously smaller than $\log 3$, the rate in the case $\epsilon = 0$, when **all** sequences of length n have to be encoded with fixed block length.

In the following paragraphs we will give two proofs for the Source Coding Theorem for the DMS, both of which make use of "typical" sequences. Remember that always $\exp\{x\} = 2^x$.

2.2.1 First Proof of the Source Coding Theorem

Definition 6 A sequence $x^n \in \mathcal{X}^n$ is (n, P, δ)-entropy-typical, if

$$\mid -\frac{1}{n} \log P^n(x^n) - H(P) \mid \le \delta.$$

We also write (P, δ)-entropy-typical if it is clear that the length is n. Similarly abbreviations (n, δ)-entropy-typical or (δ)-entropy-typical are used if the meaning is clear from the context.

The set of all (n, P, δ)-entropy-typical sequences is denoted as $E(n, P, \delta)$, so

$$E(n, P, \delta) = \{x^n \in \mathcal{X}^n :\mid -\frac{1}{n} \log P^n(x^n) - H(P) \mid \le \delta\}.$$

The proof of the theorem now consists of two parts. In the first part it is demonstrated that $E(n, P, \delta)$ is an (n, ϵ)-compression with size about $n H(P)$. In the second part it is shown that it is not possible to find an essentially better compression.

Recall that a sequence $x^n \in \mathcal{X}^n$ can be regarded as the realization of a random variable $X^n = (X_1, X_2, \ldots, X_n)$, where the X_t's are independent and identically distributed. Now for each $t \in \{1, \ldots, n\}$ we define the random variable

$$Y_t \triangleq - \log P(X_t).$$

Clearly, these random variables are again independent and identically distributed. The expected value of Y_t is

$$EY_t = - \sum_{x \in \mathcal{X}} \text{Prob}(X_t = x) \cdot \log P(x) = - \sum_{x \in \mathcal{X}} P(x) \cdot \log P(x) = H(P).$$

Further observe that

$$\log P^n(x^n) = \sum_{t=1}^{n} \log P(x_t) = -\sum_{t=1}^{n} Y_t.$$

Now we know from the weak law of large numbers that

$$P^n\big(E(n, P, \delta)\big) = \text{Prob}\left\{ \mid \frac{1}{n} \sum_{t=1}^{n} Y_t - H(P) \mid \le \delta \right\} \xrightarrow[n \to \infty]{} 1.$$

For all $x^n \in E(n, P, \delta)$ we have by definition

$$\exp\{-n(H(P) + \delta)\} \le P^n(x^n) \le \exp\{-n(H(P) - \delta)\}.$$

Hence

$$\mid E(n, P, \delta) \mid \le \exp\{n \cdot (H(P) + \delta)\}$$

and $\overline{\lim}_{n \to \infty} \frac{1}{n} \log \mid E(n, P, \delta) \mid \le H(P) + \delta$.

So $\overline{\lim}_{n \to \infty} \frac{1}{n} \log K(n, \epsilon) \le \overline{\lim}_{n \to \infty} \frac{1}{n} \log |E(n, P, \delta)| \le H(P)$, since we can choose δ arbitrarily small.

On the other hand, let A be an arbitrary (n, ϵ)-compression. When we choose n sufficiently large, then almost all probability is concentrated on $E(n, P, \delta)$, since $\lim_{n \to \infty} P^n(E(n, P, \delta)) = 1$. So for large n
$P^n(A \cap E(n, P, \delta)) \ge \frac{1-\epsilon}{2}$. Hence

$$|A| \ge \mid A \cap E(n, P, \delta) \mid = \sum_{x^n \in A \cap E(n,P,\delta)} 1$$

$$\ge \sum_{x^n \in A \cap E(n,P,\delta)} P^n(x^n) \cdot \exp\{n(H(P) - \delta)\}$$

$$\ge \frac{1-\epsilon}{2} \exp\{n \cdot (H(P) - \delta)\}.$$

Taking logarithm and dividing by n on both sides yields

$$\frac{1}{n} \log |A| \ge H(P) - \delta + \frac{1}{n} \log\left(\frac{1 - \epsilon}{2}\right).$$

Since this inequality holds for all $\delta > 0$, we have

$$\lim_{n \to \infty} \frac{1}{n} \log |A| \ge H(P),$$

and the theorem is proved.   □

Remark Let us summarize our findings on the entropy-typical sequences. From the weak law of large numbers it follows that asymptotically all probability is concentrated on the entropy-typical sequences. Further observe that all the sequences $x^n \in E(n, P, \delta)$ essentially have the same probability $P^n(x^n) = \exp\{-n \cdot H(P)\}$. There may be sequences in $\mathcal{X}^n \setminus E(n, P, \delta)$ which individually are more probable than each of the entropy-typical sequences, but in total the probability of these sequences is negligibly small.

2.2.2 Typical Sequences

Definition 7 Let $x^n \in \mathcal{X}^n$ be a sequence of length n over $\mathcal{X}$. We denote by $(x^n|x)$ the frequency of the letter x in x^n. The $|\mathcal{X}|$-tuple $((x^n|x))_{x \in \mathcal{X}}$ is named **absolute empirical** of x^n. Division by n yields a probability distribution P_{x^n}, where $P_{x^n}(x) = \frac{1}{n}(x^n|x)$ for all $x \in \mathcal{X}$ is the relative frequency of the letter x in x^n. This probability distribution is denoted as **empirical distribution** (ED) of x^n.

ED's are known under several names (see remark at the end of the lecture).

Definition 8 The set of all ED's for a given block length n over the alphabet $\mathcal{X}$ is denoted as $\mathcal{P}(n, \mathcal{X})$.

Lemma 3 $|\mathcal{P}(n, \mathcal{X})| \leq (n + 1)^{|\mathcal{X}|}$.

Of course, each letter can occur at most n times in x^n. So in each position of the absolute empirical there must be an entry $(x^n|x) \in \{0, \dots, n\}$ and Lemma 3 follows easily. It can be verified that the exact size of $\mathcal{P}(n, \mathcal{X})$ is $\binom{n+|\mathcal{X}|-1}{|\mathcal{X}|-1}$. But we are only interested in the asymptotic behaviour and with Lemma 3 it is obvious that $|\mathcal{P}(n, \mathcal{X})|$ is polynomial in n.

Definition 9 Let $P \in \mathcal{P}(n, \mathcal{X})$ be an ED. The set of all sequences $x^n \in \mathcal{X}^n$ with ED $P_{x^n} = P$ is denoted as *empirical class* $T_P^n = \{x^n \in \mathcal{X}^n : P_{x^n} = P\}$. The sequences in T_P^n are called **typical sequences**.

Example Let $\mathcal{X} = \{0, 1\}$. The ED of a sequence x^n is determined by the number of 0's in x^n, since $(x^n|0) + (x^n|1) = n$. Let for example $n = 5$ and $x^n = 00101$. Then $(x^n|0) = 3$ and $(x^n|1) = 2 = 5 - 3$. The absolute empirical of x^n here is $(3, 2)$, the ED $\left(\frac{3}{5}, \frac{2}{5}\right)$.

For general $n \in \mathbb{N}$ we have

$$\mathcal{P}(n, \{0, 1\}) = \left\{(0, 1), (\frac{1}{n}, \frac{n-1}{n}), (\frac{2}{n}, \frac{n-2}{n}), \dots (\frac{n-1}{n}, \frac{1}{n}), (1, 0)\right\}.$$

So the empirical classes are just the levels in the poset $\mathcal{P}(\{1, \dots, n\})$ and the typical sequences to $P = (\frac{i}{n}, \frac{n-i}{n})$ are those sequences with exactly i0's and $(n - i)$ 1's.

Observe that the set of entropy-typical sequences $E(n, P, \delta)$ is the union of ED classes, since permutations of the x_i's in $x^n = (x_1, \ldots, x_n)$ do not change the probability $P^n(x^n) = \prod_{x \in \mathcal{X}} P(x)^{(x^n|x)}$.

The following four lemmas will later be used in the second proof of the Coding Theorem for the DMS. They involve two (not necessarily different) ED's P and Q in $P(n, \mathcal{X})$.

Lemma 4 $|T_Q^n| \leq \exp\{n \cdot H(Q)\}$

Proof For each typical sequence $x^n \in T_Q^n$ it is

$$Q^n(x^n) = \prod_{x \in \mathcal{X}} Q(x)^{(x^n|x)} = \prod_{x \in \mathcal{X}} Q(x)^{n \cdot Q(x)} = \exp\{-n \cdot H(Q)\}.$$

Now

$$1 = \sum_{x^n \in \mathcal{X}^n} Q^n(x^n) \geq \sum_{x^n \in T_Q^n} Q^n(x^n) = |T_Q^n| \cdot \exp\{-n \cdot H(Q)\}.$$

Multiplication with $\exp\{n \cdot H(Q)\}$ on both sides yields the desired result. $\qquad \square$

Lemma 5 $P^n(T_Q^n) \leq \exp\{-n \cdot D(Q||P)\}$, *where* $D(Q||P)$ *is the relative entropy (also called I-divergence or Kullback-Leibler information) defined in this chapter.*

Proof

$$P^n(T_Q^n) = \sum_{x^n \in T_Q^n} P^n(x^n) = \sum_{x^n \in T_Q^n} \prod_{x \in \mathcal{X}} P(x)^{n \cdot Q(x)}$$

$$= |T_Q^n| \cdot \prod_{x \in \mathcal{X}} P(x)^{n \cdot Q(x)}$$

$$\leq \exp\{n \cdot H(Q)\} \cdot \exp\{n \cdot \sum_{x \in \mathcal{X}} Q(x) \cdot \log P(x)\} \text{ (Lemma 4)}$$

$$= \exp\{n \cdot (H(Q) + \sum_{x \in \mathcal{X}} Q(x) \cdot \log P(x)\}$$

$$= \exp\{n \cdot \sum_{x \in \mathcal{X}} Q(x) \cdot (-\log Q(x) + \log P(x))\}$$

$$= \exp\{-n \cdot D(Q||P)\}. \qquad \square$$

Lemma 6 $P^n(T_P^n) \geq P^n(T_Q^n)$

Proof

$$P^n(T_Q^n) = |T_Q^n| \cdot \prod_{x \in \mathcal{X}} P(x)^{n \cdot Q(x)}$$

$$= \frac{n!}{\prod\limits_{x \in \mathcal{X}} (nQ(x))!} \prod_{x \in \mathcal{X}} P(x)^{n \cdot Q(x)},$$

since $|T_Q^n|$ is the multinomial coefficient

$$\binom{n}{nQ(1), \ldots, nQ(a)} = \frac{n!}{\prod\limits_{x \in \mathcal{X}} (nQ(x))!} \text{ for } \mathcal{X} = \{1, \ldots, a\}.$$

In order to prove Lemma 6 it is sufficient to show that $\frac{P^n(T_Q^n)}{P^n(T_P^n)} \leq 1$. So

$$\frac{P^n(T_Q^n)}{P^n(T_P^n)} = \frac{\prod\limits_{x \in \mathcal{X}} (nP(x))!}{\prod\limits_{x \in \mathcal{X}} (nQ(x))!} \cdot \prod_{x \in \mathcal{X}} P(x)^{n \cdot (Q(x) - P(x))}$$

$$= \prod_{x \in \mathcal{X}} \frac{(nP(x))!}{(nQ(x))!} \cdot P(x)^{n \cdot (Q(x) - P(x))}.$$

Now for all non-negative integers $r, s \in \mathbb{N}$ it holds $\frac{r!}{s!} \leq r^{r-s}$, since for $r > s \frac{r!}{s!} = r(r-1) \cdots \cdot (s+1) \leq r^{r-s}$ and for $r < s \frac{r!}{s!} = (s(s-1) \cdots (r+1))^{-1} \leq r^{r-s}$.
Hence with $r = n \cdot P(x)$ and $s = n \cdot Q(x)$ we have

$$\frac{P^n(T_Q^n)}{P^n(T_P^n)} \leq \prod_{x \in \mathcal{X}} (n \cdot P(x))^{n \cdot (P(x) - Q(x))} \cdot P(x)^{n(Q(x) - P(x))}$$

$$= \prod_{x \in \mathcal{X}} n^{n \cdot (P(x) - Q(x))} \cdot P(x)^0 = n^{n \cdot \sum\limits_{x \in \mathcal{X}} (P(x) - Q(x))}$$

$$= n^{n \cdot \left(\sum\limits_{x \in \mathcal{X}} P(x) - \sum\limits_{x \in \mathcal{X}} Q(x) \right)} = n^{n \cdot 0} = 1. \qquad \square$$

Observe that Lemma 6 is only valid for empirical classes but not for individual sequences. Let for example

$P = \left(\frac{2}{3}, \frac{1}{3}\right)$ and $Q = (1, 0)$ be two different ED's on $\mathcal{X} = \{0, 1\}$. Then a typical sequence $x^n \in T_P^n$ consists of $\frac{2}{3} \cdot n$ 1's and $\frac{1}{3} \cdot n$ 0's, hence $P^n(x^n) = \left(\frac{2}{3}\right)^{\frac{2}{3}n} \cdot \left(\frac{1}{3}\right)^{\frac{1}{3}n}$. The typical sequence $y^n \in T_Q^n$ is the all-one sequence, which has probability $P^n(y^n) = \left(\frac{2}{3}\right)^n > P^n(x^n)$.

Lemma 7 $(n + 1)^{-|\mathcal{X}|} \exp\{n \cdot H(Q)\} \leq |T_Q^n| \leq \exp\{n \cdot H(Q)\}$

Proof The second inequality is just the statement of Lemma 4. Hence it remains to be proved that $(n+1)^{-|\mathcal{X}|} \exp\{n \cdot H(Q)\} \leq |T_Q^n|$. Now $\exp\{-n \cdot H(Q)\}|T_Q^n| = Q^n(T_Q^n)$, since all typical sequences have the same probability.

But $Q^n(T_Q^n) \geq (n + 1)^{-|\mathcal{X}|}$, since by the Lemmas 3 and 6

$$(n + 1)^{|\mathcal{X}|} \cdot Q^n(T_Q^n) \geq \sum_{P \in \mathcal{P}(n, \mathcal{X})} Q^n(T_Q^n) \geq \sum_{P \in \mathcal{P}(n, \mathcal{X})} Q^n(T_P^n)$$

$$= \sum_{P \in \mathcal{P}(n, \mathcal{X})} \sum_{x^n \in T_P^n} Q^n(x^n) = \sum_{x^n \in \mathcal{X}^n} Q^n(x^n) = 1$$

and hence $|T_Q^n| \geq (n + 1)^{-|\mathcal{X}|} \exp\{n \cdot H(Q)\}$. $\qquad\qquad\square$

Summarizing the results of the preceding lemmas the entropy can be combinatorially interpreted as a measure for the cardinality of empirical classes. More exactly, with Lemma 7 $|T_Q^n|$ is asymptotically $\exp\{n \cdot H(Q)\}$, since $(n + 1)^{|\mathcal{X}|}$ is polynomial in n.

The relative entropy can be regarded as a measure for errors.

Up to now we only considered typical sequences for probability distributions arising from relative frequencies, which we call ED. For general probability distributions the following definition is more appropriate.

Definition 10 $x^n \in \mathcal{X}^n$ is (n, P, δ)-typical, if

$$| (x^n | x) - n \cdot P(x) | \leq \delta \cdot n \quad \text{for all } x \in \mathcal{X}.$$

So for $\delta = 0$ we again have the typical sequences. As pointed out above this notion can be applied to any probability distribution. Another advantage is that we are free to choose the parameter δ. For example δ might be a constant. When Chebyshev's inequality and the weak law of large numbers are involved usually $\delta = \frac{c}{\sqrt{n}}$ for a constant c is chosen. The set of (n, P, δ)-typical sequences[1] is denoted as $T_{P,\delta}^n$ (or $T_{X,\delta}^n$).

---

[1] As for entropy-typical sequences we use analogous abbreviations for typical sequences in terms of n, P, and δ.

2.2.3 Second Proof of the Source Coding Theorem

For this proof we will use Lemmas 4 to 7 from the preceding section. Further we need from Chap. 8 that for any two probability distributions P and Q

1. Pinsker's inequality: $D(P||Q) \geq \frac{1}{2\ell n2} ||P - Q||_1^2$,
2. $|H(P) - H(Q)| \leq -\alpha \cdot \log \frac{\alpha}{|\mathcal{X}|}$ for $||P - Q||_1 \leq \alpha \leq \frac{1}{2}$, where
   $||P - Q||_1 = \sum_{x \in \mathcal{X}} |P(x) - Q(x)|$.

From these two inequalities follows immediately

Corollary 1 *If* $|H(P) - H(Q)| \geq -\alpha \cdot \log \frac{\alpha}{|\mathcal{X}|}$, *then* $D(P||Q) \geq \frac{\alpha^2}{2\ell n2}$.

Now we have all the preliminaries for the second proof of the Source Coding Theorem. Again we will introduce a special (n, ϵ)-compression with the required properties and then show that we cannot do better.

First we show that for all

$$\beta \triangleq -\alpha \cdot \log \frac{\alpha}{|\mathcal{X}|}$$

$$\lim_{n \to \infty} P^n \left(\bigcup_{\substack{Q \in \mathcal{P}(n, \mathcal{X}) \\ |H(Q) - H(P)| < \beta}} T_Q^n \right) = 1,$$

since with Lemma 5

$$\sum_{Q \in \mathcal{P}(n, \mathcal{X}) : |H(Q) - H(P)| \geq \beta} P^n(T_Q^n)$$

$$\leq (n + 1)^{|\mathcal{X}|} \max_{Q : |H(Q) - H(P)| \geq \beta} \exp\{-n \cdot D(P||Q)\}$$

$$= (n + 1)^{|\mathcal{X}|} \exp\{-n \cdot \min_{Q : |H(Q) - H(P)| \geq \beta} D(Q||P)\}$$

$$\leq (n + 1)^{|\mathcal{X}|} \exp\{-n \frac{\alpha^2}{2\ell n2}\} \xrightarrow[n \to \infty]{} 0.$$

Hence asymptotically all the probability is concentrated on Q-typical sequences with $|H(Q) - H(P)| \leq \beta$. These sequences essentially have the same probability $\exp\{-n \cdot H(P)\}$. So with Lemma 4

$$\left| \bigcup_{Q \in \mathcal{P}(n, \mathcal{X}): |H(Q)-H(P)| \leq \beta} T_Q^n \right| \leq (n+1)^{|\mathcal{X}|} \exp\left\{n \cdot \max_{Q: |H(Q)-H(P)| \leq \beta} H(Q)\right\}$$

$$\leq (n+1)^{|\mathcal{X}|} \exp\{n \cdot (H(P) + \beta)\}.$$

Choosing ϵ appropriately this union of empirical classes is an (n, ϵ)-compression, hence

$$K(n, \epsilon) \leq (n+1)^{|\mathcal{X}|} \exp\{n \cdot (H(P) + \beta)\} \text{ for all } \beta > 0$$

and again we have

$$\overline{\lim_{n \to \infty}} \frac{1}{n} \log K(n, \epsilon) \leq H(P).$$

Conversely, suppose that we are given an (n, ϵ)-compression A_n with $P^n(A_n) \geq 1 - \epsilon$ and $|A_n| = K(n, \epsilon)$.

Now we define

$$A_n'(\beta) \triangleq A_n \cap \bigcup_{Q: |H(Q)-H(P)| \leq \beta} T_Q^n$$

Obviously $|A_n| \geq |A_n'(\beta)|$ and again for n sufficiently large $P^n(A_n'(\beta)) \geq \frac{1-\epsilon}{2}$. By the pigeon-hole principle there exists an ED Q with $|H(P) - H(Q)| \leq \beta$ and $P^n(A_n \cap T_Q^n) \geq \frac{1-\epsilon}{2} \cdot P^n(T_Q^n)$. Since all the sequences in T_Q^n are equiprobable, it is

$$|A_n \cap T_Q^n| \geq \frac{1-\epsilon}{2} \cdot |T_Q^n| \geq \frac{1}{(n+1)^{|\mathcal{X}|}} \exp\{n \cdot H(Q)\} \cdot \frac{1-\epsilon}{2},$$

so

$$|A_n| = K(n, \epsilon) \geq \frac{1}{(n+1)^{|\mathcal{X}|}} \exp\{n \cdot (H(P) - \beta)\} \cdot \frac{1-\epsilon}{2}.$$

Since this holds for all $\beta > 0$ and since $\frac{1}{n} \cdot \log\left(\frac{1-\epsilon}{2}\right)$ tends to 0 for $n \to \infty$

$$\lim_{n \to \infty} \frac{1}{n} \log K(n, \epsilon) = H(P). \qquad \square$$

2.2.4 A General Entropy-Setsize Relation

The Source Coding Theorem for the DMS states that for discrete memoryless sources asymptotically all probability is concentrated on a set of size $H(P^n)$. In this paragraph arbitrary probability distributions $P = (P(1), P(2), \dots)$ on the positive integers are considered. The question here is: How much do we have to "blow up" the entropy $H(P)$ in order to store a set with a given probability? It turns out that for all $d \geq 1$ and all probability distributions with prescribed entropy $H(P)$ it is possible to find a set $J \subset \mathbb{N}$ with $|J| = \lceil 2^{H(P) \cdot d} \rceil + 1$ and $P(J) \geq 1 - \frac{1}{d}$.

Definition 11 Let $\mathcal{P}(\mathbb{N})$ denote the set of all probability distributions on the positive integers. For any $P = (P(1), P(2), \dots) \in \mathcal{P}(\mathbb{N})$ and for all $d \geq 1$ we define

$$\kappa(d, P) \triangleq \max \left\{ \sum_{j \in J} P(j) : J \subset \mathbb{N}, |J| = \lceil 2^{H(P) \cdot d} \rceil + 1 \right\}$$

and

$$\kappa(d) \triangleq \min_{P \in \mathcal{P}(\mathbb{N})} \kappa(d, P).$$

Theorem 6 (Ahlswede 1990 included in [1]) *For all $d \geq 1$*

$$\kappa(d) = 1 - \frac{1}{d}.$$

Proof The proof consists of two parts. First we shall show that $\kappa(d) \geq 1 - \frac{1}{d}$ by constructing a new probability distribution P'' with $H(P'') \leq H(P)$ for which the statement holds. Central here is the Schur-concavity of the entropy. In a second part it is demonstrated that for a certain type of distributions it is not possible to obtain a better result.

In the following let $L \triangleq \lceil 2^{H(P) \cdot d} \rceil + 1$ and let $c \triangleq P(L)$, where the probabilities are ordered non-increasingly $P(1) \geq P(2) \geq P(3) \geq \cdots$.

Now a new probability distribution P' on $\mathbb{N}$ is defined by

$$P'(i) \triangleq \begin{cases} c, & \text{for } i = 2, \dots, L \\ a \triangleq \sum_{i=1}^{L} P(i) - (L-1) \cdot c, & \text{for } i = 1 \\ P(i), & \text{for } i > L + 1 \end{cases}$$

Of course, $P'(1) \geq P'(2) \geq P'(3) \geq \cdots$, and obviously $\sum_{i=1}^{k} P(i) \leq \sum_{i=1}^{k} P'(i)$ for every $k \in \mathbb{N}$. Hence the probability distribution $P = (P(1), P(2), \dots)$ is majorized by $P' = (P'(1), P'(2), \dots)$, so $P \preceq P'$. Recall that the entropy is Schur-concave, i.e. for $P \preceq P'$ it is $H(P') \leq H(P)$. By monotonicity we can further conclude that

$$\kappa(d, P') \leq \kappa(d, P), \text{ since} \lceil 2^{H(P') \cdot d} \rceil + 1 \leq \lceil 2^{H(P) \cdot d} \rceil + 1 = L \text{ and } \sum_{i=1}^{L} P(i) = \sum_{i=1}^{L} P'(i).$$

Furthermore, again the Schur-concavity is exploited to define for a suitable $T > L$ a probability distribution P'' by

$$P''(i) \triangleq \begin{cases} P'(i), & \text{for } i = 1, \ldots, L \\ c, & \text{for } L \leq i \leq T - 1 \\ b \leq c, & \text{for } i = T \\ 0, & \text{for } i > T. \end{cases}$$

With the same argumentation as above, $H(P'') \leq H(P')$ and $\kappa(d, P'') \leq \kappa(d, P')$.

Therefore it suffices to prove $\kappa(d, P) \geq 1 - \frac{1}{d}$ only for probability distributions of the form $P = (a, c, \ldots, c, b, 0, \ldots)$ with $a \geq c \geq b$.

Since $L = \lceil 2^{H(P) \cdot d} \rceil + 1 \geq 2$, obviously $\kappa(d, P) \geq a + c \geq a + b$. If $a + b \geq 1 - \frac{1}{d}$ we are done, and otherwise we have

$$d(1 - a - b) > 1.$$

Then $c = \frac{1-a-b}{T-2}$ and

$$H(P) = -a \cdot \log a - b \cdot \log b - (T - 2) \cdot \frac{1-a-b}{T-2} \cdot \log \left(\frac{1-a-b}{T-2} \right)$$

$$= -a \cdot \log a - b \cdot \log b - (1 - a - b) \cdot \log(1 - a - b)$$

$$+ (1 - a - b) \cdot \log(T - 2)$$

$$\geq (1 - a - b) \cdot \log(T - 2)$$

and hence

$$\lceil 2^{H(P) \cdot d} \rceil + 1 \geq \lceil 2^{d \cdot (1-a-b) \cdot \log(T-2)} \rceil + 1$$

$$\geq T - 1 + 1 = T.$$

This implies $\kappa(d, P) = 1 \geq 1 - \frac{1}{d}$ in this case.

Conversely, fix d and consider only distributions $Q = (a, c, c, \ldots, c, 0, \ldots)$, that is, with $b = c$ in the notation above and also with $a > 1 - \frac{1}{d}$.

Observe that now $c = \frac{1-a}{T-1}$. Then

$$\kappa(d, Q) \leq a + \lceil 2^{H(Q) \cdot d} \rceil \cdot c$$

$$= a + \lceil 2^{[-a \cdot \log a - (T-1) \cdot c \cdot \log c] \cdot d} \rceil \cdot c$$

$$\leq a + 2^{[-a \cdot \log a - (1-a) \cdot \log \frac{1-a}{T-1}] \cdot d} \cdot \frac{1-a}{T-1} + \frac{1-a}{T-1}$$

$$= a + 2^{-a \cdot \log a - (1-a) \cdot \log(1-a)} \cdot (T - 1)^{d(1-a)-1} \cdot (1 - a) + \frac{1-a}{T-1},$$

which tends to a for $T \to \infty$, since $a > 1 - \frac{1}{d}$ and hence $d(1 - a) - 1 < 0$.

So $\kappa(d) \leq \inf_Q \kappa(d, Q)$ and since this is true for all $a > 1 - \frac{1}{d}$, the result follows. $\square$

Remark Empirical distributions play an important role in Statistics and so do random variables. We find it therefore appealing to introduce in addition to the abbreviation RV for a random variable the abbreviation ED for an empirical distribution. This symmetry in notation is also convenient because for entropy there are the two notions $H(X)$ and $H(P)$ in use depending on properties of entropy considered. The situation is similar for mutual information (defined in the next lecture) and other information measures. Also for sets $T^n_{X,\delta}$ and $T^n_{P,\delta}$, which are equal if $P = P_X$. For ED's there are other names in use. In combinatorial (or algebraic) Coding Theory mostly for $((x^n|x))_{x \in \mathcal{X}}$ the name composition is used, a name, also occurring in probabilistic Coding Theory, for instance in [3]. Very popular is the name type introduced in the book [4]. An obvious drawback is that this word is used for many, many classifications of objects and structures in science. In particular in Game Theory it has been introduced by Harsanyi [6] a long time ago for other properties of distributions (see Handbook of Game Theory, pp. 669–707). We therefore avoid it. One of the oldest, perhaps the oldest name used, is Komplexion, due to Boltzmann.

2.3   Lecture on Rate Distortion Theory

We are given a DMS $(X_t)_{t=1}^{\infty}$ with the X_t's taking values in a finite alphabet $\mathcal{X}$. A word $x^n \in \mathcal{X}^n$ will be encoded using an encoding function $f : \mathcal{X}^n \to \mathbb{N}$, e.g., it might be prepared for a storage medium. The word will be reproduced over an alphabet $\hat{\mathcal{X}}$ using a decoding function $g : \mathbb{N} \to \hat{\mathcal{X}}^n$. We call the random variable $\hat{X}^n \triangleq g(f(X^n))$ the *reproduction* (or *estimate*) of X^n. A function $d : \mathcal{X} \times \hat{\mathcal{X}} \to \mathbb{R}_+$ will serve as a measure for distortion of a single letter. The distortion measure for words of length n is of sum-type, i.e., it is just the sum of the single letters' distortions $d(x^n, \hat{x}^n) = \sum_{t=1}^{n} d(x_t, \hat{x}_t)$. Hence, the following definition makes sense.

Definition 12 The estimate $\hat{X}^n$ has **fidelity** D, if the average expected distortion $\sum_{t=1}^{n} \mathbb{E}d(X_t, \hat{X}_t) \leq D \cdot n$.

Shannon gave some examples. Think for example of $\mathcal{X}$ as the latin alphabet $\{a, b, c, \ldots, z\}$. Then the reproduction alphabet might be as follows:

$$\hat{\mathcal{X}} = \{a, b, c, \ldots, z\} \cup \{\overset{?}{a}, \overset{?}{b}, \overset{?}{c}, \ldots \overset{?}{z}\} \cup \{?\}, \text{ where "}\overset{?}{a}\text{" means "probably } a\text{", etc.,}$$

and "?" means that the letter cannot be reproduced.

A possible distortion measure is the *Hamming distance* (in the case $\mathcal{X} = \hat{\mathcal{X}}$). Other important measures for distortion on $\mathcal{X} = \hat{\mathcal{X}} = \{1, \ldots, a\}$ are the *Lee distance* $d_L : \mathcal{X} \times \hat{\mathcal{X}} \to \mathbb{R}_+$, with $d_L(x, \hat{x}) = \max\{|x - \hat{x}|, a - |x - \hat{x}|\}$, and the *Taxi metric* $d_T : \mathcal{X} \times \hat{\mathcal{X}} \to \mathbb{R}_+$, defined by $d_T(x, \hat{x}) = |x - \hat{x}|$. These two measures are more refined than the Hamming distance, since the distortion is not as large for letters "close" to x (observe that all the above functions are metrics) as for letters at a large distance from x.

Let $||f||$ denote the cardinality of the range of f. Then the *rate* of f is defined as $\frac{\log ||f||}{n}$. It might be interpreted as the storage space used per letter.

Definition 13 The **rate distortion function** $R(D)$ is defined as

$$R(D) \triangleq \min\{R : \exists (f, g) \text{ with rate}(f) \leq R \text{ and fidelity } D\}.$$

Observe that it suffices to find encoding and decoding function (f_n, g_n) for some $n \in \mathbb{N}$ in the definition, since we then can achieve rate R and fidelity D by encoding in blocks of length n using f_n as encoding function.

The quantity $I(X \wedge Y | Z) = H(X|Z) - H(X|Y, Z)$, called *conditional mutual information*, has found a coding theoretic interpretation for two-way channels, and multiple-access channels, and is an important quantity for many other problems in multi-user communication. Furthermore, it is used in the Rate Distortion Theorem. Some of the inequalities used in the proof of the following theorem are proven in the appendix of this book.

It turns out that the rate distortion function can also be expressed in terms of entropy, namely

Theorem 7 (Rate Distortion Theorem, Shannon, 1948 [11]) *For every $D \geq 0$ and all distortion measures $d : \mathcal{X} \times \hat{\mathcal{X}} \to \mathbb{R}_+$*

$$R(D) = A(D) \triangleq \min_{W : \mathcal{X} \to \hat{\mathcal{X}}, \mathbb{E}d(X, \hat{X}) \leq D} I(X \wedge \hat{X}),$$

where X is the generic random variable of the source and $\hat{X}$ is the output variable of the channel W.

2.3.1 Proof of the Converse

We shall first prove the "converse": $R(D) \geq A(D)$. Therefore let f_n, g_n and $\hat{\mathcal{X}}^n$ be given with $\mathbb{E}d(X^n, \hat{X}^n) = D \cdot n$. Now

$$n \text{ rate } (f_n) = \log ||f_n|| \geq \log ||g_n(f_n)|| \geq \log(\hat{X}^n) \geq H(\hat{X}^n)$$
$$\geq I(X^n \wedge \hat{X}^n) \left(= H(\hat{X}^n) - H(\hat{X}^n | X^n)\right)$$
$$\geq \sum_{t=1}^{n} I(X_t \wedge \hat{X}_t)$$

by Dobrushin's inequality. So we know that

$$R(D) \geq \frac{1}{n} \sum_{t=1}^{n} I(X_t \wedge \hat{X}_t) \geq \frac{1}{n} \sum_{t=1}^{n} A(D_t),$$

where $D_t \triangleq \mathbb{E}d(X_t, \hat{X}_t)$ and hence $nD = \sum_{t=1}^{n} D_t$.

Now assume that X_t and $\hat{X}_t$ are connected by the channel W_t for every $t = 1, \ldots, n$. Hence the distribution of $\hat{X}_t$ is $P_t \cdot W_t$, when P_t is the distribution of X_t.

Since all the X_t's have the same distribution, say P, they can be described by the generic random variable X and

$$\frac{1}{n} \sum_{t=1}^{n} I(X_t \wedge \hat{X}_t) = \frac{1}{n} \sum_{t=1}^{n} I(X \wedge \hat{X}_t) \geq I(X \wedge \hat{X}),$$

where $\hat{X}$ has the distribution $P \cdot \left(\frac{1}{n} \sum_{t=1}^{n} W_t\right)$, by convexity of the mutual information.

Hence $A(D)$ is also convex, i.e. $\frac{1}{n} \sum_{t=1}^{n} A(D_t) \geq A(D)$, and we are done, since fidelity D is assumed by the channel $W \triangleq \frac{1}{n} \sum_{t=1}^{n} W_t$.

2.3.2 Proof of the Direct Part

In order to prove the "direct part", $R(D) \leq A(D)$, observe that there is a canonical partition on $\mathcal{X}^n$ defined by the sets $D_u = \{x^n \in \mathcal{X}^n : g(f(x^n)) = u\}$ for all $u \in \hat{\mathcal{X}}^n$, which are estimates for some $x^n \in \mathcal{X}^n$. So $\mathcal{X}^n$ is partitioned into sets $D_u, u \in \hat{\mathcal{X}}^n$, the elements of which are all those words $x^n \in \mathcal{X}^n$ that are reproduced as u.

The idea now is to find a covering of $\mathcal{X}^n$ consisting of sequences generated by typical sequences of $\hat{\mathcal{X}}^n$. From this covering we obtain a partition by dividing the intersection of the sets of the covering in some way, such that those sets are finally disjoint.

Observe that a code corresponds to a maximal packing, whereas we are now searching for a minimal covering. It turns out that not only a code but also a covering is achievable by Shannon's method, which makes use of typical sequences $u \in T_{\hat{X}}^n$ over one alphabet (here $\hat{X}$, in Channel Coding Theory typical sequences over X are encoded) and the sets generated by them over the other alphabet. We shall now reformulate the problem in terms of hypergraphs. We are given the hypergraph

$$\mathcal{H} \triangleq \left(T_{X,\delta}^n, (T_{X|\hat{X},\hat{\delta}}^n, (u))_{u \in T_{\hat{X}}^n}\right), \hat{\delta} = \frac{1}{\sqrt{n}}, \delta = \hat{\delta}|\mathcal{X}|,$$

in which we are looking for a minimal covering, i.e., a subset $\mathcal{C}$ of the hyper-edges, such that all vertices are contained in one hyper-edge $E \in \mathcal{C}$.

First we shall demonstrate that this method achieves fidelity D. More exactly, we show that for a typical sequence $u \in T_{\hat{X}}^n$ and a word $x^n \in T_{X|\hat{X},\hat{\delta}}^n(u)$, the difference

$|\frac{1}{n}d(x^n, u) - \mathbb{E}d(X, \hat{X})|$ becomes arbitrarily small.

$| d(x^n, u) - n\,\mathbb{E}d(X, \hat{X}) |$

$$= | d(x^n, u) - n \sum_{x \in \mathcal{X}} \sum_{\hat{x} \in \hat{\mathcal{X}}} P_{X\hat{X}}(x, \hat{x})d(x, \hat{x}) |$$

$$= | \sum_{x \in \mathcal{X}} \sum_{\hat{x} \in \hat{\mathcal{X}}} (x^n, u|x, \hat{x}) \cdot d(x, \hat{x}) - n \sum_{x \in \mathcal{X}} \sum_{\hat{x} \in \hat{\mathcal{X}}} \hat{P}(\hat{x})w(x|\hat{x}) \cdot d(x, \hat{x}) |$$

$$\leq \sum_{x \in \mathcal{X}} \sum_{\hat{x} \in \hat{\mathcal{X}}} d(x, \hat{x}) \cdot | (x^n, u|x, \hat{x}) - n \cdot \hat{P}(\hat{x})w(x|\hat{x}) |$$

$$= \sum_{x \in \mathcal{X}} \sum_{\hat{x} \in \hat{\mathcal{X}}} d(x, \hat{x}) \cdot | (x^n, u|x, \hat{x}) - (u|\hat{x})w(x, \hat{x}) |$$

$$\leq n\hat{\delta} \cdot d_{\max} \cdot |\mathcal{X}| \cdot |\hat{\mathcal{X}}|,$$

where $d_{\max} \triangleq \max\limits_{x, \hat{x}} d(x, \hat{x})$, since by definition of the generated sequences $| (x^n, u|x, \hat{x}) - (u|\hat{x})w(x, \hat{x}) | \leq n\hat{\delta}$. So by an appropriate choice of $\hat{\delta}$, e.g. $\hat{\delta} \triangleq \frac{1}{\sqrt{n}}$, it can be assured that $| \frac{1}{n}d(x^n, u) - \mathbb{E}d(x, \hat{x}) |$ becomes arbitrarily small for large n.

Hence, using this method, there are no problems with the fidelity and we can concentrate on the rate. The question is: How many typical sequences $u \in T_{\hat{X}}^n$ are needed to cover the elements of $T_{X, \delta}^n$ by generated sequences from $T_{X|\hat{X}, \hat{\delta}}^n(u)$ for some u?

Or in terms of hypergraphs: What is the minimum size of a covering for the hypergraph $\big(T_{X, \delta}, (T_{X|\hat{X}, \hat{\delta}}^n(u))_{u \in T_{\hat{X}}^n}\big)$?

Central for the answer is the following Covering Lemma.

Lemma 8 (Covering Lemma) *For a hypergraph $(\mathcal{V}, \mathcal{E})$ with $\min\limits_{v \in \mathcal{V}} \deg(v) \geq \sigma$ there exists a covering $\mathcal{C} \subset \mathcal{E}$ with $|\mathcal{C}| \leq |\mathcal{E}|\sigma^{-1} \log |\mathcal{V}| + 1$.*

Proof We select at random hyper-edges $E^{(1)}, \ldots, E^{(k)}$ according to the uniform distribution. The probability that some point $v \in \mathcal{V}$ is not contained in one of these edges is

$$\text{Prob}\big(v \notin \bigcup_{i=1}^{k} E^{(i)}\big) = \text{Prob}(v \notin E^{(1)}) \cdot \text{Prob}(v \notin E^{(2)}) \cdot \cdots \cdot \text{Prob}(v \notin E^{(k)})$$

$$= \left(\frac{|\mathcal{E}| - \deg(v)}{|\mathcal{E}|}\right)^k,$$

since $|\mathcal{E}|$ is the total number of edges, $|\mathcal{E}| - \deg(v)$ is the number of edges not containing v, and hence $\frac{|\mathcal{E}| - \deg(v)}{|\mathcal{E}|}$ is the probability that v is not contained in $E^{(i)}$ for every $i = 1, \ldots, k$. Since $\min_{v \in \mathcal{V}} \deg(v) \geq \sigma$ we can further estimate

$$\mathrm{Prob}\left(v \notin \bigcup_{i=1}^{k} E^{(i)}\right) \leq \left(\frac{|\mathcal{E}| - \sigma}{|\mathcal{E}|}\right)^{k} = \left(1 - \frac{\sigma}{|\mathcal{E}|}\right)^{k}.$$

If $E^{(1)}, \ldots, E^{(k)}$ is not a covering of $\mathcal{V}$, then there exists a point $v \in \mathcal{V}$ not contained in one of these edges. The probability for this event is upper bounded by

$$\mathrm{Prob}\left(\exists\, v \in \mathcal{V} : v \notin \bigcup_{i=1}^{k} E^{(i)}\right) \leq \left(1 - \frac{\sigma}{|\mathcal{E}|}\right)^{k} \cdot |\mathcal{V}|.$$

Now observe that if $\left(1 - \frac{\sigma}{|\mathcal{E}|}\right)^{k} \cdot |\mathcal{V}| < 1$, then the probability for the existence of a covering is positive, and we can conclude that in this case a covering really exists. Taking the logarithm on both sides we obtain the required condition for the size of the covering $k \cdot \log\left(1 - \frac{\sigma}{|\mathcal{E}|}\right) + \log|\mathcal{V}| < 0$, or equivalently $\log|\mathcal{V}| < -k \cdot \log\left(1 - \frac{\sigma}{|\mathcal{E}|}\right)$. Because of the well-known identity $x - 1 \geq \log x$ (for all $x \in \mathbb{R}_+$) we obtain

$$\log|\mathcal{V}| \leq k \frac{\sigma}{|\mathcal{E}|}.$$

This finally yields the condition

$$|\mathcal{C}| \geq |\mathcal{E}| \cdot \sigma^{-1} \cdot \log|\mathcal{V}|$$

for the existence of a covering. This condition is already fulfilled for the smallest integer $|\mathcal{C}| \leq |\mathcal{E}| \cdot \sigma^{-1} \cdot \log|\mathcal{V}| + 1$. $\qquad\square$

The Covering Lemma can now be applied to prove the direct part of the rate distortion theorem. We shall again consider the hypergraph $\mathcal{H} = (\mathcal{V}, \mathcal{E}) = \left(T_{X,\delta}^{n}, \left(T_{X|\hat{X},\hat{\delta}}^{n}(u)\right)_{u \in T_{\hat{X}}^{n}}\right)$.

By Lemmas 49 and 50 (lecture on list reduction) we have the following estimations:

$$|\mathcal{V}| \leq \exp\{H(X) \cdot n + g_1(\delta)n\}$$
$$|\mathcal{E}| \leq \exp\{H(\hat{X}) \cdot n\}$$
$$\deg(v) \geq \exp\{H(\hat{X}|X) \cdot n - g_2(\hat{\delta}, \delta)n\} \text{ for all } v \in \mathcal{V}.$$

Here $g_1(\delta)$ and $g_2(\delta, \hat{\delta})$ tend to 0 as $n \to \infty$.

Application of the Covering Lemma yields

$$|\mathcal{C}| \leq |\mathcal{E}| \cdot \sigma^{-1} \cdot \log |\mathcal{V}| + 1$$
$$\leq \exp\{H(\hat{X}) \cdot n - H(\hat{X}|X)n + g_2(\delta, \hat{\delta}) \cdot n\} \cdot \left(H(X) \cdot n + g_1(\delta)n\right) + 1$$
$$= \exp\{I(X \wedge \hat{X}) \cdot n + g_2(\delta, \hat{\delta}) \cdot n\} \cdot \left(H(X) \cdot n + g_1(\delta) \cdot n\right) + 1$$

and hence $\lim_{n\to\infty} \frac{1}{n} \log |\mathcal{C}| \leq I(X \wedge \hat{X})$ for every channel $W : \mathcal{X} \to \hat{\mathcal{X}}$ from which $R(D) \leq A(D)$ follows. $\square$

References

1. R. Ahlswede, B. Verboven, On identification via multi-way channels with feedback. IEEE Trans. Inf. Theory **37**(5), 1519–1526 (1991)
2. R.B. Ash, *Information Theory* (Interscience Publishers, New York, 1965)
3. R.E. Blahut, *Theory and Practice of Error Control Codes* (Addison-Wesley Publishing Company, Advanced Book Program, Reading, MA, 1983)
4. I. Csiszár, J. Körner, *Coding Theorems for Discrete Memoryless Systems, Probability and Mathematical Statistics* (Academic Press, New York, 1981)
5. R.M. Fano, *The Transmission of Information, Technical Report No. 65, Research Laboratory of Electronics, M.I.T.* (Cambridge, Mass, 1949)
6. J.C. Harsanyi, Game and decision theoretic models in ethics. in *Handbook of Game Theory with Economic Applications*, vol. 1, 1st edn, ed. by R.J. Aumann, S. Hart (1992), pp. 669–707
7. D.A. Huffman, A method for the construction of minimum-redundancy codes. Proc. Inst. Radio Eng. **40**(9), 1098–1101 (1952)
8. L.G. Kraft, A device for quantizing, grouping, and coding amplitude modulated pulses. *MS Thesis Electrical Engineering Department* (Institute of Technology, Cambridge, 1949)
9. B. McMillan, Two inequalities implied by unique decipherability. IEEE Trans. Inf. Theor. **2**(4), 115–116 (1956)
10. F.M. Reza, *An Introduction to Information Theory* (McGraw-Hill, New York, 1961)
11. C.E. Shannon, A mathematical theory of communication. Bell Sys. Tech. Jour. **27**, 398–403 (1948)
12. D. Welsh, *Codes and Cryptography* (Oxford Science Publications, The Clarendon Press, Oxford University Press, New York, 1988)

Chapter 3
The Entropy as a Measure of Uncertainty

3.1 Lecture on the Entropy Function for Finite-valued Random Variables

3.1.1 Axiomatic Characterization of the Entropy Function

Up to now we had an *operational* access to the entropy function, i.e., the entropy was involved into the solution of a mathematical problem. More specifically, the entropy turned out to be a measure for data compression. In this lecture we will take a different approach and interpret the entropy as a measure of uncertainty of an experiment with n possible outcomes, where each outcome will take place with a certain probability. The approach will be *axiomatic*, i.e., some "reasonable" conditions which a measure of uncertainty should possess are postulated. These conditions yield a set of functional equalities from which it can be derived that a measure of uncertainty is necessarily of the form $K \cdot H(P)$, where K is a constant factor. Several systems of axioms have been formulated. We will present the axioms of Shannon [1], Khinchin [2] and Faddeev [3] (see also the book [4]). For technical reasons in this paragraph the notation p_i instead of $P(i)$ is used.

It is understood throughout that $H(p_1, \ldots, p_n)$ is defined only for a complete set of probabilities, i.e., a set of non-negative numbers whose sum equals one.

Theorem 8 (Shannon 1948 [1]) *Let $H(p_1, \ldots, p_n)$ be a function with the following properties*

(i) *H is continuous in p_i, $i = 1, \ldots, n$.*
(ii) *If $p_i = \frac{1}{n}$ for $i = 1, \ldots, n$, then H is monotonically increasing in n.*
(iii) *(Grouping axiom) For $P = (p_1, \ldots, p_n)$ let $Q = (q_1, \ldots, q_s)$ be defined by $q_j \triangleq \sum_{i=n_{j-1}+1}^{n_j} p_i$ for $j = 1, \ldots, s$ and $0 = n_0 < n_1 < \cdots < n_s = n$. Then*

$$H(P) = H(Q) + \sum_{j=1}^{s} q_j \, H\left(\frac{p_{n_{j-1}+1}}{q_j}, \ldots, \frac{p_{n_j}}{q_j}\right)$$

A. Ahlswede et al. (eds.), *Storing and Transmitting Data*,
Foundations in Signal Processing, Communications and Networking 10,
DOI: 10.1007/978-3-319-05479-7_3, © Springer International Publishing Switzerland 2014

Then $H(p_1, \ldots, p_n)$ is necessarily of the form $-K \cdot \sum_{i=1}^{n} p_i \cdot \log p_i$, where K is a constant factor.

Before proving the theorem let us briefly discuss the three axioms for a measure of uncertainty. The first two properties seem to be quite reasonable. Small changes in the probabilities should only have little influence on the uncertainty and the uncertainty about an experiment with n possible outcomes should be larger than the uncertainty about an experiment with $m < n$ outcomes. In the extreme case $n = 1$, for example, the result of the experiment is known before, and hence there is no uncertainty. The third axiom, however, is not evident; several other postulates have been proposed here. It may be illustrated by the following example. Assume we are searching for a person in a hotel. Let $p_{n_{j-1}+i}, i \leq n_j - n_{j-1}$ denote the probability that the person can be found in room No. i on floor No. j. Then q_j is the probability that the person is on the jth floor. $H\left(\frac{p_{n_{j-1}+1}}{q_j}, \ldots, \frac{p_{n_j}}{q_j}\right)$ is the remaining uncertainty, when we already know that the person can be found on the jth floor.

Proof of the theorem Define $A(n) \triangleq H(\frac{1}{n}, \ldots, \frac{1}{n})$ and let $n = s^m$ for some s, $m \in \mathbb{N}$. With $Q = (\frac{1}{s}, \ldots, \frac{1}{s})$, where $\frac{1}{s} = q_j = \sum_{(j-1)s^{m-1}+1}^{js^{m-1}} \frac{1}{s^m}$ for $j = 1, \ldots, s$, the grouping axiom can be applied to obtain

$$A(s^m) = A(s) + s \cdot \frac{1}{s} \cdot A(s^{m-1}) = A(s) + A(s^{m-1}).$$

Iterating this procedure yields

$$A(s^m) = m \cdot A(s).$$

Analogously, we obtain for some $t, \ell \in \mathbb{N}$,

$$A(t^\ell) = \ell \cdot A(t).$$

For all $\ell \in \mathbb{N}$ there exists a number $m \in \mathbb{N}$, such that $s^m \leq t^\ell < s^{m+1}$. From this we can conclude that $\frac{m \cdot \log s}{\ell \cdot \log s} \leq \frac{\ell \cdot \log t}{\ell \cdot \log s} < \frac{(m+1) \cdot \log s}{\ell \cdot \log s}$, and hence

$$\frac{m}{\ell} \leq \frac{\log t}{\log s} < \frac{m}{\ell} + \frac{1}{\ell}.$$

From the monotonicity (Axiom (ii)) follows that $A(s^m) \leq A(t^\ell) \leq A(s^{m+1})$ and further $m \cdot A(s) \leq \ell A(t) \leq (m+1) \cdot A(s)$, and with the same computation as above

$$\frac{m}{\ell} \leq \frac{A(t)}{A(s)} \leq \frac{m}{\ell} + \frac{1}{\ell}.$$

Combining the two results yields

$$\left| \frac{A(t)}{A(s)} - \frac{\log t}{\log s} \right| \le \frac{1}{\ell}.$$

Fixing s and letting ℓ tend to infinity, finally we obtain

$$A(t) = \frac{A(s)}{\log s} \cdot \log t = K \cdot \log t$$

where $K \triangleq \frac{A(s)}{\log s}$ is constant, since s is fixed.

So the theorem is proved for the uniform distribution. For distributions $P = (p_1, \ldots, p_n)$, where the p_i's are rational numbers, it can be assumed w.l.o.g. that all the p_i's have the same divisor, hence $p_i = \frac{n_i}{\sum_{i=1}^n n_i}$ for $i = 1, \ldots, n$. Now the grouping axiom is applied to P and the uniform distribution $\left(\frac{1}{\sum_{i=1}^n n_i}, \ldots, \frac{1}{\sum_{i=1}^n n_i} \right)$ to obtain $K \cdot \log \left(\sum_{i=1}^n n_i \right) = H(p_1, \ldots, p_n) + \sum_{i=1}^n p_i \cdot K \log n_i$. So

$$H(p_1, \ldots, p_n) = K \left(\sum_{i=1}^n p_i \cdot \log \left(\sum_{j=1}^n n_j \right) - \sum_{i=1}^n p_i \cdot \log n_i \right)$$

$$= K \left(- \sum_{i=1}^n p_i \cdot \log \frac{n_i}{\sum_{j=1}^n n_j} \right)$$

$$= -K \sum_{i=1}^n p_i \cdot \log p_i.$$

Finally, since H is required to be continuous (Axiom (i)),

$H(p_1, \ldots, p_n) = -K \cdot \sum_{i=1}^n p_i \cdot \log p_i$ also holds, if the p_i's are real numbers.

So it is demonstrated that a function $H(p_1, \ldots, p_n)$, for which the three axioms of the theorem hold, must be of the form $H(p_1, \ldots, p_n) = -K \sum_{i=1}^n p_i \cdot \log p_i$. On the other hand, it is easily verified that $-K \sum_{i=1}^n p_i \cdot \log p_i$ is continuous in $p_i, i = 1, \ldots, n$ and monotonely increasing in n for $P = (\frac{1}{n}, \ldots, \frac{1}{n})$. The grouping axiom holds, since for $P = (p_1, \ldots, p_n)$ and $Q = (q_1, \ldots, q_s)$, with $q_j = \sum_{n_{j-1}+1}^{n_j} p_i, j = 1, \ldots, s, 0 = n_0 < n_1 < \cdots < n_s = n$,

$$H(Q) + \sum_{j=1}^s q_j H \left(\frac{p_{n_{j-1}+1}}{q_j}, \ldots, \frac{p_{n_j}}{q_j} \right)$$

$$= -K \cdot \left(\sum_{j=1}^s q_j \cdot \log q_j + \sum_{j=1}^s q_j \sum_{i=n_{j-1}+1}^{n_j} \frac{p_i}{q_j} \cdot \log \frac{p_i}{q_j} \right)$$

$$= -K \cdot \left(\sum_{j=1}^{s} \left(q_j \cdot \log q_j + \sum_{i=n_{j-1}+1}^{n_j} p_i \cdot \log p_i - \log q_j \cdot \sum_{i=n_{j-1}+1}^{n_j} p_i \right) \right)$$

$$= -K \cdot \left(\sum_{i=1}^{n} p_i \cdot \log p_i \right) = H(P).$$ $\square$

In the above theorem Shannon's axiomatic characterization of the entropy as a measure of uncertainty is presented. Khinchin [2] proposed the following set of axioms, which also determine the entropy up to a constant factor:

(i') $H(p, 1 - p)$ is continuous in p.
(ii') $H(p_1, \ldots, p_n)$ is symmetric.
(iii') $H(p_1, \ldots, p_n) \le H(\frac{1}{n}, \ldots, \frac{1}{n})$.
(iv') $H(p_1, \ldots, p_n, 0) = H(p_1, \ldots, p_n)$.
(v') (Grouping axiom)

$$H(q_{11}, \ldots, q_{1m_1}, q_{21}, \ldots, q_{2m_2}, \ldots, q_{n1}, \ldots, q_{nm_n})$$

$$= H(p_1, \ldots, p_n) + \sum_{i=1}^{n} p_i \cdot H\left(\frac{q_{i1}}{p_i}, \ldots, \frac{q_{im_i}}{p_i} \right),$$

where

$$p_i = q_{i1} + \cdots + q_{im_i}.$$

Essential here is the replacement of the monotonicity by the weaker property "symmetry". Khinchin's axioms were later refined by Faddeev.

Theorem 9 (Faddeev 1956 [3]) *Let $H(p_1, \ldots, p_n)$ be a function with the following properties*

(i'') $H(p, 1 - p)$ *is continuous for $0 \le p \le 1$.*
(ii'') $H(p_1, \ldots, p_n)$ *is symmetric for all $n \in \mathbb{N}$.*
(iii'') *If $p_n = q_1 + q_2 > 0$, then*

$$H(p_1, \ldots, p_{n-1}, q_1, q_2) = H(p_1, \ldots, p_n) + p_n \cdot H\left(\frac{q_1}{p_n}, \frac{q_2}{p_n} \right).$$

Then $H(p_1, \ldots, p_n) = -K \cdot \sum_{i=1}^{n} p_i \cdot \log p_i$ with a constant factor $K \in \mathbb{R}$.

In order to prove the theorem four lemmas are required.

Lemma 9 $H(1, 0) = 0$

Proof From (iii'') it follows $H(\frac{1}{2}, \frac{1}{2}, 0) = H(\frac{1}{2}, \frac{1}{2}) + \frac{1}{2}H(1, 0)$. Further (ii'') and (iii'') yield $H(\frac{1}{2}, \frac{1}{2}, 0) = H(0, \frac{1}{2}, \frac{1}{2}) = H(0, 1) + H(\frac{1}{2}, \frac{1}{2})$. So $\frac{1}{2}H(1, 0) = H(1, 0)$ and, of course, $H(1, 0) = 0$. $\square$

Lemma 10 $H(p_1, \ldots, p_n, 0) = H(p_1, \ldots, p_n)$

Proof Because of the symmetry we can assume w.l.o.g. that $p_n > 0$. $H(p_1, \ldots, p_n, 0) = H(p_1, \ldots, p_n) + p_n \cdot H(1, 0) = H(p_1, \ldots, p_n)$ with (iii") and Lemma 9. $\qquad\square$

Lemma 11

$$H(p_1, \ldots, p_{n-1}, q_1, \ldots, q_m) = H(p_1, \ldots, p_n) + p_n \cdot H\left(\frac{q_1}{p_n}, \ldots, \frac{q_m}{p_n}\right),$$

where $p_n = q_1 + \cdots + q_m > 0$.

Proof Induction on m:

For $m = 2$ the statement of Lemma 11 is just property (iii"). From Lemma 10 it is clear that we only have to consider the case when $q_i > 0, i = 1, \ldots, m$. Suppose that there is an m such that the assertion is true for all n. Then using condition (iii), we have

$$H(p_1, \ldots, p_{n-1}, q_1, \ldots, q_{m+1})$$

$$= H(p_1, \ldots, p_{n-1}, q_1, p') + p' \cdot H\left(\frac{q_2}{p'}, \ldots, \frac{q_{m+1}}{p'}\right)$$

$$= H(p_1, \ldots, p_n) + p_n \cdot H\left(\frac{q_1}{p_n}, \frac{p'}{p_n}\right) + p' \cdot H\left(\frac{q_2}{p'}, \ldots, \frac{q_{m+1}}{p'}\right)$$

where $p' = q_2 + \cdots + q_{m+1}$. Further,

$$H\left(\frac{q_1}{p_n}, \ldots, \frac{q_{m+1}}{p_n}\right) = H\left(\frac{q_1}{p_n}, \frac{p'}{p_n}\right) + \frac{p'}{p_n} \cdot H\left(\frac{q_2}{p'}, \ldots, \frac{q_{m+1}}{p'}\right).$$

Substituting this into the preceding equation, we obtain the assertion of the lemma for $m + 1$. $\qquad\square$

Lemma 12 $H(q_{11}, \ldots, q_{1m_1}, q_{21}, \ldots, q_{2m_2}, \ldots, q_{n1}, \ldots, q_{nm_n}) = H(p_1, \ldots, p_n) + \sum_{i=1}^n p_i \cdot H\left(\frac{q_{i1}}{p_i}, \ldots, \frac{q_{im_i}}{p_i}\right)$, *where $p_i = q_{i1} + \cdots + q_{im_i}$ for $i = 1, \ldots, n$.*

Proof By Lemma 11

$$H(q_{11}, \ldots, q_{1m_1}, q_{21}, \ldots, q_{2m_2}, \ldots, q_{n1}, \ldots, q_{nm_n})$$

$$= p_n \cdot H\left(\frac{q_{n1}}{p_n}, \ldots, \frac{q_{nm_n}}{p_n}\right) + H\left(q_{11}, \ldots, q_{1m_1}, \ldots, q_{n-1,1}, \ldots, q_{n-1,m_{n-1}}, p_n\right).$$

Iteration of this procedure yields the desired result. $\qquad\square$

Proof of the theorem Let $F(n) = H(\frac{1}{n}, \ldots, \frac{1}{n})$ for $n \geq 2$ and $F(1) = 0$ (since $F(1) = H(1) = H(1, 0) = 0$). From the general grouping axiom (Lemma 12) it can be concluded that $F(m \cdot n) = F(m) + F(n)$, letting $m_1 = \cdots = m_n = m$. Further, applying Lemma 11 yields $H(\frac{1}{n}, \ldots, \frac{1}{n}) = H(\frac{1}{n}, \frac{n-1}{n}) + \frac{n-1}{n} \cdot H(\frac{1}{n-1}, \ldots, \frac{1}{n-1})$. Now $H(\frac{1}{n}, \frac{n-1}{n}) = F(n) - \frac{n-1}{n} F(n)$, and since $\lim_{n \to \infty} H(\frac{1}{n}, \frac{n-1}{n}) = H(0, 1) = 0$ because of 1″, we have $\lim_{n \to \infty} (F(n) - \frac{n-1}{n} F(n-1)) = 0$ and hence $\lim_{n \to \infty} (F(n) - F(n-1)) = 0$. Summarizing, $F : \mathbb{N} \to \mathbb{R}$ is a multiplicative function with $F(n \cdot m) = F(n) + F(m)$

$$\lim_{n \to \infty} (F(n) - F(n-1)) = 0.$$

In the following lemma it will be shown that $F(n)$ is necessarily of the form $c \cdot \log n$, where c is a constant factor. With this knowledge we can proceed along the lines in the proof of Shannon's theorem to obtain the desired result.   □

Lemma 13 *Let* $f : \mathbb{N} \to \mathbb{R}$ *be a multiplicative function with*

(i) $f(n \cdot m) = f(n) + f(m)$
(ii) $\lim_{n \to \infty} (f(n+1) - f(n)) = 0$

Then $f(n) = c \cdot \log n$ *for some constant factor* $c \in \mathbb{R}$.

This Lemma was originally stated by Erdös and rediscovered by Faddeev. The following proof is due to Rényi. The proof makes use of another lemma in which an additional condition on the function is introduced.

Lemma 14 *Let* $g : \mathbb{N} \to \mathbb{R}$ *be a multiplicative function with*

(i) $g(n \cdot m) = g(n) + g(m)$.
(ii) $\lim_{n \to \infty} (g(n+1) - g(n)) = 0$.
(iii) *There exists an* $N > 1$ *with* $g(N) = 0$.

Then $\lim_{n \to \infty} \frac{g(n)}{\log n} = 0$.

Before proving Lemma 14 we will first demonstrate that Lemma 13 follows easily from Lemma 14. Therefore define

$$g(n) \triangleq f(n) - \frac{f(N) \cdot \log n}{\log N}$$

for a fixed $N > 1$ and a function f with the properties required in Lemma 13. Then g obviously has the properties (ii) and (iii) from Lemma 14 and also (i) is fulfilled, since

$$g(n \cdot m) = f(n \cdot m) - \frac{f(N) \cdot \log(n \cdot m)}{\log N}$$

$$= f(n) - \frac{f(N) \cdot \log n}{\log N} + f(m) - \frac{f(N) \cdot \log m}{\log N} = g(n) + g(m).$$

Now by Lemma 14

$$0 = \lim_{n \to \infty} \frac{g(n)}{\log n} = \lim_{n \to \infty} \left(\frac{f(n)}{\log n} - \frac{f(N)}{\log N} \right), \quad \text{and hence}$$

$\lim_{n \to \infty} \frac{f(n)}{\log n} = \frac{f(N)}{\log N} \triangleq c$ for some constant $c \in \mathbb{R}$.

Since N could have been arbitrarily chosen, we have $\frac{f(N')}{\log N'} = c$ for all $N' > 1$, and hence Lemma 13 is proved.

Proof of Lemma 14 First a function $G : \mathbb{N} \cup \{-1, 0\} \to \mathbb{R}$ is defined by

$$G(-1) \triangleq 0, \, G(k) \triangleq \max_{N^k \le n < N^{k+1}} |g(n)| \quad \text{for } k = 0, 1, 2, \ldots$$

Furthermore, for $k = 0, 1, 2, \ldots$ let

$$\delta_k \triangleq \max_{N^k \le n < N^{k+1}} |g(n+1) - g(n)|$$

Since $\lim_{n \to \infty} (g(n+1) - g(n)) = 0$, for every $\varepsilon > 0$ we find an $n(\varepsilon)$ such that $|g(n+1) - g(n)| \le \varepsilon$ for $N^k > n \ge n(\varepsilon)$ and

$$\lim_{k \to \infty} \delta_k = 0.$$

Further observe that for $N^k \le n < N^{k+1}$ $\frac{g(n)}{\log n} \le \frac{G(k)}{k \cdot \log N}$. Hence it is sufficient to prove that $\lim_{k \to \infty} \frac{G(k)}{k \cdot \log N} = 0$, or better, since N is fixed, $\lim_{k \to \infty} \frac{G(k)}{k} = 0$. Therefore choose n arbitrary and k such that $N^k \le n < N^{k+1}$ and set $n' \triangleq N \cdot \lfloor \frac{n}{N} \rfloor$.

The telescoping sum now is

$$g(n) = g(n') + \sum_{\ell=n'}^{n-1} (g(\ell+1) - g(\ell)).$$

Hence

$$|g(n)| \le |g(n')| + \sum_{\ell=n'}^{n-1} |g(\ell+1) - g(\ell)| \le g(n') + N\delta_k.$$

By the properties (i) and (iii) required in Lemma 14 now

$$g(n') = g \left(N \cdot \left\lfloor \frac{n}{N} \right\rfloor \right) = g(N) + g \left(\left\lfloor \frac{n}{N} \right\rfloor \right) = g \left(\left\lfloor \frac{n}{N} \right\rfloor \right),$$

and so

$$|g(n)| \le g \left(\left\lfloor \frac{n}{N} \right\rfloor \right) + N\delta_k.$$

This inequality especially holds for n with $|g(n)| = G(k)$, and since $N^{k-1} \leq \lfloor \frac{n}{N} \rfloor < N^k$ we obtain

$$G(k) \leq G(k-1) + N\delta_k$$

and further for every $m \in \mathbb{N}$

$$\sum_{k=0}^{m} G(k) \leq \sum_{k=0}^{m} G(k-1) + N \cdot \sum_{k=0}^{m} \delta_k.$$

We can conclude that

$$G(m) = \sum_{k=0}^{m} \big(G(k) - G(k-1)\big) \leq N \sum_{k=0}^{m} \delta_k$$

and hence

$$\frac{G(m)}{m} \leq N \frac{\sum_{k=0}^{m} \delta_k}{m}.$$

Since $\lim_{k \to \infty} \delta_k = 0$, also $\lim_{m \to \infty} \frac{1}{m} \sum_{k=0}^{m} \delta_k = 0$.

Hence $\lim_{m \to \infty} \frac{G(m)}{m} = 0$ and we are done.   □

3.1.2 Formal Properties of the Entropy Function

Usually a RV is a function $X : \Omega \to \mathbb{R}$, where (Ω, Q) is a probability space with sample space Ω and a probability distribution Q on Ω. For a set $I \subset \mathbb{R}$ we denote by

$$P_X(I) \triangleq \mathrm{Prob}(X \in I) = Q\big(\{\omega : X(\omega) \in I\}\big).$$

For $I = (-\infty, t)$ we shortly write $F_X(t) \triangleq P_X(-\infty, t)$ and obtain the distribution (function).

For our purposes it is more appropriate to consider a discrete RV $X : \Omega \to \mathcal{X}$ for some discrete $\mathcal{X}$. The notation $P_X(x)$ now is used for $P_X(x) \triangleq \mathrm{Prob}(X = x) = Q\big(\{\omega : X(\omega) = x\}\big)$.

Especially, we can choose $\Omega = \mathcal{X}$ and $Q = P_X$.

Accordingly, P_{XY} is defined for a RV $(X, Y) : \Omega \times \Omega \to \mathcal{X} \times \mathcal{Y}$ defined on the product space $\Omega \times \Omega$. In this case the projections on the components $\mathcal{X}$ and $\mathcal{Y}$ will be described by the RV's $X : \Omega \times \Omega \to \mathcal{X}$ and $Y : \Omega \times \Omega \to \mathcal{Y}$ with $P_X(x) = \sum_{y \in \mathcal{Y}} P_{XY}(x, y)$ and $P_Y(y) = \sum_{x \in \mathcal{X}} P_{XY}(x, y)$.

For the entropy $H\big((X, Y)\big)$ of the RV (X, Y) we will shortly write $H(X, Y)$, hence

$$H(X, Y) = -\sum_{x \in \mathcal{X}} \sum_{y \in \mathcal{Y}} P_{XY}(x, y) \cdot \log P_{XY}(x, y).$$

$H(X, Y)$ is sometimes denoted as the *joint entropy* of X and Y.

Definition 14 The **conditional entropy** $H(X|Y)$ is defined as

$$H(X|Y) = H(X, Y) - H(Y).$$

The conditional entropy can be interpreted as the remaining uncertainty on the realization of the RV (X, Y), when the outcome of Y is already known. After this heuristic approach we shall now give a formal description of the conditional entropy in terms of probabilities. Therefore we define for all $y \in \mathcal{Y}$ the conditional entropy of X given y

$$H(X|Y = y) \triangleq -\sum_{x \in \mathcal{X}} P_{X|Y}(x|y) \cdot \log P_{X|Y}(x|y),$$

where $P_{X|Y}(x|y) = \frac{P_{XY}(x,y)}{P_Y(y)}$ is the usual conditional probability.

Theorem 10

$$H(X|Y) = \sum_{y \in \mathcal{Y}} P_Y(y) \cdot H(X|Y = y)$$

Proof

$$H(X|Y) = H(X, Y) - H(Y)$$

$$= -\sum_{x \in \mathcal{X}} \sum_{y \in \mathcal{Y}} P_{XY} \cdot \log P_{XY}(x, y) + \sum_{y \in \mathcal{Y}} P_Y(y) \cdot \log P_Y(y)$$

$$= -\sum_{x \in \mathcal{X}} \sum_{y \in \mathcal{Y}} P_{XY}(x, y) \cdot \log \left(P_{X|Y}(x|y) \cdot P_Y(y) \right)$$

$$+ \sum_{y \in \mathcal{Y}} P_Y(y) \cdot \log P_Y(y)$$

$$= -\sum_{x \in \mathcal{X}} \sum_{y \in \mathcal{Y}} P_{XY}(x, y) \cdot \log P_{X|Y}$$

$$- \sum_{x \in \mathcal{X}} \sum_{y \in \mathcal{Y}} P_{XY}(x, y) \cdot \log P_Y(y) + \sum_{y \in \mathcal{Y}} P_Y(y) \cdot \log P_Y(y)$$

$$= -\sum_{x \in \mathcal{X}} \sum_{y \in \mathcal{Y}} P_{XY}(x, y) \cdot \log P_{X|Y}(x|y) \tag{3.1}$$

$$\text{(since } P_Y(y) = \sum_{x \in \mathcal{X}} P_{XY}(x, y))$$

$$= - \sum_{x \in \mathcal{X}} \sum_{y \in \mathcal{Y}} P_{X|Y}(x|y) \cdot P_Y(y) \cdot \log P_{X|Y}(x|y)$$

$$= \sum_{y \in \mathcal{Y}} P_Y(y) \cdot \left(- \sum_{x \in \mathcal{X}} P_{X|Y}(x|y) \cdot \log P_{X|Y}(x|y)\right)$$

$$= \sum_{y \in \mathcal{Y}} P_Y(y) \cdot H(X|Y = y). \qquad \square$$

Lemma 15 $H(X, Y) \leq H(X) + H(Y)$ *with equality, exactly if X and Y are independent.*

Proof Since $P_X(x) = \sum_{y \in \mathcal{Y}} P_{XY}(x, y)$ and $P_Y(y) = \sum_{x \in \mathcal{X}} P_{XY}(x, y)$ we have

$$H(X) + H(Y) = - \sum_{x \in \mathcal{X}} P_X(x) \cdot \log P_X(x) - \sum_{y \in \mathcal{Y}} P_Y(y) \cdot \log P_Y(y)$$

$$= - \sum_{\substack{x \in \mathcal{X} \\ y \in \mathcal{Y}}} P_{XY}(x, y)\big(\log P_X(x) + \log P_Y(y)\big)$$

$$= - \sum_{\substack{x \in \mathcal{X} \\ y \in \mathcal{Y}}} P_{XY}(x, y) \cdot \log\big(P_X(x) \cdot P_Y(y)\big).$$

On the other hand $H(X, Y) = - \sum_{\substack{x \in \mathcal{X} \\ y \in \mathcal{Y}}} P_{XY}(x, y) \cdot \log\big(P_{XY}(x, y)\big)$.

Now

$$H(X) + X(Y) - H(X, Y) = \sum_{\substack{x \in \mathcal{X} \\ y \in \mathcal{Y}}} P_{XY}(x, y) \cdot \log \frac{P_{XY}(x, y)}{P_X(x) \cdot P_Y(y)}$$

$$= D(P_{XY} \| P_X \cdot P_Y).$$

But we know that for all probability distributions P and Q $D(P\|Q) \geq 0$ and $D(P\|Q) = 0$, exactly if $P = Q$. Applied to our problem this yields the desired result, since $P_{XY} = P_X \cdot P_Y$, exactly if X and Y are independent. $\qquad \square$

Remark If we replace the grouping axiom for a measure of uncertainty by Lemma 15, we obtain a further function fulfilling the new postulates, namely the Rényi-entropy (of order $\alpha > 0$) defined by

$$H_\alpha(X) \triangleq \frac{1}{1 - \alpha} \sum_{x \in \mathcal{X}} P_X(x)^\alpha.$$

It can be shown that $\lim_{\alpha \to 1} H_\alpha(X) = H(X)$. In the Appendix it is shown that Shannon's entropy $H(X)$ is the logarithm of a geometric mean, whereas the Rényi-entropy is the (weighted) logarithm of an arithmetic mean.

Lemma 16 *For RV's $X_1, X_2, \ldots, X_n$*

$$H(X_1, \ldots, X_n) \leq H(X_1) + \cdots + H(X_n)$$

with equality exactly if $X_1, \ldots, X_n$ are independent.

Proof Letting $X = X_1$ and $Y = (X_2, \ldots, X_n)$ we obtain by the preceding lemma that $H(X_1, \ldots, X_n) \leq H(X_1) + H(X_2, \ldots, X_n)$ with equality if X_1 and $(X_2, \ldots, X_n)$ are independent. Iteration of this procedure finally yields the result. $\square$

Lemma 17 $H(X|Y) \leq H(X)$ with equality, if and only if X and Y are independent.

Proof By definition of $H(X|Y)$ and Lemma 15 we have

$$H(X|Y) = H(X, Y) - H(Y) \leq H(X) + H(Y) - H(Y) = H(X). \qquad \square$$

Lemma 18 $H(X|YZ) \leq H(X|Z)$ with equality, if and only if X and Y are conditionally independent, i.e., $P_{XY|Z} = P_{X|Z} \cdot P_{Y|Z}$.

Proof The idea of the proof is that for all $z \in \mathcal{Z}$ by Lemma 17 $H(X|Y, Z = z) \leq H(X|Z = z)$. Averaging then gives

$$\sum_{z \in \mathcal{Z}} P_Z(z) \cdot H(X|Y, Z = z) = H(X|YZ) \leq H(X|Z)$$

$$= \sum_{z \in \mathcal{Z}} P_Z(z) \cdot H(X|Z = z).$$

The exact computation follows. By (3.1) in the proof of Theorem 10

$$H(X|Z) - H(X|YZ) = -\sum_{x \in \mathcal{X}} \sum_{z \in \mathcal{Z}} P_{XZ}(x, z) \cdot \log P_{X|Z}(x|z)$$

$$+ \sum_{x \in \mathcal{X}} \sum_{y \in \mathcal{Y}} \sum_{z \in \mathcal{Z}} P_{XYZ}(x, y, z) \cdot \log P_{X|YZ}(x|yz)$$

$$= -\sum_{z \in \mathcal{Z}} P_Z(z) \cdot \left(\sum_{x \in \mathcal{X}} P_{X|Z}(x|z) \cdot \log P_{X|Z}(x|z) \right.$$

$$\left. - \sum_{x \in \mathcal{X}} \sum_{y \in \mathcal{Y}} P_{XY|Z}(x, y|z) \cdot \log P_{X|YZ}(x|y, z) \right)$$

$$= -\sum_{z \in \mathcal{Z}} P_Z(z) \cdot \sum_{x \in \mathcal{X}} \sum_{y \in \mathcal{Y}} P_{XY|Z}(x, y|z) \cdot \left(\log P_{X|Z}(x, z) - \log P_{X|YZ}(x|y, z) \right)$$

$$= -\sum_{z \in \mathcal{Z}} P_Z(z) \cdot \sum_{x \in \mathcal{X}} \sum_{y \in \mathcal{Y}} P_{XY|Z}(x, y|z) \cdot \log \frac{P_{X|Z}(x|z) \cdot P_{Y|Z}(y|z)}{P_{XY|Z}(x, y|z)}$$

$$= \sum_{z \in \mathcal{Z}} P_Z(z) \cdot D\left(P_{XY|Z} \| P_{X|Z} \cdot P_{Y|Z} \right) \geq 0$$

with equality, exactly if $P_{XY|Z} = P_{X|Z} \cdot P_{Y|Z}$. $\square$

Remark The conditional independence $P_{XY|Z} = P_{X|Z} \cdot P_{Y|Z}$ is also denoted as Markov property, since it is characteristic for Markov chains. So (X, Z, Y) form a Markov chain, symbolized by

$$X \multimap Z \multimap Y,$$

hence $P_{Y|ZX} = P_{Y|Z}$, i.e., the realization of Y only depends on the last state Z and not on the further past. An easy calculation now yields the Markov property.

$P_{XZY} = P_Z \cdot P_{XY|Z} = P_Z \cdot P_{X|Z} \cdot P_{Y|XZ} = P_X \cdot P_{X|Z} \cdot P_{Y|Z}$, since (X, Z, Y) is a Markov chain. Division by P_Z gives the Markov property.

Lemma 19 *Let P be a probability distribution and W a doubly-stochastic matrix. For the probability distribution $P' = P \cdot W$ we have: $H(P') \geq H(P)$ with equality, exactly if W is a permutation matrix.*

Lemma 20 *Let X and Y be RV's with $Y = f(X)$ for some function f. Then $H(Y) \leq H(X)$ with equality, exactly if f is bijective.*

Proof $H(X, f(X)) = H(X) + H(f(X)|X)$. $H(f(X)|X) = 0$, however, since there is no uncertainty about the function value $f(x)$, when x is known. Hence

$$H(X) = H(X, f(X)) = H(f(X)) + H(X|f(X))$$

and $H(X|f(X)) \geq 0$ with equality, if we can conclude from the function value $f(x)$ to x. This is only the case if f is bijective. $\square$

3.1.3 Entropy and Information, Inequalities

Some General Remarks

In the first two chapters we saw that entropy and information occur as basic parameters in coding problems. The theorems proved demonstrate the importance of these notions. Shannon always felt that these coding theoretic interpretations are their main justification. Nevertheless he did propose an axiomatic characterisation of entropy, which was shown by Faddeev [3] to really define entropy within a multiplicative constant. The intuition behind those axioms originates from an attempt to measure numerically the "uncertainty" in a probabilistic experiment. Those are the axioms

1. $H\left(\frac{1}{M}, \ldots, \frac{1}{M}\right)$ is a monotonically increasing function of M.
2. $H\left(\frac{1}{ML}, \ldots, \frac{1}{ML}\right) = H\left(\frac{1}{M} + \cdots + \frac{1}{M}\right) + H\left(\frac{1}{L} + \cdots + \frac{1}{L}\right)$ for $M, L = 1, 2, \ldots$.
3. $H(p, 1 - p)$ is a continuous function of p.
4. $H(p_1, \ldots, p_M) = H(p_1 + \cdots + p_r, p_{r+1} + \cdots + p_M) = (p_1 + \cdots + p_r)$
$$H\left(p_1 \bigg/ \sum_{i=1}^{r} p_i, \ldots, p_r \bigg/ \sum_{i=1}^{r} p_i\right) + (p_{r+1} + \cdots + p_M)$$
$$H\left(p_{r+1} \bigg/ \sum_{i=r+1}^{M} p_i, \ldots, p_M \bigg/ \sum_{i=r+1}^{M} p_i\right) (r = 1, 2, \ldots, M - 1).$$

Axiom 4 is called the grouping axiom and it has been considered to be the least "intuitive" one by many authors. In an almost infinite series of papers those axioms have been weakened and recently B. Forte succeeded in showing that Axiom 4 can be replaced by

Axiom 4': $H(XY) \leq H(X) + H(Y)$ for all finite RV's X and Y.

With this brief sketch we leave now any further discussions about foundations. We do not discuss the importance of reflection on intuitive ideas relating to those notions and we shall try to suggest intuitive interpretations, whereever we find those appropiate. Information Theory, like many disciplines in Applied Mathematics, vitally depends on a fruitful tension between the attempt to mathematical rigor on one side and the wish for meaningful answers to questions coming from the real world on the other side. At any time we have to be willing to review our mathematical assumptions.

Otherwise we really have to take the criticism about mathematicians attributed to Goethe (and given in loose translation) "If you have a problem in real life and ask a mathematician for a solution, then he formulates and reformulates your problem many times and finally also gives you a solution. But this solution has nothing to do anymore with the question you originally asked."

With respect to Information Theory we think that it is fair to say that mathematicians have responded to the many questions raised by communication engineers and have also come up with mathematical theorems, which in turn have sharpened the understanding of engineering problems.

In the present paragraph we introduce entropy and information quantities, which occur in coding problems, and we attempt to give a fairly complete list of their properties. Beyond this we shall discuss quantities, which have not yet found a coding theoretic interpretation, but might suggest new coding problems. Proofs are always included for those results which are needed in this book. Section 3.1.2 contains classical inequalities from real analysis, which are important for Information Theory.

Entropy, Conditional Entropy, and their Properties

We recall that the entropy $H(X)$ of a finite-valued RV X was defined as

$$H(X) = -\sum_{x \in \mathcal{X}} \text{Prob}(X = x) \log \text{Pr}(X = x). \qquad (3.2)$$

We write $P(x)$ instead of $\text{Pr}(X = x)$ and also use the convention $\|X\|$ for the cardinality of the range of a function. For a (general) RV $Z = (X, Y)$ it is natural to define

$$H(Z) = H(X, Y) = -\sum_{x \in \mathcal{X}} \sum_{y \in \mathcal{Y}} \text{Pr}(X = x, Y = y) \log \text{Pr}(X = x, Y = y) \quad (3.3)$$

and abbreviate $\text{Pr}(X = x, Y = y)$ as $P(x, y)$.

The *conditional entropy of X given Y* can be introduced by

$$H(X|Y) = H(X, Y) - H(Y). \tag{3.4}$$

In the sequel a simple inequality is needed, which is an immediate consequence of (3.3) and stated as lemma, because of its frequent use.

Lemma 21 *For two probability vectors* $P = (p_1, \ldots, p_a)$ *and* $Q = (q_1, \ldots, q_a)$

$$-\sum_{i=1}^{a} p_i \log p_i \leq -\sum_{i=1}^{a} p_i \log q_i$$

with equality iff $p_i = q_i$ *for all* i.

The result can also be derived from the inequality $\log x \leq x - 1$ with equality if $x = 1$.

We are now ready to state

Lemma 22 *The function* $H(X) = H(P)$ *is strictly concave (3.2) in* $P = (p_1, \ldots, p_a)$ *and satisfies* $0 \leq H(X) \leq \log a$, *with equality to o iff* $P_i = 1$ *for some* i *and with equality to* $\log a$ *iff* $p_i = \frac{1}{a}$ *for all* i.

Proof Since $z(t) = -t \log t$ is strictly concave, $H(P)$ inherits this property and

$$H(P) = a \sum_{i=1}^{a} \frac{1}{a} z(p_i) \leq a \, z \left(\frac{1}{a} \sum p_i \right) (\text{"="} \text{ if } p_i = \frac{1}{a} \text{ for all } i)$$

$$= a \, z \left(\frac{1}{a} \right) = \log a.$$

The remaining assertions are obvious.

Theorem 11 $H(Y|X) \leq H(Y)$.

Proof This is a consequence of Lemma 20.

Mutual Information and Conditional Mutual Information

The function

$$R(p, w) = H(q) - \sum_{j} q_j \, H\big(w(\cdot|j)\big), q = p \cdot w,$$

plays an important role in channel coding, in fact, $\max\limits_{p} R(p, w)$ gave the exact value of the channel capacity for a DMC. This quantity therefore deserves special attention. Clearly,

$$R(p, w) = -\sum_x \sum_y P(x)w(y|x) \log \frac{w(y|x)}{q(y)} = -\sum_x \sum_y P(x, y) \log \frac{P(x, y)}{P(x)P(y)},$$

where $P(x, y) = P(x)w(y|x)$.

We have

Lemma 23 $R(p, w)$ *is concave in p and convex in w.*

Proof $R(p, w) = H(p \cdot w) - \sum p_i H(w(\cdot|i))$ and since $p \cdot w$ and $\sum p_i H(w(\cdot|i))$ are linear in p and H is a concave function concavity in p follows.

As a consequence of the arithmetic-geometric mean inequality one can derive

$$\sum r_i \log \frac{r_i}{u_i} \geq (r_1 + \cdots + r_m) \log \frac{r_1 + \cdots + r_m}{u_1 + \cdots + u_m} \quad r_i, u_i \geq 0.$$

In many cases concerning the study of connections between random variables our standard RV language is preferable:

$$R(p, w) = H(Y) - H(Y|X).$$

We denote this quantity also by $I(X \wedge Y)$. Clearly, $H(Y) \geq H(Y|X)$ and thus $I(X \wedge Y) \geq 0$. This can be shown by using

$$I(X \wedge Y) = D(p(x, y) \| p(x)p(y)).$$

Interpreting $H(X)$ as the uncertainty in a random experiment (X, p) and $H(X|Y)$ as the remaining average uncertainly about X after Y has been observed, one might now say that the difference of the uncertainty about X before and after having observed Y. Y is the "information" gained through this observation. This information is 0 iff X and Y are independent. That there is some meaning to this heuristic will become clearer in Section .... This language is sometimes a useful guide, but of course by no means a substitute for a mathematically rigorous theory.

Of course, $I(X \wedge Y|Z) \geq 0$. Equality holds iff $H(X|Z) = H(X|Y, Z)$ iff X and Y are independent conditionally on Z iff X, Z, Y form a Markov chain in that order.

We use the convient notation $X \to Z \to Y$ iff the RV's form a Markov chain in this order. It is also convenient to write $I(X_1, \ldots, X_n \wedge Y_1, \ldots, Y_n | Z_1 Z_2 \ldots Z_n)$ instead of $I((X_1, \ldots, X_n) \wedge (Y_1, \ldots, Y_n)|(Z_1, \ldots, Z_n))$. The following elementary formula was first formulated by Kolmogorov and rediscovered and first used in coding theory in [5].

$$I(X \wedge YZ) = I(X \wedge Y) + I(X \wedge Z|Y) \tag{3.5}$$

and holds, because

$$H(X) - H(X|YZ) = H(X|Y) - H(X|YZ) + H(X) - H(X|Y).$$

In general there is no inequality between $I(X \wedge Y|Z)$ and $I(X \wedge Y)$. If Z is independent of (X, Y), then equality holds. However, already if Z is independent of one variable Y, say, always

$$I(X \wedge Y|Z) \geq I(X \wedge Y). \tag{3.6}$$

This fact was used in [5, 6] and matters because independent sources were considered. The verification of (3.6) is straightforward:

$$I(X \wedge Y|Z) = H(Y) - H(Y|X, Z) \geq H(Y) - H(Y|X).$$

Also, if $X \leftrightarrow Y \leftrightarrow Z$, then by (3.5)

$$I(X \wedge YZ) = I(X \wedge Y). \tag{3.7}$$

Since

$$I(X \wedge YZ) = H(X) - H(X|YZ) \geq H(X) - H(X|Z) = I(X \wedge Z)$$

we have proved the so called *data processing "theorem"*, whose significance is explained in Sect. 8.2

Lemma 24 (Data-processing) *If* $X \rightarrow Y \rightarrow Z$, *then* $I(X \wedge Y) \geq I(X \wedge Z)$ *and* $I(Y \wedge Z) \geq I(X \wedge Z)$.

An important special case:

$$U = f(X), V = g(Y) \text{ and hence } U \rightarrow X \rightarrow Y; U \rightarrow Y \rightarrow V.$$

Therefore, $I(X \wedge Y) \geq I(U \wedge Y) \geq I(U \wedge V) = I\big(f(X) \wedge g(Y)\big)$.

Often this result is stated in terms of partitions: partitioning of the $\mathcal{Y}$ space (or $\mathcal{X}$ space) decreases information.

We leave it to the reader to verify

$$I\big(f(X) \wedge Y|Z\big) \leq I(X \wedge Y|Z)$$

and also the generalized Kolmogorov rule

$$I(XY \wedge Z|U) = I(X \wedge Z|U) + I(Y \wedge Z|UX).$$

Two Basic Inequalities for Mutual Information in Case of Memory

Lemmas 23 and 25 shows a certain duality in the behaviour of $R(p, w)$ as function of p and w with respect to convexity properties. There is a similar duality between "source and channel" with respect to memory.

Theorem 12 *Let $X^n = (X_1, \ldots, X_n)$ and $Y^n = (Y_1, \ldots, Y_n)$ have a joint distribution. Then*

(i) $I(X^n \wedge Y^n) \geq \sum_{t=1}^{n} I(X_t \wedge Y_t)$ *if the X_t's are independent. (Dobrushin's inequality [7, 8])*

(ii) $I(X^n \wedge Y^n) \leq \sum_{t=1}^{n} I(X_t \wedge Y_t)$ *if Y^n and X^n are connected by a memoryless channel, that is,* $\Pr(Y^n = y^n | X^n = x^n) = \prod_{t=1}^{n} w_t(y_t | x_t)$.

Proof We prove (i) inductively and use the identity

$$I(XY \wedge Z) + I(X \wedge Y) = I(X \wedge YZ) + I(Y \wedge Z),$$

which is easily verified with the help of (3.7).

Assume that

$$I(X^{n-1}, Y^{n-1}) \geq \sum_{t=1}^{n-1} I(X_t, Y_t).$$

Then

$$
\begin{aligned}
I(X^n \wedge Y^n) &= I(X^{n-1} X_n \wedge Y^{n-1} Y_n) \\
&= I(X^{n-1} \wedge X_n Y^{n-1} Y_n) + I(X_n \wedge Y^{n-1} Y_n) - I(X^{n-1} \wedge X_n) \\
&\geq I(X^{n-1} \wedge Y^{n-1}) + I(X_n \wedge Y_n)
\end{aligned}
$$

as a consequence of the data-processing theorem and the fact that X^{n-1} and X_n are independent.

The following steps lead to (ii):

$$
\begin{aligned}
I(X^n \wedge Y^n) &= H(Y^n) - H(Y^n | X^n) \\
&= H(Y^n) - \sum \Pr(X^n = x^n) H(Y^n | X^n = x^n) \\
&= H(Y^n) - \sum_{t-1}^{n} H(Y_t | X_t) \leq \sum_{t-1}^{n} H(Y_t) - \sum_{t=1}^{n} H(Y_t | X_t) \\
&= \sum_{t=1}^{n} I(Y_t \wedge X_t).
\end{aligned}
$$

Shannon's Lemma about the Capacity Formula

Lemma 25 *A p^x maximizes $R(p, w) = \sum_{x,y} P(x) w(y|x) \log \frac{w(y|x)}{q(y)}$, if and only if for some sonstant K*

$$\sum_y w(y|x) \log \frac{w(y|x)}{q^*(y)} = K, \ \text{if } P(x) > 0$$

$$\leq K, \ \text{if } P(x) = 0.$$

Such a K equals C. There is a unique q corresponding to the maximizing p's.

The following proof of Shannon's lemma ([1]) is due to Topsøe ([9, 10]).
Proof For any distribution q^* on $\mathcal{Y}$ one has

$$\sum P(x)w(y|x) \log \frac{w(y|x)}{q(y)} + \sum q(y) \log \frac{q(y)}{q^*(y)} = \sum P(x)w(y|x) \cdot \log \frac{w(y|x)}{q^*(y)}$$

$$(3.8)$$

and $\sum q(y) \log \frac{q(y)}{q^*(y)} \geq 0$ with equality iff $q \equiv q^*$ by the lemma.

Let $\{p_\ell(\cdot) : \ell = 1, \ldots, L\}$ be a set of PD's on $\mathcal{X}$ and $s = (s_1, \ldots, s_L)$ a PD
on $\{1, \ldots, L\}$. Also, set $q_L = p_L \cdot w$, $p_0 = \sum s_\ell p_\ell$, and $q_0 = \sum s_\ell q_\ell$. For any
distribution q^* on $\mathcal{Y}$ we have

$$\sum s_\ell p_\ell(x)w(y|x) \log \frac{w(y|x)}{q_\ell(y)} + \sum s_\ell q_\ell(y) \log \frac{q_\ell(y)}{q^*(y)}$$

$$= \sum s_\ell p_\ell(x)w(y|x) \log \frac{w(y|x)}{\sum s_\ell q_\ell(y)} + \sum s_\ell q_\ell(y) \log \frac{\sum s_\ell q_\ell(y)}{q^*(y)}.$$

If we choose $q^* = \sum s_\ell q_\ell$, then we see that $\sum s_\ell R(p_\ell, w) \leq R(p_0, w)$ with
equality iff q_ℓ does not depend on ℓ for those ℓ with $s_\ell > 0$. Hence we also have the
uniqueness of the maximizing q's.

The conditions of the lemma are sufficient for $\overline{P}$ to be optimal, because it implies

$$\sum P(x)w(y|x) \log \frac{w(y|x)}{\overline{P}(y)} \leq K$$

and therefore by equation (3.8) $R(p, w) \leq K$ for all p.

Assume now that $\overline{p}$ is optimal. We show necessity of the lemma by a variational
argument.

Define $C_x = \sum w(y|x) \log \frac{w(y|x)}{q(y)}$, $x \in \mathcal{X}$, then $C = \sum \overline{P}(x)C_x$.

Assume there exists an x_0 with $C_{x_0} > C$ and let δ_{x_0} be the PD concentrated on
x_0. For $0 \leq \alpha \leq 1$ we put $p_\alpha = \alpha \delta_0 + (1 - \alpha)\overline{p}^*$ and denote the corresponding
output distribution on $\mathcal{Y}$ by q_α.

Now we make for the first time use of (3.17) with $q^* = q_\alpha$ and obtain

$$(1-\alpha)C + (1-\alpha) \sum \overline{P}(y) \log \frac{\overline{P}(y)}{q_\alpha(y)} + \alpha \sum_y w(y|x_0) \log \frac{w(y|x_0)}{q_\alpha(y)} = R(p_\alpha, w).$$

Therefore,

$$R(p_\alpha, w) \geq (1 - \alpha)C + \alpha \sum w(y|x_0) \log \frac{w(y|x_0)}{q_\alpha(y)}$$

and since $C_{x_0} > C$ we can find a positive α such that $\sum w(y|x_0) \log \frac{w(y|x_0)}{q_\alpha(y)} > C$ holds.

The inequality $R(p_\alpha, w) > C$ contradicts the definition of C and the proof is complete.

3.2 Lecture on General Entropy

3.2.1 Introduction

By Kraft's inequality a uniquely decipherable code over a b letter alphabet $\mathcal{Y}$ with codeword length $l_1, \ldots, l_N$ exists iff

$$\sum_{i=1}^{N} b^{-l_i} \leq 1. \tag{3.9}$$

To compare different codes and pick out an optimum code it is customary to examine the mean length, $\sum l_i p_i$, and to minimize this quantity. This is a good procedure if the cost of using a sequence of length l_i is directly proportional to l_i. However, there may be occasions when the cost is another function of l_i.

Example If there is an "end of message" symbol an optimal code for a probability distribution $p_1 \geq p_2 \geq \ldots$ is given by

$$0, 1, \ldots, b, 00, 01, \ldots, 0b, 10, \ldots, 1b, \ldots, bb, 000, \ldots \tag{3.10}$$

(it is only neccessary that the code c is an injective function, which is a weaker condition than uniquely decipherable). In order to get bounds for the mean length of such a code we introduce a concave function $\phi : [0, \infty) \to \mathbb{R}$. Now if for the function ϕ and the set of codeword length (3.10) the inequality

$$\sum_{i=1}^{N} b^{-\phi(l_i)} \leq 1 \tag{3.11}$$

holds, then by Kraft's inequality a uniquely decipherable code with codeword length $\phi(l_i), \ldots, \phi(l_N)$ exists and by the Noiseless Coding Theorem

$$H(P) \leq \sum_{i=1}^{N} p_i \phi(l_i) < H(P) + 1 \tag{3.12}$$

holds, where $H(P)$ is the Shannon entropy. Now using the assumtion that ϕ is concave we get

$$\sum_{i=1}^{N} p_i \phi(l_i) \leq \phi(\sum_{i=1}^{N} p_i l_i). \qquad (3.13)$$

If we assume in addition that ϕ is strictly monotonic than ϕ has an inverse ϕ^{-1} and we get

$$\phi^{-1}(H(P)) \leq \sum_{i=1}^{n} p_i l_i. \qquad (3.14)$$

Thus, in some circumstances, it might be more appropriate to choose a code which minimizes the quantity

$$L(P, \vec{l}, \phi) \triangleq \phi^{-1}\left(\sum_{i=1}^{N} p_i \phi(l_i)\right). \qquad (3.15)$$

For a code C we will also use the notation $L(P, C, \phi)$.

3.2.2 Properties of the Generalized Entropy

The minimum of (3.15) will be called generalized entropy and denoted by

$$H(P, \phi) \triangleq \inf_{C} L(P, C, \phi) \qquad (3.16)$$

where the length distributions of the code C has to satisfy (3.11).
   Some properties are

   (i) $H(P, \phi)$ is a symmetric function of $p_1, \ldots, p_N$.
   (ii) If $P_1 = (p_1, \ldots, p_N)$ and $P_2 = (p_1, \ldots, p_N, 0)$, then $H(P_1, \phi) = H(P_2, \phi)$.
   (iii) For $P = (1, 0, \ldots, 0)$ $H(P, \phi) = 0$.
   (iv) For $P = (p_1, \ldots, p_N)$ $H(P, \phi) \leq log_b N$.

3.2.3 Rényi's Entropy

In [11] Campbell has introduced a quantity which resembles the mean length. Let a code length of order t be defined by

$$l(t) = \frac{1}{t} log_b\left(\sum_{i=1}^{N} p_i b^{tl_i}\right) \qquad (0 < t < \infty) \qquad (3.17)$$

(We have defined $\phi(x) = b^{tx}$).

An application of l'Hospital's rule shows that

$$l(0) = \lim_{t \to \infty} l(t) = \sum_{i=1}^{N} p_i l_i. \tag{3.18}$$

For large t,

$$\sum_{i=1}^{N} p_i b^{tl_i} \approx p_j b^{tl_j}, \tag{3.19}$$

where l_j is the largest of the numbers $l_1, \ldots, l_N$. Thus,

$$l(\infty) = \lim_{t \to \infty} l(t) = \max_{1 \le i \le N} l_i. \tag{3.20}$$

Moreover (Beckenbach and Bellman, 1961, see p.16), $l(t)$ is a monotonic non-decreasing function of t. Thus $l(0)$ is the conventional measure of mean length and $l(\infty)$ is the measure which would be used if the maximum length were of prime importance.

Note also that if all l_i are the same, say $l_i = l$, then $l(t) = l$. This is a reasonable property for any measure of length to possess.

Before proceeding to the coding theorem we need a definition. Rényi (1961, 1962) has introduced the entropy of order α, defined by

$$H_\alpha(P) \triangleq \frac{1}{1-\alpha} log_b \left(\sum_{i=1}^{N} p_i^\alpha \right) \qquad (\alpha \neq 1). \tag{3.21}$$

L'Hospital's rule shows that

$$H_1(P) \triangleq \lim_{\alpha \to 1} H_\alpha(P) = -\sum_{i=1}^{N} p_i log_b p_i. \tag{3.22}$$

Thus $H_1(P)$ is the ordinary Shannon entropy.

We can view Rényi's entropy as a generalization of Shannon's entropy.

The entropy of order α behaves in much the same way as H_1. For example, H_α is a continuous and symmetric function of $p_1, \ldots, p_N$. If $p_i = \frac{1}{N}$ for each i, then $H_\alpha = log_D N$. In addition, if X and Y are two independent sources, then

$$H_\alpha(X, Y) = H_\alpha(X) + H_\alpha(Y). \tag{3.23}$$

Properties of this sort can be used to give an axiomatic characterization of H_α in fashion similar to the well known axiomatic characterization of H_1. Axiomatic characterizations of H_α have been studied by Rényi (1961, 1962), Aczél and Daróczy

(1963a, 1963b), and Daróczy (1963) and Campbell (1965). The theorem of Campbell
[11] can be regarded as giving an alternative characterization of H_α in the same way
that the noiseless coding theorem provides an alternative characterization of H_1.

An inequality relating H_α and $l(t)$ is provided by the

Lemma 26 *Let* $n_1, \ldots, n_N$ *satisfy* (3.9). *Then*

$$l(t) \geq H_\alpha, \tag{3.24}$$

where $\alpha = 1/(t+1)$.

Proof If $t = 0$ and $\alpha = 1$ the result is given by Feinstein (1958) in his proof of the
neiseless coding theorem.

If $t = \infty$ and $\alpha = 0$ we have $l(\infty) = \max n_i$ and $H_0 = log_D N$. If n_i satisfy
(3.9) we must have

$$D^{-n_i} \leq \frac{1}{N}$$

for at least one value of i and hence for the maximum n_i. It follows easily that
$\max n_i \geq log_D N$.

Now let $0 < t < \infty$. By Hölder's inequality,

$$\left(\sum_{i=1}^{N} x_i^p \right)^{\frac{1}{p}} \left(\sum_{i=1}^{N} y_i^q \right)^{\frac{1}{q}} \leq \sum_{i=1}^{N} x_i y_i \tag{3.25}$$

where $\frac{1}{p} + \frac{1}{q} = 1$ and $p < 1$. Note that the direction of Hölder's inequality is the
reverse of the usual one for $p < 1$ (Beckenbach and Bellman, 1961, see p. 19). In
(3.25) becomes

$$\left(\sum_{i=1}^{N} p_i D^{t n_i} \right)^{-\frac{1}{t}} \left(\sum_{i=1}^{N} p_i^\alpha \right)^{\frac{1}{1-\alpha}} \leq \sum_{i=1}^{N} D^{-n_i}.$$

Therefore

$$\left(\sum_{i=1}^{N} p_i D^{t n_i} \right)^{\frac{1}{t}} \geq \frac{\left(\sum_{i=1}^{N} p_i^\alpha \right)^{\frac{1}{1-\alpha}}}{\sum_{i=1}^{N} D^{-n_i}} \geq \left(\sum_{i=1}^{N} p_i^\alpha \right)^{\frac{1}{1-\alpha}}, \tag{3.26}$$

where the last inequality follows from the assumtion that (3.9) is satisfied. If we
take logarithms of the first and third members of (3.26) we have the statement of the
lemma.   □

An easy calculation shows that we have equality in (3.24) and (3.26) and (3.9) is satisfied if

$$D^{-n_i} = \frac{p_i^\alpha}{\sum_{j=1}^N p_j^\alpha},$$

or

$$n_i = -\alpha \log_D p_i + \log_D \left(\sum_{j=1}^N p_j^\alpha \right). \tag{3.27}$$

Thus, if we ignore the additional constraint that each n_i should be an integer, it is seen that the minimum possible value of $l(t)$ is H_α, where $\alpha = \frac{1}{t+1}$. Moreover, as $p_i \to 0$, the optimum value of n_i is asymptotic to $\frac{-\log_D p_i}{t+1}$, so that the optimum length is less than $-\log_D p_i$ for $t > 0$ and sufficiently small p_i.

We can now proceed to prove a coding theorem for a noiseless channel with independent input symbols. Let a sequence of input symbols be generated independently, where each symbol is governed by the probability distribution $(p_1, \ldots, p_N)$. Consider a typical input sequence of length M, say $s = (a_1, \ldots, a_M)$. The probability of s is

$$P(s) = p_{i_1} \cdots p_{i_M} \tag{3.28}$$

if $a_1 = x_{i_1}, \ldots, a_M = x_{i_M}$. Let $n(s)$ be the length of the code sequence for s in some uniquely decipherable code. Then the length of order t for the M-sequences is

$$l_M(t) = \frac{1}{t} \log_D \sum P(s) D^{tn(s)} \qquad (0 < t < \infty), \tag{3.29}$$

where the summation extends over the N^M sequences s. The entropy of order α of this product space is

$$H_\alpha(M) = \frac{1}{1-\alpha} \log_D Q, \tag{3.30}$$

where

$$Q = \sum P(s)^\alpha. \tag{3.31}$$

It follows directly from (3.28) that

$$Q = \left(\sum_{i=1}^N p_i^\alpha \right)^M \tag{3.32}$$

and hence that

$$H_\alpha(M) = M H_\alpha. \tag{3.33}$$

Now let $n(s)$ be an integer which satisfies

$$-\alpha \log_D P(s) + \log_D Q \leq n(s) < 1 - \alpha \log_D P(s) + \log_D Q. \qquad (3.34)$$

As we remarked earlier in connection with (3.27), if every $n(s)$ equals the left member of (3.34) that

$$P(s)^{-\alpha t} Q^t \leq D^{tn(s)} < D^t P(s)^{-\alpha t} Q^t. \qquad (3.35)$$

If we multiply each member of (3.35) by $P(s)$, sum over all s, and use the fact that $\alpha t = 1 - \alpha$, we get

$$Q^{1+t} \leq \sum P(s) D^{tn(s)} < D^t Q^{1+t}.$$

Now we take logarithms, divide by t, and use the relations $1 + t = \frac{1}{\alpha}$ and $\alpha t = 1 - \alpha$. From (3.29) and (3.30) we have

$$H_\alpha(M) \leq l_M(t) < H_\alpha(M) + 1. \qquad (3.36)$$

Finally, if we divide by M and use (3.33), we have

$$H_\alpha \leq \frac{l_M(t)}{M} < H_\alpha + \frac{1}{M}. \qquad (3.37)$$

The quantity $\frac{l_M(t)}{M}$ can be called the average code length of order t per input symbol. By choosing M sufficiently large the average length can be made as close to H_α as desired. Thus we have proved most of

Theorem 13 *Let $\alpha = \frac{1}{1+t}$. By encoding sufficiently long sequences of input symbols it is possible to make the average code length of order t per input symbol as close to H_α as desired. It is not possible to find a uniquely decipherable code whose average length of order t is less than H_α.*

The second half of the theorem follows directly from (3.24) and (3.33). If $t = 0$ this is just the ordinary coding theorem. If $t = \infty$ the above proof is not quite correct but the theorem is still true. In this case we choose each $n(s)$ to satisfy

$$log_D N^M \leq n(s) < 1 + log_D N^M.$$

Then since $H_0 = log_D N$ and $l_M(\infty) = \max(n(s))$, we have

$$H_0 \leq \frac{l_M(\infty)}{M} < H_0 + \frac{1}{M}.$$

thus the theorem follows as before.

There are further entropy measures not covered by the class of generalized entropies. Therefore we introduce a unification of entropy measures denoted as (h, ϕ)-entropies.

3.2.4 (h, ϕ)-Entropies

Salicrú et al. (1993) presented a unification of entropy measures by means of (h, ϕ)-entropy measures. In our context the (h, ϕ)-entropy associated to a probability distribution P is given by

$$H_\phi^h(P) = h\left(\sum_{i=1}^{N} \phi(p_i)\right) \tag{3.38}$$

where either $\phi : [0, \infty) \to \mathbb{R}$ is concave and $h : \mathbb{R} \to \mathbb{R}$ is increasing or $\phi : [0, \infty) \to R$ is convex and $h : \mathbb{R} \to \mathbb{R}$ is decreasing. In the following table we shall suppose that ϕ is concave and h is increasing and we present some examples of (h, ϕ)-entropy measures.

$\phi(x)$	$h(x)$	(h, ϕ)-entropy
$-x\log x$	x	Shannon (1948)
x^r	$(1-r)^{-1}\log x$	Rényi (1961)
x^{r-m+1}	$(m-r)^{-1}\log x$	Varma (1966)
$x^{\frac{r}{m}}$	$m(m-r)\log x$	Varma (1966)
$(1-s)^{-1}(x^s - x)$	x	Havrda and Charvat (1967)
$x^{1/t}$	$(2^{t-1} - 1)^{-1}(x^t - 1)$	Arimoto (1971)
$x\log x$	$(1-s)^{-1}(exp_2((s-1)x) - 1)$	Sharma and Mittal (1975)
x^r	$(1-s)^{-1}(x^{\frac{s-1}{r-1}-1})$	Sharma and Mittal (1975)
$x^r\log x$	$-2^{r-1}x$	Taneja (1975)
$x^r - x^s$	$(2^{1-r} - 2^{1-s})^{-1}x$	Sharma and Taneja (1975)
$(1 + \lambda x)(log(1 + \lambda x))$	$(1 + 1/\lambda)log(1 + \lambda) - \frac{x}{\lambda}$	Ferreri (1980)

The convexity in the applications of all af them play a key role. An advantage of the unification is obvious, i.e., one can draw the common property of those entropies and find a common formulation to their applications.

3.3 Lecture on the Entropy Function for Continuous Random Variables

3.3.1 Differential Entropy and Shannon's Entropy Power Inequalities

We will generalize the notion of Shannon's entropy function introduced by the formula

$$H(P) = -\sum_{x \in \mathcal{X}} P(x) \log P(x),$$

where $\mathcal{X}$ is a finite set and $\{P(x), x \in \mathcal{X}\}$ is a PD on the set $\mathcal{X}$.

Recall that we write for $H(P)$ also $H(X)$, if X is a RV with distribution $P_X = P$.

We explained earlier this ambiguous notation, which does not result in confusion, if one gets used to it. Studying independent properties of RV's the notation $H(X)$ is more convenient, and studying for instance continuity problems in the distribution P the notation $H(P)$ is more convenient.

Now we consider instead of discrete RV's continuous RV's, which have a probability density function (abbreviated PDF) $f : \mathbb{R} \to \mathbb{R}_+$.

Definition 15 The function

$$h(f) = - \int f(x) \ln f(x) dx,$$

where $f(x), x \in (-\infty, \infty)$, is a PDF, is called the differential entropy.

For example, if $f(x) = 1/a$ for $x \in [0, a]$ and $f(x) = 0$, otherwise, where $a > 0$ is a given constant, then

$$h(f) = - \int\limits_0^a \frac{1}{a} \ln \frac{1}{a} dx = \ln a.$$

Again it is convenient to use two notations, $h(f)$ and $h(X)$ in this case.

Definition 16 For two continuous RV's X and Y the function

$$h(X|Y) = h(X, Y) - h(Y)$$

is called the conditional differential entropy.

Definition 17 For two densities f and g the function

$$D(f\|g) = \int f(x) \ln \frac{f(x)}{g(x)} dx$$

is called the relative differential entropy and the function

$$I(X; Y) = D(P_{XY}|P_X \times P_Y) \geq 0$$

is the continuous mutual information.

Note that $I(X; Y)$ has the destinction of retaining its fundamental significance as a measure of discrete mutual information, because it is the supremum of discrete mutual information defined for partitions.

Lemma 27 *The following properties hold:*

(i) *Rescaling : for all $c > 0$,*

$$h(cX) = h(X) + \ln c,$$

i.e., contrary to the ordinary enropy function, differential entropy can be negative and it is not invariant under multiplication by a constant.

(ii) *Invariance to the shifts :*
$$h(f + c) = h(f),$$

i.e., one can assign $c = -\mathbb{E}X$ and do not change $h(f)$.

(iii) *Transformations: for a continuous vector valued RV $\vec{X}$ and a matrix A*

$$h(A\vec{X}) = h(\vec{X}) + \log(det A),$$

and in general for a bijective transformation from $\vec{X}$ to $T\vec{X} = \vec{Y}$ the Jacobian of T replaces $det A$.

Lemma 28 *The following properties hold*

(i) *For two densities f and g*
$$D(f\|g) \geq 0,$$

with equality if $f = g$ almost everywhere.

(ii) *For continuous RV's X and Y*

$$I(X; Y) \geq 0 \quad and \quad h(X|Y) \leq h(X)$$

with equalities iff X and Y are independent.

(iii) *The chain rule for differential entropy holds as in the discrete case*

$$h(X_1, \ldots, X_n) = \sum_{i=1}^{n} h(X_i|X_1, \ldots, X_{i-1}) \leq \sum_{i=1}^{n} h(X_i),$$

with equality iff $X_1, X_2, \ldots, X_n$ are independent.

An important operation on the space of the density functions is known as a *convolution*: if f_1 and f_2 are two densities, then their convolution, $f = f_1 * f_2$, is defined by

$$f(z) = \int_{-\infty}^{\infty} f_1(x) f_2(z - x) dx.$$

Note that for independent RV's X and Y, we have $f_{X+Y} = f_X * f_Y$.

The result formulated below plays a key role in many approaches to multi-user problems. This result was conjectured by Shannon in 1949. The first step of the proof was proposed by de Bruijn and the proof was completed by Stam [12].

Theorem 14 (Entropy power inequality, EPI) *Suppose that X and Y are independent RV's having densities f and g. Then*

$$e^{2h(X+Y)} \geq e^{2h(X)} + e^{2h(Y)} \tag{3.39}$$

with equality iff X and Y are Gaussian.

The fact that the equation in (3.39) is valid for independent Gaussian RV's can be easily checked. Let

$$G_\sigma(x) = \frac{1}{\sqrt{2\pi\sigma^2}} e^{-\frac{x^2}{2\sigma^2}},$$

then

$$
\begin{aligned}
h(G_\sigma) &= - \int_{-\infty}^{\infty} \frac{1}{\sqrt{2\pi\sigma^2}} e^{-\frac{x^2}{2\sigma^2}} \left[-\ln\sqrt{2\pi\sigma^2} - \frac{x^2}{2\sigma^2} \right] dx \\
&= \ln\sqrt{2\pi\sigma^2} + \sigma^2 \frac{1}{2\sigma^2} \\
&= \frac{1}{2}\ln 2\pi e\sigma^2.
\end{aligned}
$$

Hence,

$$e^{2h(G_\sigma)} = 2\pi e\sigma^2.$$

If X and Y are independent Gaussian RV's, then their sum is also a Gaussian RV with density $G_{(\sigma_X^2 + \sigma_Y^2)^{1/2}}$ and

$$\exp\{2h(G_{(\sigma_X^2+\sigma_Y^2)^{1/2}})\} = 2\pi e(\sigma_X^2 + \sigma_Y^2) = \exp\{2h(G_{\sigma_X})\} + \exp\{2h(G_{\sigma_Y})\}.$$

Note that the EPI is scale invariant, i.e., we can multiply both sides of (3.39) by $e^{2\ln c}$, where $c > 0$, and obtain

$$e^{2h(cX+cY)} \geq e^{2h(cX)} + e^{2h(cY)}.$$

Furthermore, the following statement is true.

Lemma 29

$$\max_{X:Var(X)=\sigma^2} h(X) = h(G_\sigma).$$

Proof Note that

$$\int_{-\infty}^{\infty} f(x) \ln \frac{1}{G_\sigma(x)} dx = \ln \sqrt{2\pi\sigma^2} + \int_{-\infty}^{\infty} f(x) \frac{x^2}{2\sigma^2} dx = h(G_\sigma),$$

where w.l.o.g. (the differential entropy is invariant to the shifts) we assumed that $\mathbb{E}(X) = 0$. Using the well-known inequality for the divergence between f and G_σ we obtain

$$0 \le D(f \parallel G_\sigma) = \int_{-\infty}^{\infty} f(x) \ln \frac{f(x)}{G_\sigma(x)} dx = h(G_\sigma) - h(f). \qquad \square$$

Let us check that (3.39) is true when

$$f_X(x) = \begin{cases} 1/a & \text{if } x \in [0, a] \\ 0 & \text{otherwise,} \end{cases}$$

$$f_Y(y) = \begin{cases} 1 & \text{if } y \in [0, 1] \\ 0 & \text{otherwise,} \end{cases}$$

where $a \ge 1$ is given. Setting $Z = X + Y$ we get

$$f_Z(z) = \begin{cases} z/a & \text{if } z \in [0, 1) \\ 1/a & \text{if } z \in [1, a] \\ (1 + a - z)/a & \text{if } z \in (a, 1 + a] \\ 0 & \text{otherwise.} \end{cases}$$

Then

$$h(f_X) = \ln a, \quad h(f_Y) = 0,$$

and

$$h(f_Z) = -\int_0^1 \frac{z}{a} \ln \frac{z}{a} dz - \int_1^a \frac{1}{a} \ln \frac{1}{a} dz - \int_a^{1+a} \frac{1 + a - z}{a} \ln \frac{1 + a - z}{a} dz$$

$$= \frac{2I}{a} + \ln a$$

where

$$I = -\int_0^1 z \ln z \, dz.$$

It is easy to see that

$$-\int z \ln z\, dz = -\frac{1}{2}z^2 \ln z + \frac{1}{4}z^2.$$

Since

$$\lim_{z \to 0} z^2 \ln z = 0,$$

we obtain $I = 1/4$, and the EPI inequality becomes

$$\frac{1}{a} + \ln a^2 \geq \ln(a^2 + 1),$$

or

$$1 \geq \max_{a \geq 1} a \cdot \ln(1 + a^{-2}) = \ln 2,$$

where the last equation follows from the observation that the maximum is attained for $a = 1$.

The entropy maximizing property of Lemma 29 extends to the multivariate normal

Lemma 30 *Let the continuous random vector* $\vec{X}$ *have also mean and covariance* $K = \mathbb{E}\vec{X}\vec{X}^*$, *where* $\vec{X}^*$ *is the transposed vector to* $\vec{X}$, *i.e.,* $K_{ij} = \mathbb{E}X_i X_j$, $1 \leq i\, m j \leq n$, *then*

$$h(\vec{X}) \leq \frac{1}{2} \ln(2\pi e)^n det(K),$$

with equality iff $f(\vec{X}) = \phi_K(\vec{X})$, *where*

$$\phi_K(\vec{X}) = \frac{1}{(2\pi)^{\frac{n}{2}} det(K)^{\frac{1}{2}}} e^{-\frac{1}{2}\vec{X}^* K^{-1} \vec{X}}$$

is the multivariate Gaussian density with mean $\vec{0}$ *and covariance matrix* K.

Proof Note by simple calculation that

$$-\int \phi_K \ln \phi_K = \frac{1}{2} \ln(2\pi e)^n det(K). \tag{3.40}$$

Let $g(\vec{X})$ be any density satisfying

$$\int g(\vec{X}) X_i X_j d\vec{X} = K_{ij} \quad \text{for all } i, j,$$

then

$$0 \leq D(g\|\phi_K) = \int g \ln \frac{g}{\phi_K} = -h(g) - \int g \ln \phi_K$$
$$= -h(g) - \int \phi_K \ln \phi_K = -h(g) + h(\phi_K),$$

where the equation $\int g \ln \phi_K = \int \phi_K \ln \phi_K$ follows from the fact that g and ϕ_K have the same moments of the quadratic form $\ln \phi_K(\vec{X})$. $\qquad\square$

3.3.2 Probabilistic Notion of Entropy

The notion of entropy in Thermodynamics was introduced by R. Clausius (1876), who formulated the Second Law of Thermodynamics—*the entropy of a closed system is steadily increasing*. The notion of entropy and the Second Law have been put in another light by L. Boltzmann (1896–1898) who showed that *the entropy is proportional to the logarithm of the state probability,* and proved on this basis that *the increase of entropy for a closed system is not sure, but has a very great probability provided that the system is staying in a state with small probability.* Boltzmann also introduced the so-called H-function. We will present the results of the paper by J. Balatoni and A. Rényi [13] and consider the general notion of entropy in the sense of Probability Theory and discuss its connections with the notion of entropy in Statistical Mechanics, and H-function (we also strongly recommend [14, 15] for reading).

The entropy function

$$H(X) = - \sum_x P(x) \log P(x),$$

introduced by C. Shannon, has several properties that allow to consider it as a measure of the uncertaincy in the value of a RV X. These properties are as follows:

(i) if $P(x) = 1$ for some x, then $H(X) = 0$.

(ii) if $f : \mathcal{X} \leftrightarrow \mathcal{X}$ is a one-to-one mapping, then $H(f(X)) = H(X)$ and if there are at least two values, x_k and x_l such that $P(x_k), P(x_l) > 0$ and $f(x_k) = f(x_l)$, then $H(f(X)) < H(X)$.

(iii) $H(X, Y) \leq H(X) + H(Y)$ with the equation iff the RVs X and Y are statistically independent.

(iv) $H(X)$ continuously depends on the distribution of RV X;

(v) if n is the maximal number of all possible values of X, then $H(X) \leq \log n$ with the equation iff the distribution P is uniform.

(vi) if $X_1, \ldots, X_k$ are RVs such that $\Pr\{X_i = x\}\Pr\{X_j = x\} = 0$ for all $i \neq j$ and $x \in \mathcal{X}$, and if a RV X coincides with X_i with probability q_i, then

$$H(X) = \sum_{i=1}^{n} q_i H(X_i) - \sum_{i=1}^{n} q_i \log q_i.$$

(vii) if $A_1, \ldots, A_k$ are possible outcomes of a random experiment where A_i has probability p_i, then $X = -\log p_k$ may be regarded as a RV and $\mathbb{E}X = H(X)$.

Repeating the experiment N times we get that a sequence containing N_i outcomes $A_i, i = 1, ..., k$, has probability $P = p_1^{N_1}, ..., p_k^{N_k}$. Thus,

$$\mathbb{E} \log P = -N H_0,$$

where

$$H_0 = -\sum_{i=1}^{k} p_i \log p_i$$

denotes the entropy of the experiment. Furthermore,

$$Var(\log P) \le \sum_{i=1}^{k} Var(N_i) \log^2 p_i = N \sum_{i=1}^{k} p_i (1 - p_i) \log^2 p_i.$$

Consequently, $\log P$ makes, as a rule, oscillations of order $\sqrt{N}$ around the value $-N H_0$ and

$$P \sim 2^{-N H_0}.$$

3.3.3 Connections Between Entropy in the Sense of Statistical Mechanics, Boltzmann's H-Function, and Entropy in the Sense of Probability Theory

Let us consider the model of a perfect gas. If a mono-atomic gas consisting of N atoms, each of mass m, has temperature T and the gas is enclosed in a vessel of volume V, then—after leaving the gas to itself for a sufficiently long time—in consequence of collisions between atoms a state will evolve in which a distribution of the atoms in the vessel is nearly uniform and the velocities of the atoms approximately follow the Maxwell distribution, i.e., velocity components of the atoms are mutually independent and each has approximately normal distribution with expectation 0 and standard deviation $\sigma = \sqrt{kT/m}$, where k is Boltzmann's constant. Thus the position of a gas atom "chosen" at random may be regarded, as a 3-dimensional random vector with uniform distribution in all of the vessels, while the velocity components of the atom may be looked upon as random variables that are independent of each other and the position of the atom and have normal distribution.

We will use the following facts.

(i) If $\overrightarrow{X} = (X_1, ..., X_n)$ is the n-dimensional random vector having probability density function $f(\mathbf{x})$, then its entropy is given by the formula

$$H(\overrightarrow{X}) = -\int f(\mathbf{x}) \log f(\mathbf{x}) \mathbf{dx},$$

where $\mathbf{dx} = dx_1 \ldots dx_n$ and the integral is extended over the n-dimensional space. If, in particular, $\vec{X}$ has uniform distribution in an n-dimensional domain of volume V, then

$$H(\vec{X}) = \log V.$$

(ii) If $X_1, \ldots, X_n$ are independent RV's with absolutely continuous distribution, then the n-dimensional entropy of their joint distribution is equal to the sum of the one-dimensional entropies of the separate variables, i.e.,

$$H(\vec{X}) = \sum_{i=1}^{n} H(X_i).$$

(iii) If X has Gaussian distribution, then

$$H(X) = \frac{1}{2} \log(2\pi e \sigma^2).$$

Let $\vec{X} = (X_1, X_2, X_3)$ denote the vector consisting of the velocity coordinates and $\vec{Y} = (Y_1, Y_2, Y_3)$ denote the position vector of an atom. Then the atom is characterized by the 6-dimensional random vector $\vec{Z} = (\vec{X}, \vec{Y})$ and

$$H(\vec{Z}) = H(X_1) + H(X_2) + H(X_3) + H(\vec{Y}) = \frac{3}{2} \log(2\pi e \sigma^2) + \log V.$$

Taking into account that $\sigma = \sqrt{kT/m}$, we find

$$H(\vec{Z}) = \frac{3}{2} \log T + \log V + \frac{3}{2} \log \frac{2\pi e k}{m}.$$

Since each atom has this amount of entropy, the entropy of the whole gas Γ, provided that there are altogether N atoms present, equals

$$H(\Gamma) = N \left[\frac{3}{2} \log T + \log V + \frac{3}{2} \log \frac{2\pi e k}{m} \right]. \tag{3.41}$$

The definition of entropy in Thermodynamics, as it is well-known, leads, apart from the selection of the base of the logarithms and the factor k, to the same result, i.e., if S denotes the thermodynamical entropy, then

$$S = kN \left[\frac{3}{2} \log T + \log V + c \right],$$

where the value of the constant c depends neither on volume nor on temperature. This formula can be proved in the following way. We write

$$dS = \frac{dQ}{T},$$

where dQ signifies the quantity of heat transferred to the gas; $dQ = dE + pdV$, where E denotes the kinetic energy, p the pressure, and V the volume of the gas. Making use of the formula $E = (3/2)kTN$ well-known from the Kinetic Theory of Gases, and the gas law $pV = NkT$, we obtain

$$dQ = \frac{3}{2}NkdT + NkT\frac{dV}{V}.$$

Therefore,

$$dS = \frac{3}{2}Nk\frac{dT}{T} + Nk\frac{dV}{V},$$

$$S = Nk\left[\frac{3}{2}\log T + \log V + c\right].$$

Consequently, since in Thermodynamics the additive constant can be selected arbitrarily, the thermodynamical entropy is, apart from a constant factor, equal to the differential entropy of the gas.

Boltzmann defined the so-called H-function (note that the term H-functional would be more correct) by the formula

$$H = \int f(x)\log f(x)dx.$$

This function is identical with -1 times the probabilistic entropy.

Let us now examine entropy in the sense of Statistical Mechanics, where the Boltzmann formula

$$S = k\log W = -H + C$$

is valid. W stands for the probability of a momentary state and H is the H-function belonging to the state. Note that it is necessary to distinguish between the two notions of entropy: entropy as a RV on the one hand, and maximal (or probabilistic) entropy, as a *constant* characterizing the system on the other.

Let us divide the phase space into n cells. If the probability that a particle enters the ith cell is equal to $1/n$ and there are N particles, then the probability that the ith cell has N_i particles, $i = 1, ..., n$, is expressed by the polynomial distribution

$$\frac{N!}{N_1!...N_n!}\frac{1}{n^N},$$

provided that

$$\sum_{i=1}^{n} N_i = N, \quad \sum_{i=1}^{n} N_i E_i = E,$$

where E_i is the energy level of the ith cell and E is the total energy of the system. This means that the probability of the state of the system is expressed as

$$W = c \frac{N!}{N_1!...N_n!},$$

where c does not depend on N_1, ..., N_n. Let us determine the approximate value of $\log W$ when the numbers N_1, ..., N_n are very large. Let $f_i = N_i/N$. Then according to the Stirling's approximation formula, and

$$\log W \approx -N \sum_{i=1}^{N} f_i \log f_i + \log c. \tag{3.42}$$

Thus, the logarithm of W is, apart from an additive constant, proportional to the entropy of the empirical distribution. The most probable state is the one for which $-\sum_{i=1}^{N} f_i \log f_i$ is maximal, where the maximization domain contains $f_1, ..., f_N$ such that

$$\sum_{i=1}^{n} f_i = f, \quad \sum_{i=1}^{n} f_i E_i = \frac{E}{N}. \tag{3.43}$$

Using the method of Lagrange multipliers, we obtain that the maximum is attained for

$$f_i = A e^{-\beta E_i},$$

where the constants A and β should be determined by the Eq. (3.43).

Let the system under consideration be a mono-atomic perfect gas enclosed in a vessel of volume V and consisting of atoms of mass m; we have the relation $E/N = (3kT)/2$ which can in this connection be regarded as the definition of temperature. Assuming that cells of equal probability in the phase space correspond to regions of equal volume in space and to regions of equal volume in the space of velocity components we find, approximately replacing the sums by integrals, that the constants must satisfy the following conditions:

$$1 = V A \int_{-\infty}^{\infty} \int_{-\infty}^{\infty} \int_{-\infty}^{\infty} \exp\{-m\beta(x_1^2 + x_2^2 + x_3^2)/2\} dx_1 dx_2 dx_3,$$

$$\frac{3kT}{2} = \frac{1}{2} m V A \int_{-\infty}^{\infty} \int_{-\infty}^{\infty} \int_{-\infty}^{\infty} (x_1^2 + x_2^2 + x_3^2)$$

$$\times \exp\{-m\beta(x_1^2 + x_2^2 + x_3^2)/2\} dx_1 dx_2 dx_3.$$

Calculating the integrals we get

$$1 = V A \left(\frac{2\pi}{m\beta} \right)^{3/2},$$

$$\frac{3kT}{2} = \frac{1}{2} m V A \left(\frac{2\pi}{m\beta} \right)^{3/2} \frac{3}{m\beta}.$$

Thus,

$$A = \frac{(2\pi m/k)^{3/2}}{V T^{3/2}}, \quad \beta = \frac{1}{kT}.$$

Hence, using (3.42), we obtain the following approximation for the entropy of a perfect gas,

$$H = \log W \approx \frac{3kT}{2} = -N \sum_{i=1}^{n} f_i (\log A - \beta E_i + \log c)$$

$$= -N \log A + \beta E + \log c$$

$$= N \left(\frac{3}{2} \log T + \log V \right) + C,$$

where all terms independent of T and V are included into C. Thus, we arrived at the expression (3.41)!

3.3.4 Determinantal Inequalities via Differential Entropy

We assume in this section that K is a non-negative definite symmetric $n \times n$-matrix. In our main chapter on inequalities we saw that convexity is a prime issue for basic analytical and also entropy inequalities. Perhaps it is not susprising that the latter kept some properties of their origin and yield proofs for determinantal inequalities where proofs based on convexity arguments were given before.

We begin with the simplest inequality used for Gaussian channels with feedback [16].

Theorem 15 *If K_1 and K_2 are nonnegative definite symmetric matrices, then*

$$det(K_1 + K_2) \geq det(K_1)$$

Proof Let $\vec{X}$, $\vec{Y}$ be independent random vectors with density functions ϕ_{K_1} and ϕ_{K_2}. Then $\vec{X} + \vec{Y}$ has density $\phi_{K_1+K_2}$, and hence

$$\frac{1}{2} \ln(2\pi e)^n det(K_1 + K_2) = h(\vec{X} + \vec{Y}) \geq h(\vec{X} + \vec{Y} | \vec{Y})$$

$$= h(\vec{X} | \vec{Y}) = h(\vec{X}) = \frac{1}{2} \ln(2\pi e)^n det(K_1) \qquad \square$$

Theorem 16 (Hadamard) *If $K = (K_{ij})$ is a nonnegative definite symmetric matrix, then*

$$det(K) \leq \prod_{1 \leq i \leq n} K_{ii}$$

with equality iff $K_{ij} = 0$ for $i \neq j$.

Proof Let random vector $\vec{X} = (X_1, X_2, \ldots, X_n)$ have density ϕ_K, then

$$\frac{1}{2} \ln(2\pi e)^n det(K) = h(X_1, X_2, \ldots, X_n)$$

$$\leq \sum_{i=1}^{n} h(X_i) = \sum_{i=1}^{n} \frac{1}{2} \ln 2\pi e det(K_{ii})$$

with equality iff $X_1, X_2, \ldots, X_n$ are independent, i.e., $K_{ij} = 0, i \neq j$. $\qquad\square$

Theorem 17 (Ky Fan) $\ln det(K)$ *is concave.*

Proof Let $\vec{X}_1$ and $\vec{X}_2$ be Gaussian distributed n-vectors with densities ϕ_{K_1} and ϕ_{K_2}, and let RV θ satisfy

$$Pr(\theta = 1) = \lambda, \quad Pr(\theta = 2) = 1 - \lambda, \quad 0 \leq \lambda \leq 1.$$

Let $\vec{X}_1, \vec{X}_2$, and θ be independent and let $Z = X_\theta$. Then Z has covariance $K_Z = \lambda K_1 + (1 - \lambda)K_2$. However, Z is not multivariate Gaussian and therefore (by Lemma 30)

$$\frac{1}{2} \ln(2\pi e)^n det(\lambda K_1 + (1 - \lambda)K_2) \geq h(Z) \geq h(Z|\theta) \quad \text{(by Lemma 28 (ii))}$$

$$= \lambda \frac{1}{2} \ln(2\pi e)^n det(K_1) + (1 - \lambda) \frac{1}{2} \ln(2\pi e)^n det(K_2)$$

Thus

$$det(\lambda K_1 + (1 - \lambda)K_2) \geq det(K_1)^\lambda det(K_2)^{1-\lambda},$$

as claimed. $\qquad\square$

Finally we use the deeper EPI and derive again Minkowski's inequality

$$det(K_1 + K_2)^{\frac{1}{n}} \geq det(K_1)^{\frac{1}{n}} > det(K_2)^{\frac{1}{n}} \tag{3.44}$$

Proof Let $\vec{X}_1, \vec{X}_2$ be independent with densities ϕ_{K_1} and ϕ_{K_2}. Then $\vec{X}_1 + \vec{X}_2$ has density $\phi_{K_1+K_2}$ and the EPI yields

$$(2\pi e)\det(K_1 + K_2)^{\frac{1}{n}} = e^{(2/n)h(\vec{X}_1+\vec{X}_2)}$$

$$\geq e^{(2/n)h(\vec{X}_1)} + e^{(2/n)h(\vec{X}_2)}$$

$$= (2\pi e)\det(K_1)^{\frac{1}{n}} + (2\pi e)\det(K_2)^{\frac{1}{n}}. \qquad \square$$

Further Reading

1. J. Aczél, Z. Daróczy, Charakterisierung der entropien positiver Ordnung und der Shannon Chen Entropie. Acla Math. Acad. Sci. Hung. **14**, 95–121 (1963)
2. E.F. Beckenbach, R. Bellman, *Inequalities* (Springer, Berlin, 1961)
3. R.E. Bellman, Notes on matrix theory IV: an inequality due to Bergstrøm. Amer. Math. Monthly **62**, 172–173 (1955)
4. N. Blachman, The convolution inequality for entropy powers. IEEE Trans. Inf. Theory **IT-11**, 267–271 (1965)
5. B.S. Choi, T. Cover, An information-theoretic proof of Burg's maximum entropy spectrum. Proc. IEEE **72**, 1094–1095 (1984)
6. T. Cover, A. El Gamal, An information theoretic proof of Hadamard's inequality. IEEE Trans. Inf. Theory **IT-29**, 930–931 (1983)
7. I. Csiszár, *I*-Divergence geometry of probability distributions and minimization problems. Ann. Probab. **3**(1), 146–158 (1975)
8. I. Csiszár, Why least squares and maximum entropy? An axiomatic approach to interference for linear inverse problems. Ann. Statist. **19**(4), 2033–2066 (1991)
9. I.Csiszár, New axiomatic results on inference for inverse problems. Studia Sci. Math. Hungar **26**, 207–237 (1991)
10. Z. Daróczy, Über die gemeinsame Charakterisierung der zu den nicht vollständigen Verteilungen gehörigen Entropien von Shannon und Rényi". Z. Wahrscheinlichkeitstheorie u. Verw. Gebiete **1**, 381–388 (1963)
11. Ky Fan, On a theorem of Weyl concerning the eigenvalues of linear transformations, II. Proc. Natl. Acad. Sci. USA **36**, 31–35 (1950)
12. Ky Fan, Some inequalities concerning positive-definite matrices. Proc. Cambridge Philos. Soc. **51**, 414–421 (1955)
13. A. Feinstein, *Foundations of Information Theory* (McGraw-Hill, New York, 1958)
14. S. Kullback, *Information Theory and Statistics* (Willey, New York, 1959)
15. A. Marshall, I. Olkin, *Inequalities: Theory of Majorization and Its Applications* (Academic Press, New York, 1979)
16. A. Marshall, I. Olkin, A convexity proof of Hadamard's inequality. Amer. Math. Monthly **89**, 687–688 (1982)
17. H. Minkowski, Diskontunuitätsbereich für arithmetische Äquivalenz. J. für Math. **129**, 220–274 (1950)
18. L. Mirsky, On a generalization of Hadamard's determinantal inequality due to Szasz. Arch. Math. **VIII**, pp. 274–275 (1957)
19. A. Oppenheim, Inequalities connected with definite hermitian forms. J. Lond. Math. Soc. **5**, 114–119 (1930)
20. A. Rényi, On measures of entropy and information, *Proceedings of the Berkeley Symposium on Mathematical Statistics and Probability*, Vol. 1 (University of California Press, Berkeley, 1961), pp. 547–561

21. A. Rényi, *Wahrscheinlichkeitsrechnung Mit einem Anhang über Informationstheory* (VEB Deutscher Verlag der Wissenschaften, Berlin, 1962)
22. A. Rényi, On the amount of information in a frequency-count, 1964. *Selected Papers of Alfred Rényi*, Vol. 3 (Akadémiai Kiadó, Budapest, 1976), pp. 280–283
23. A. Rényi, On the amount of missing information and the Neymann-Pearson lemma, 1966. *Selected Papers of Alfred Rényi*, Vol. 3 (Akadémiai Kiadó, Budapest, 1976), pp. 432–439
24. A. Rényi, On the amount of information in a random variable concerning an event, 1966. *Selected Papers of Alfred Rényi*, Vol. 3 (Akadémiai Kiadó, Budapest, 1976), pp. 440–443
25. A. Rényi, Statistics and information theory, 1967. *Selected Papers of Alfred Rényi*, Vol. 3 (Akadémiai Kiadó, Budapest, 1976), pp. 449–456
26. A. Rényi, On some basic problems of statistics from the point of view of information theory, 1965–1967. *Selected Papers of Alfred Rényi*, Vol. 3 (Akadémiai Kiadó, Budapest, 1976), pp. 479–491
27. A. Rényi, Information and statistics, 1968. *Selected Papers of Alfred Rényi*, Vol. 3 (Akadémiai Kiadó, Budapest, 1976), pp. 528–530
28. J.E. Shore, R.W. Johnson, Axiomatic derivation of the principle of maximum entropy and principle of minimum of cross-entropy. IEEE Trans. Inf. Theory **26**, 26–37 (1980)

References

1. C.E. Shannon, A mathematical theory of communication. Bell Syst. Tech. J. **27**, 623–656 (1948)
2. A.Y. Khinchin, The concept of entropy in probability theory, (in Russian), Uspekhi Mat. Nauk, 8(3), *translation in "Mathematical Foundations of Information Theory"* (Dover Publications Inc, New York, 1953). 1958, pp. 3–20
3. D.K. Faddeev, On the concept of entropy of a finite probabilistic scheme. Uspehi Mat. Nauk **11**(1), 227–231 (1956)
4. J. Aczél, Z. Daróczy, *On measures of information and their characterizations, Mathematics in Science and Engineering*, vol. 115 (Academic Press, New York-London, 1975)
5. R. Ahlswede, Multi-way communication channels, in *Proceedings of 2nd International Symposium on Information Theory, Thakadsor, Armenian SSR*, vol. 23–52 (Akademiai Kiado, Budapest, 1973) Sept. 1971
6. R. Ahlswede, The capacity region of a channel with two senders and two receivers. Ann. Probab. **2**(5), 805–814 (1974)
7. R.L. Dobrushin, General formulation of shannon's basic theorem in information theory, (in russian) uspehi mat. Nauk **14**(6), 3–104 (1959)
8. R.L. Dobrushin, *Arbeiten zur Informationstheorie IV* (VEB Deutscher Verlag der Wissenschaften, Berlin, 1963)
9. F. Topsøe, A new proof of a result concerning computation of the capacity for a discrete channel. Z. Wahrscheinlichkeitstheorie und Verw. Gebiete **22**, 166–168 (1972)
10. F. Topsøe, *Informationstheorie* (Teubner, Stuttgart, 1973)
11. L.L. Campbell, A coding theorem and Rényi' s entropy. Inf. Control **8**(4), 423–429 (1965)
12. A. Stam, Some inequalities satisfied by the quantities of information of fisher and shannon. Inf. Control **2**, 101–112 (1959)
13. J. Balatoni, A. Rényi, On the notion of entropy, 1956. *Selected Papers of Alfred Rényi*, Vol. 1, 558–585 (Akadémiai Kiadó, Budapest, 1976)
14. A. Rényi, On the amount of information concerning an unknown parameter in a space of observations, 1964. *Selected Papers of Alfred Rényi*, Vol. 3 (Akadémiai Kiadó, Budapest, 1976), pp. 272–279

15. A. Rényi, On some problems of statistics from the point of view of information theory, 1969.
    Selected Papers of Alfred Rényi, Vol. 3 (Akadémiai Kiadó, Budapest, 1976), pp. 560–576
16. S. Pombra, T. Cover, Gaussian feedback capacity. IEEE Trans. Inf. Theory **35**(1), 37–43 (1989)

Chapter 4
Universal Coding

4.1 Lecture on Universal Codeword Sets and Representations of the Integers

4.1.1 Minimal Codes for Positive Integers

We will deal with the source $(\mathbb{N}, P)$ whose set of messages is $\mathbb{N} = \{1, 2, ...\}$ and the probability distribution is given by $P = \{P(j), j \in \mathbb{N}\}$. We also assume that

$$P(1) \geq P(2) \geq P(3) \geq \cdots$$

Note that our considerations can be extended to a more general case of countable sources defined as follows.

Definition 18 A countable source is a countable set of messages $\mathcal{X}$ and a probability distribution function $P : \mathcal{X} \to (0, 1]$ that assigns a probability $P(m) > 0$ to each message $m \in \mathcal{X}$.

We can number the elements of a countable set $\mathcal{X}$ in order of non-increasing probabilities, i.e.,

$$\mathcal{X} = \{m_1, m_2, m_3, ...\}, \quad P(m_1) \geq P(m_2) \geq P(m_3) \geq \cdots$$

and use the methods developed for positive integers meaning the bijection $j \leftrightarrow m_j$, $j \in \mathbb{N}$.

Definition 19 Let B be a finite set of symbols and let B^* be the set of all finite sequences of symbols, where each symbol belongs to B. A $|B|$-ary code for the source $(\mathbb{N}, P)$ is a set $C \subseteq B^*$ and a one-to-one function $f : \mathbb{N} \to C$, which assigns a distinct codeword to each integer j such that $P(j) > 0$.

A. Ahlswede et al. (eds.), *Storing and Transmitting Data*,
Foundations in Signal Processing, Communications and Networking 10,
DOI: 10.1007/978-3-319-05479-7_4, © Springer International Publishing Switzerland 2014

1. The code (C, f) is uniquely decipherable iff every concatenation of a finite number of codewords is a distinct sequence in B^* so that

$$f^* : \mathbb{N}^* \to C^* \implies (f^*)^{-1} : C^* \to \mathbb{N}^*,$$

where $f^*(\mathbf{j}) = (f(j_1), f(j_2), ...) \in C^*$ for all $\mathbf{j} = (j_1, j_2, ...) \in \mathbb{N}^*$ such that $P(j_1), P(j_2), ... > 0$.

2. A code is saturated iff adding any new sequence $c' \in B^* \backslash C$ to C gives a code which is not uniquely decipherable.

3. A code is effective iff there is an effective procedure that can decide whether a given sequence in B^* is a codeword in C or not. For if C is not effective no algorithm can distinguish between C and B^*, while if C is effective an algorithm can partition a given sequence in B^* into segments in all possible ways and test the segments for membership in C.

Let us index the elements of C in order of non-decreasing length, and define the length function

$$L(j) = |c_j|,$$

where $|c_j|$ denotes the length of the sequence c_j, $j = 1, ..., |C|$, i.e.,

$$L(1) \le L(2) \le L(3) \le \cdots$$

The following results relate the properties of the function L to properties of the set C.

Theorem 18 *1. Let $C \subseteq B^*$ be a uniquely decipherable set with length function L. Then*

(i) *L satisfies the Kraft inequality*

$$\sum_{j=1}^{|C|} |B|^{-L(j)} \le 1; \tag{4.1}$$

(ii) *if there is equation in (4.1), then C is saturated;*

(iii) *if C is effective, then L is an effectively computable function.*

2. Let L be a non-decreasing function satisfying (4.1). Then

(i) *there is a prefix set $C \subseteq B^*$ with length function L;*

(ii) *if the prefix set $C \subseteq B^*$ in 2(i) is saturated, then every sequence in B^* is either the prefix of some sequence in C or has some sequence in C as prefix;*

(iii) *if C is finite and 2(ii) is valid, then there is equality in (4.1);*

(iv) *if L is effectively computable, the prefix set C in 2(ii) is effective.*

Proof The statement 1(i) is standard and 2(ii) follows from the observation that adding a sequence C' to C gives a term in (4.1) and violates the inequality. The statement 1(iii) follows since generating B^* in order of non-decreasing length and testing for membership in C gives an algorithm that computes L.

A prefix set can be represented as a set of paths of a non-uniform tree. If L satisfies (4.1), it is possible to derive an algorithm constructing the paths of length k when all shorter paths are known, and 2(i) follows. The property 2(ii) holds since if some $c' \in B^*$ is neither the prefix of any member of C nor has any member of C as a prefix, it can be added to C and the code is not saturated. The same argument proves 2(iii). The property 2(iv) is evident.   □

It is evident that no one-to-one code from $\mathcal{X}$ to C has a smaller average length than the encoding does $m_j \to c_j$, which assigns longer codewords to less probable messages and has the minimal codeword length

$$\sum_{j \in \mathbb{N}} P(j) \cdot |c_j| \tag{4.2}$$

determined by C and P alone. Such a code from $\mathbb{N}$ to C is called *minimal*. The average codeword length (4.2) of a minimal code does not depend on the details of the set C, but only on its length function L and is just the average value of the function L averaged with respect to the given distribution P, denoted by

$$\mathbb{E}_P(L) = \sum_{j \in \mathbb{N}} P(j) \cdot L(j).$$

By Theorem 18, if any decipherable code for a source $(\mathbb{N}, P)$ is minimal, there is a prefix code with the same codeword length, and the prefix set is effective if the uniquely decipherable set is also. There is, therefore, no advantage in either average codeword length or effective decipherability to be gained by using a uniquely decipherable set that is not a prefix set. Theorem 19 gives well-known lower and upper bounds to $\mathbb{E}_P(L)$ for uniquely decipherable and prefix sets.

Theorem 19 *Let B be a finite set and let $(\mathbb{N}, P)$ be a source with entropy*

$$H(P) = -\sum_{j \in \mathbb{N}} P(j) \cdot \log P(j)$$

computed using the logarithmic base $|B|$.

(i) *If C is a uniquely decipherable set with length function L, then*

$$\mathbb{E}_P(L) \geq \begin{cases} 0, & \text{if } H(P) = 0, \\ \max\{1, H(P)\}, & \text{if } 0 < H(P) < \infty. \end{cases}$$

(ii) *There is a prefix set $C_P \subseteq B^*$ with length function L_P given by*

$$L_P(j) = \lceil - \log P(j) \rceil, \quad j = 1, ..., |C_P| \tag{4.3}$$

and minimal average codeword length

$$\mathbb{E}_P(L_P) = \begin{cases} 0, & \text{if } H(P) = 0, \\ 1 + H(P), & \text{if } 0 < H(P) < \infty. \end{cases}$$

Proof Note that $H(P) = 0$ iff $P(m) > 0$ only for one $m \in \mathbb{N}$ (it means that $P(m) = 1$). Then the null string λ of no symbols is in B^* and will do for the single codeword, with length $|\lambda| = 0$. The set $C(\{\lambda\})$ satisfies the prefix condition by default. For $H(P) > 0$ there must be more than one message. Then λ is not a codeword since $\lambda c = c$ for any $c \in C$ so $\{c, \lambda\}$ is not a uniquely decipherable set. Thus all codeword lengths (and their average) must be at least 1. The upper bound $1 + H(P)$ in (ii) is best possible in the sense that for every $H > 0$ and $\varepsilon > 0$ there is a distribution P with $H(P) = H$ and with $\mathbb{E}_P(L) = 1 + H - \varepsilon$, for every L that satisfies the Kraft inequality. $\square$

Let us summarize our discoveries.

Proposition 1 *For a source $(\mathbb{N}, P)$ with entropy $H(P) = H$ such that $0 < H < \infty$ and for a uniquely decipherable code with the length function L, let*

$$R_P(L|H) = \frac{\mathbb{E}_P(L)}{\max\{1, H\}} \tag{4.4}$$

denote the ratio of the average codeword length to its minimal possible value. Then

$$R_P(L|H) \geq R_P(L_P|H) \tag{4.5}$$

and

$$R_P(L_P|H) \leq \begin{cases} 1 + H, & \text{if } 0 < H \leq 1, \\ 1 + 1/H, & \text{if } 1 < H < \infty, \end{cases} \tag{4.6}$$

$$R_P(L_P|H) \leq 2, \quad \text{for all } 0 < H < \infty,$$

$$\lim_{H \to 0} R_P(L_P|H) \leq 1,$$

$$\lim_{H \to \infty} R_P(L_P|H) \leq 1.$$

Remark The relationships (4.5) and (4.6) give the (fixed to variable-to-length) *noiseless coding theorem* for a stationary memoryless source selected from the set $\mathbb{N}$, selecting successive messages with statistical independence from the fixed probability distribution P. Let $P_n : \mathbb{N}^n \to (0, 1)$ be the probability distribution on n-tuples of messages. Then $H = H(P) > 0$ and $H(P_n) = nH$. Hence, encoding $(\mathbb{N}^n, P_n)$ into the set C_{P_n} of Theorem 19 gives

$$\lim_{H \to \infty} R_{P_n}(L_{P_n} | H(P_n)) = \lim_{H \to \infty} R_{P_n}(L_{P_n} | nH) = \lim_{H \to \infty} \left[1 + \frac{1}{nH}\right] = 1,$$

so that for any $\varepsilon > 0$ choosing $n > (\varepsilon H(P))^{-1}$ gives an average codeword length less than $(1 + \varepsilon)H(P_n)$.

4.1.2 Universal Codeword Sets

The characteristics described in the previous section need not be attained for the source $(\mathbb{N}, P)$ unless the codeword set chosen is C_P of which the length function L_P is specifically designed to match the distribution P of that particular source (see (4.3)). We will consider a different situation in which a single set C is used for the encoding of any countable source with entropy H, for $0 < H < \infty$.

Let $f : \mathbb{N} \to C$ map the positive integers into the members of the set C in order of non-decreasing length, so that setting $f(j) = c_j$ gives the kind of indexing with $|c_j| \le |c_{j+1}|$, $j \in \mathbb{N}$. If C is a uniquely decipherable set in B^* given f, then the function f is called a $|B|$-ary *representation of the integers*; otherwise f is called a $|B|$-ary *encoding of the integers*. The length function of the set C will be denoted by L so that $L(j) = |c_j|$. and the average codeword length will be denoted by $\mathbb{E}_P(L)$.

Definition 20 Given a code (C, f) with length function L let

$$F(H) = \max_{P:H(P)=H} \mathbb{E}_P(L),$$

where the maximum is taken over all *PDs* P such that $H(P) = H$, and let

$$K = \max_{0 < H < \infty} \frac{F(H)}{\max\{1, H\}}.$$

(i) The code is universal iff $K < \infty$.
(ii) The code is asymptotically optimal (abbreviated as a.o.) iff

$$\lim_{H \to \infty} \frac{F(H)}{H} = 1$$

provided that this limit exists.

We restrict our attention to the binary case ($B = \{0, 1\}$) and begin with the *standard binary representations and encodings* using the following notations: given $j \in \mathbb{N}$, let

$$L^*(j) = 1 + \lfloor \log j \rfloor \tag{4.7}$$

and

$$j^{(l)} = j_l, \mathrm{mod}2 = \begin{cases} 0, & \text{if } j_l \text{ is even,} \\ 1, & \text{if } j_l \text{ is odd,} \end{cases} \qquad (4.8)$$

where

$$j_0 = j - 1, \qquad\qquad\qquad\qquad\qquad\qquad (4.9)$$
$$j_l = \lfloor j_{l-1}/2 \rfloor, \quad l = 1, ..., L^*(j) - 1.$$

The key tool which will be widely used in the analysis is the *Wyner inequality* [8]:

$$\mathbb{E}_P(L) \leq H(P). \qquad (4.10)$$

This inequality immediately follows from the observation

$$0 \geq \log \sum_{i=1}^{j} P(i) \geq \log(j P(j)) \geq \log j + \log P(j).$$

4.1.3 Standard Binary Representations and Encodings of Integers

Standard representations are widely used for doing arithmetic and counting operations. Decoding an arbitrary universal set into a standard representation is possible by Theorem 18 if L is computable but need not be easy. Fortunately some standard representations are universal. Unfortunately none of them are universal binary representations or are asymptotically optimal.

4.1.3.1 Unary Representation (α)

The simplest binary representation of $\mathbb{N}$ is known in Turing machine theory as *unary representation*. The vector

$$\alpha(j) = 0^{j-1}1, \quad |\alpha(j)| = j$$

is a codeword for the integer $j \in \mathbb{N}$. The codeword set is complete and has a two-state acceptor that halts on the first non-zero symbol. Unary codes are close to be optimal for some exponential distributions [6] but are not universal; for example

$$P(j) = \frac{6}{\pi^2 j^2}, \quad j \in \mathbb{N}$$

has finite entropy but $\mathbb{E}_P(L_\alpha) = \infty$.

4.1.3.2   Binary Encoding (β)

The vector

$$\beta(j) = (j^{(L^*(j)-1)}, ..., j^{(0)}), \quad |\beta(j)| = L^*(j)$$

is a codeword for the integer $j \in \mathbb{N}$. The encoding β is not a representation since it is not uniquely decipherable. ($\beta(5) = 101 = \beta(2)\beta(1)$ illustrates this statement).

4.1.3.3   Modified Binary Encoding ($\hat{\beta}$)

The vector

$$\hat{\beta}(j) = (j^{(L^*(j)-2)}, ..., j^{(0)}), \quad |\hat{\beta}| = L^*(j) - 1$$

is a codeword for the integer $j > 1$, i.e., the first symbol of a codeword $\beta(j)$ is omitted since in $\hat{\beta}(j)$ it is always equal to 1. If $j = 1$, then the codeword is defined as the null sequence: $\hat{\beta}(1) = \lambda$.

4.1.3.4   Binary Representation (τ)

The code alphabet of a "binary representation" contains three symbols: 0, 1, and $\square$, where $\square$ is an end-of-word symbol. The representation is defined as

$$\tau(j) = \beta(j)\square, \quad |\tau(j)| = L^*(j) + 1.$$

This representation is in fact ternary, and the Wyner inequality (4.10) shows that it is universal, since

$$\mathbb{E}_P(L_\tau) \le 2 + H(P)\log 3.$$

Therefore,

$$F_\tau(H) \le H \log 3 + 2.$$

However τ is not complete (for example, we can add the codeword $0\square$).

4.1.3.5   Modified Binary Representation ($\hat{\tau}$)

The representation is defined as

$$\hat{\tau}(j) = \hat{\beta}(j)\square, \quad |\hat{\tau}(j)| = L^*(j).$$

The properties of $\hat{\tau}$ are similar to the properties of the representation τ.

4.1.4 Some Non-standard Representations of Integers

As we can see from the previous section the unary encoding α is not universal, the binary encodings β and $\hat{\beta}$ are not representations, and the "binary" representations τ and $\hat{\tau}$ are ternary, so there is no standard universal binary representation. In this section we consider some "non-standard" binary representations.

4.1.4.1 k-Universal Representation (ρ_k)

For a given $k \geq 1$, let us construct a representation ρ_k as follows. Write $j \in \mathbb{N}$ in base 2^k notation, which takes $1 + \lfloor \log j \rfloor$ base 2^k symbols, and separate the 2^k-ary symbols with 0. Terminate the sequence with 1.

Formally, let us extend the definitions (7)–(9) as follows:

$$L_k^*(j) = 1 + \lfloor (\log j)/k \rfloor \tag{4.11}$$

and

$$j_k^{(l)} = j_{kl}, \mathrm{mod} 2^k, \tag{4.12}$$

where

$$j_{0k} = j - 1, \tag{4.13}$$
$$j_{lk} = \lfloor j_{l-1,k}/2^k \rfloor, \quad l = 1, ..., L_k^*(j) - 1.$$

Then

$$|\rho_k(j)| = (1 + \lfloor \log_{2^k} j \rfloor)(k + 1) \tag{4.14}$$
$$\leq \left(1 + \frac{1}{k} \log j\right)(k + 1)$$
$$= 1 + k + \left(1 + \frac{1}{k}\right) \log j.$$

The code sequence can be decoded by a $(k + 2)$-state acceptor, i.e., the decoding algorithm uses a counter, starts, reads, and prints the next k symbols, halts if the $(k + 1)$-st symbol is 1, and increases the value of the counter by one and returns to "start" if the $(k + 1)$-st symbol is 0. The formal description is as follows:

1. set the pointer at the first position;
2. print the current k input symbols and shift the pointer by k positions to the right;
3. if the pointer points to 1, go to 5;
4. shift the pointer by 1 position to the right and go to 2;
5. end: $p_k(j)$ has been printed out.

The fact that ρ_k is a universal representation follows from (4.14) and the Wyner inequality (4.10) so that

$$\mathbb{E}_P(L_k) \leq 1 + k + \left(1 + \frac{1}{k}\right) H(P).$$

Therefore,

$$F_k(H) \leq 1 + k + \left(1 + \frac{1}{k}\right) H$$

and

$$K_k \leq 2 + k + \frac{1}{k}, \quad \text{for all } H < \infty,$$

$$\lim_{H \to \infty} \frac{F_k(H)}{H} = 1 + \frac{1}{k}.$$

The representation ρ_k is complete for integers belonging to the set $\mathbb{N} \cup \{0\}$ (the codeword $0^k 1$ can be added to the set of codewords).

4.1.4.2 Extended k-Universal Representation (ρ)

The choice of increasing values of k in (4.11–4.13) gives the sequence of representations ρ_k, which tends to an asymptotically optimal representation, but no single ρ_k is itself an asymptotical optimal one.

Let us define $n = n(j) \in \mathbb{N}$ by the inequalities

$$\frac{n(n+1)}{2} \geq 1 + \lfloor \log j \rfloor \geq \frac{n(n-1)}{2} + 1. \tag{4.15}$$

Let

$$m = m(j) = \frac{n(n+1)}{2} - 1 - \lfloor \log j \rfloor.$$

Construct $\rho(j)$ by padding the binary representation of j on the left with m initial zeroes and partitioning the resulting sequence of $n(n+1)/2$ bits into n segments, the kth segment of length k. Insert 0 between each pair of segments and 1 after the last symbol. Then

$$|\rho(j)| = \frac{n(n+1)}{2} + n = n + m + 1 + \lfloor \log j \rfloor. \tag{4.16}$$

Using (4.15) we conclude that $m \leq n - 1$ and

$$8(1 + \lfloor \log j \rfloor) \geq 4n(n-1) + 8 = (2n-1)^2 + 7.$$

Thus,

$$2n \leq 1 + \sqrt{1 + 8\lfloor \log j \rfloor}$$

and we can continue (4.16) as follows:

$$|\rho(j)| \leq 2n + \log j + 1 + \sqrt{1 + 8\lfloor \log j \rfloor} \leq \log j + 1 + \sqrt{1 + 8\lfloor \log j \rfloor}.$$

Using Wyner's inequality and the convexity of the square root function in averaging this expression gives

$$\mathbb{E}_P(L_\rho) \leq 1 + H(P) + \sqrt{1 + 8H(P)}. \tag{4.17}$$

Hence,

$$F_\rho(H) = 1 + H + \sqrt{1 + 8H}$$

and

$$K_\rho \leq 5, \quad \text{for all} \ H < \infty,$$
$$\lim_{H \to \infty} \frac{F_\rho(H)}{H} = 1.$$

4.1.4.3 Compound Representation (γ)

The representation is constructed in accordance with the rule:

$$\gamma(j) = \alpha(L^*(j))\hat{\beta}(j),$$

i.e., we first write the length of a standard binary representation of j and then encode j using a modified binary encoding. Since $L^*(j) \leq 1 + \log j$, Wyner's inequality leads to

$$\mathbb{E}_P(L_\gamma) \leq 1 + 2H(P).$$

Thus,

$$F_\gamma(H) \leq 1 + 2H$$

and

$$K_\gamma \leq 3,$$
$$\lim_{H \to \infty} \frac{F_\gamma(H)}{H} \leq 2,$$

i.e., γ is not asymptotically optimal.

An acceptor first decodes $L^*(j)$ and then uses it as a counter while decoding j.

4.1.4.4 Doubly Compound Representation (δ)

The representation is constructed in accordance with the rule:

$$\gamma(j) = \gamma(|\beta(j)|)\hat{\beta}(j),$$

i.e., we first write use representation γ to encode the length of a standard binary representation of j and then encode j using a modified binary encoding. Then

$$|\delta(j)| = |\gamma(|\beta(j)|)| + |\hat{\beta}(j)|$$
$$= 1 + 2\lfloor \log |\beta(j)| \rfloor + \lfloor \log j \rfloor$$
$$= 1 + \lfloor \log j \rfloor + 2\lfloor \log(1 + \lfloor \log j \rfloor) \rfloor.$$

Note that $|\delta(j)|$ increases by 2 units whenever j increases by 1 from $2^{2^k-1} - 1$ to 2^{2^k-1}.

The Wyner inequality and the convexity of the logarithm function prove the inequalities:

$$F_\delta(H) \leq 1 + H + 2\log(1 + H),$$
$$K_\delta \leq 4,$$
$$\lim_{H \to \infty} \frac{F_\delta(H)}{H} \leq 1,$$

i.e., δ is an asymptotically optimal representation.

4.1.4.5 Penultimate Representation (ω)

The representation ω is constructed in accordance with the following algorithm:

1. write 0;
2. set $i := j$;
3. if $\lfloor \log i \rfloor = 0$, halt;
4. write $\beta(i)$ to the left of the previous string;
5. set $i := \lfloor \log i \rfloor$ and go to 3.

It is easy to see that there is an equally simple left-to-right decoding algorithm. Define $l^k(j)$ by induction

$$l^0(j) = j,$$
$$l^{i+1}(j) = l^1(l^i(j)), \quad i = 0, 1, \dots.$$

Then

$$|\beta(l^i(j))| = l^{i+1}(j) + 1$$

Table 4.1 Standard binary representations and encodings for integer $j \in \{1, 2, ...\}$, where $j^{(l)} = 0$ if j_l is even and $j^{(l)} = 0$ if j_l is odd; $j_0 = j - 1$ and $j_l = \lfloor j_{l-1}/2 \rfloor$, $l = 1, ..., \lfloor \log j \rfloor$

	Name	$L(j)$	Codeword
α	Unary representation	j	$0^{j-1}1$
β	Binary encoding	$1 + \lfloor \log j \rfloor$	$(j^{(L^*(j)-1)}, ..., j^{(0)})$
$\hat{\beta}$	Modified binary encoding	$\lfloor \log j \rfloor$	$(j^{(L^*(j)-2)}, ..., j^{(0)})$
τ	Binary representation	$2 + \lfloor \log j \rfloor$	$(j^{(L^*(j)-1)}, ..., j^{(0)})\square$
$\hat{\tau}$	Modified binary representation	$1 + \lfloor \log j \rfloor$	$(j^{(L^*(j)-2)}, ..., j^{(0)})\square$

and

$$\omega(j) = \sum_{i=1}^{k} \beta(l^{k-i}(j)) + 1 = 1 + \sum_{i=1}^{k} (l^i(j)) + 1),$$

where the summation stops with the integer k such that $l^k(j) = 1$.

The analysis of the ω representation is not as simple as the analysis of the previous methods. It can be shown ([1, 5]) that

$$F_\omega(H) \leq \frac{5}{2} H + 1$$

and $K_\omega \leq 7/2$. Since

$$\lim_{j \to \infty} \frac{l^{m+1}(j)}{l^m(j)},$$

we conclude that

$$\lim_{H \to \infty} \frac{F_\omega(H)}{H} \leq 1,$$

i.e., the representation is asymptotically optimal (Tables 4.1, 4.2, 4.3, 4.4, 4.5 and 4.6).

4.1.5 Universal Codes

Let us start with the definition of some classes of sources that extends the Definition 18 of a countable source.

Definition 21 A class $\mathcal{A}$ of stationary countable finite-entropy sources without memory includes a source $(\mathcal{X}, P)$ if

(i) $\mathcal{X} = \mathbb{N}$;
(ii) the PD P has a finite entropy, i.e., $0 < H(P) < \infty$;
(iii) the source output sequence of length n takes the value $\mathbf{m} = (m_1, ..., m_n)$ with probability

Table 4.2 Standard binary representations and encodings for integers $j = 1, ..., 16$

j	α	β	$\hat{\beta}$	τ	$\hat{\tau}$
1	1	1	λ	1□	□
2	01	10	0	10□	0□
3	$0^2 1$	11	1	11□	1□
4	$0^3 1$	100	00	100□	00□
5	$0^4 1$	101	01	101□	01□
6	$0^5 1$	110	10	110□	10□
7	$0^6 1$	111	11	111□	11□
8	$0^7 1$	1000	000	1000□	000□
9	$0^8 1$	1001	001	1001□	001□
10	$0^9 1$	1010	010	1010□	010□
11	$0^{10} 1$	1011	011	1011□	011□
12	$0^{11} 1$	1110	100	1110□	100□
13	$0^{12} 1$	1101	101	1101□	101□
14	$0^{13} 1$	1110	110	1110□	110□
15	$0^{14} 1$	1111	111	1111□	111□
16	$0^{15} 1$	10000	0000	10000□	0000□

Table 4.3 Characteristics of standard binary representations and encodings for integers

	$F(H)$	K	$\lim_{H \to \infty} F(H)/H$	Properties
α	∞	∞		Non-universal complete
β				Not a representation
$\hat{\beta}$				Not a representation
τ	$H \log 3 + 2$	$\log 3 + 2$	$\log 3$	Universal ternary not complete
$\hat{\tau}$	$H \log 3 + 1$	$\log 3 + 1$	$\log 3$	Universal ternary not complete

Table 4.4 Universal binary representations for integer $j \in \{1, 2, ...\}$, where the parameters $n(j)$ and $m(j)$ are defined in the text

	Name	$L(j)$	Codeword
ρ_k	k-universal representation	$(1 + \lfloor \log_{2^k} j \rfloor)(k+1)$	$\beta(j_k^{(L_k^*(j))-1})0...\beta(j_k^{(0)})1$
ρ	Extended k-universal representation	$n(j)(3n(j)+1)/2$	$\rho(0^{m(j)})\beta(j)$
γ	Compound representation	$1 + 2\lfloor \log j \rfloor$	$\alpha(L^*(j))\hat{\beta}(j)$
δ	Doubly compound representation	$1 + 2\lfloor \log L^*(j) \rfloor + \lfloor \log j \rfloor$	$\gamma(L^*(j))\hat{\beta}(j)$
ω	Penultimate representation	$1 + \sum_{m \le k}(l^m(j) + 1)$, where $l^0(j) = j$, $l^1(j) = \lfloor \log j \rfloor$, $l^{m+1}(j) = l^1(l^m(j))$, $k : l^k(j) = 1$	$\beta(l^{k-1}(j))...\beta(l^0(j))0$

Table 4.5 Universal binary representations for integers $j = 1, ..., 16$

j	ρ_2	ρ	γ	δ	ω
1	01 1	1 1	1	1	0
2	10 1	0 0 10 1	01 0	010 0	10 0
3	11 1	0 0 11 1	01 1	010 1	11 0
4	011 00 1	1 0 00 1	001 00	011 00	10 100 0
5	010 01 1	1 0 01 1	001 01	011 01	10 101 0
6	010 10 1	1 0 10 1	001 10	011 10	10 110 0
7	010 11 1	1 0 11 1	001 11	011 11	10 111 0
8	100 00 1	0 0 01 0 000 1	0001 000	00100 000	11 1000 0
9	100 01 1	0 0 01 0 001 1	0001 001	00100 001	11 1001 0
10	100 10 1	0 0 01 0 010 1	0001 010	00100 010	11 1010 0
11	100 11 1	0 0 01 0 011 1	0001 011	00100 011	11 1011 0
12	110 00 1	0 0 01 0 100 1	0001 100	00100 100	11 1100 0
13	110 01 1	0 0 01 0 101 1	0001 101	00100 101	11 1101 0
14	110 10 1	0 0 01 0 110 1	0001 110	00100 110	11 1110 0
15	110 11 1	01 1	0001 111	00100 111	11 1111 0
16	010000001	000010000	000010000	001010000	10100100000

Table 4.6 Characteristics of universal binary representations for integers, where $K_\infty = \lim_{H \to \infty} F(H)/H$

	$F(H)$	K	K_∞	Properties
ρ_k	$\leq \left(1 + \frac{1}{k}\right) H + 1 + k$	$\leq 2 + k + \frac{1}{k}$	$\leq 1 + \frac{1}{k}$	Complete for $\mathbb{N} \cup \{0\}$ not a.o
ρ	$\leq 1 + H + \sqrt{1 + 8H}$	≤ 5	1	Complete for $\mathbb{N} \cup \{0\}$ a.o
δ	$\leq H + 2\log(1 + H) + 1$	≤ 4	1	Complete a.o
γ	$\leq 2H + 1$	≤ 3	≤ 2	Complete for $\mathbb{N} \cup \{0\}$ not a.o
ω	$\leq (5/2)H + 1$	$\leq 7/2$	1	Complete a.o

$$P_n(\mathbf{m}) = \prod_{j=1}^{n} P(m_j).$$

A class $\mathcal{M}$ of monotonic sources is a subclass of the class of stationary countable finite-entropy sources without memory, i.e., $\mathcal{M} \subset \mathcal{A}$, such that

(iv) $1 > P(j) \geq P(j + 1)$, $j \in \mathbb{N}$.

The subset $\mathcal{A}_k \subset \mathcal{A}$ of k-ary sources satisfies (1)–(3) and the restriction

(v) $P(j) > 0$, iff $j = 1, ..., k$.

The monotonic k-ary sources satisfy (1)–(5), i.e., $\mathcal{M}_k = \mathcal{A}_k \cap \mathcal{M}$.

The source coding theorem (see Proposition 1) shows that for each $P \in \mathcal{A}$ there is a sequence of sets C_n and a sequence of codes $\mu_n : \mathcal{X}^n \to C_n$ such that the entropy performance measure

$$\sum_{\mathbf{m} \in \mathcal{X}^n} \frac{P_n(\mathbf{m}) |\mu_n(\mathbf{m})|}{H(P_n)} \tag{4.18}$$

approaches 1 as n increases. The previous considerations show that there exist single asymptotically optimal sets, for example C_ρ, such that a sequence $\mu_n'' : \mathcal{X}^n \to C_\rho$ has the same limiting performance when the codes depend on P (since μ_n maps $\mathcal{X}^n$ onto C_ρ in order of decreasing probabilities) but C_ρ does not. It is also possible to show that there is a single sequence $\mu_n'' : \mathcal{X}^n \to C_n$ having the same limiting performance for all $P \in \mathcal{A}$ [2].

Definition 22 The code is universal for a class of sources with respect to the performance measure if there is a uniform bound to the measure for all sources in the class.

In the further considerations we denote the representation of the integers by ρ and while estimating the basic quantities assume that ρ is the Extended k-Universal representation. The algorithms are also valid when ρ is any representation.

4.1.5.1 Coding Runs of Zeroes

Let us consider the class $\mathcal{M}_2$ that includes Bernoulli sources ($\mathcal{X} = \{0, 1\}$) such that 1's and 0's independently appear with probabilities $p \in (0, 1/2]$ and $q = 1 - p$, respectively.

Universal run-length coding treats the infinite source output sequence as a concatenation $\alpha(j_1)\alpha(j_2)...$ of unary encodings of a sequence $j_1 j_2...$ of integers (the lengths of runs of zeroes followed by one) and decodes and re-encodes the integers into the concatenation $\rho(j_1)\rho(j_2)...$ of their universal representations. The probability of the integer j is

$$P(j) = q^{j-1} p$$

decreasing in j, and the entropy is equal to

$$H(P) = -\sum_{j \in \mathbb{N}} q^{j-1} p \log(q^{j-1} p) = \frac{h(p)}{p} \geq 2,$$

where

$$h(p) = -q \log q - p \log p.$$

Since P is known as a function of p, the ratio

$$\frac{\mathbb{E}_P(L_\rho)}{\max\{1, H(P)\}} = \frac{\mathbb{E}_P(L_\rho) \cdot p}{h(P)}$$

can be estimated for all $p \in (0, 1/2]$ (see (4.17)).

4.1.5.2 Coding Runs of Zeroes and Ones

Let

$$\mathbf{m} = 0^{j_1} 1^{k_1} 0^{j_2} 1^{k_2} \dots$$

be the output sequence of the source belonging to $\mathcal{A}_2$. The PDs are defined as $P_0 = \{q^{j-1}p, \ j \in \mathbb{N}\}$ and $P_1 = \{p^{j-1}q, \ j \in \mathbb{N}\}$. Both are decreasing functions. Since a pair (j, k) always contains one integer from each distribution, the concatenation $\rho(j)\rho(k)$ has entropy performance measure

$$\frac{\mathbb{E}_{P_0}(L_\rho) + \mathbb{E}_{P_1}(L_\rho)}{H(P_0) + H(P_1)}.$$

Computing this ratio as a function of $p \in (0, 1)$ gives a uniform bound of order 1.6 [3].

The encoding scheme has practical interest since it works well for a memoryless source and works even better for a Markov source whose state is the last output symbol and whose conditional probabilities of staying in the same state are greater than the corresponding steady-state probabilities. *It should, therefore, do well for both line drawings and large objects in facsimile encoding.*

4.1.5.3 Universal Codes for $\mathcal{A}_2$

A source in $\mathcal{A}_2$ can be universally encoded by factoring it into $k - 1$ independent sources in $\mathcal{A}_2$ and a universal run-length encoding of their outputs.

Let $\mathcal{X} = \{1, \dots, k\}$ and let $\mathbf{m} = (j_1, \dots, j_n)$ be the first n messages. Represent each occurrence of the integer $j \in \mathcal{X}$ in $\mathbf{m}$ by the unary codeword $\alpha(j) = 0^{j-1}1$. Write $\alpha(j_s)$ as the sth column of the array A having k rows and n columns, the first symbol of each codeword occupying the first row.

Example Let $k = 4$, $\mathcal{X} = \{1, 2, 3, 4\}$ and

$$\mathbf{m} = (2, 1, 3, 2, 4, 1, 3, \dots).$$

Then we construct the sequence

$$(\alpha(2), \alpha(1), \alpha(3), \alpha(2), \alpha(4), \alpha(1), \alpha(3), \dots)$$

and store it in following array:

$$A = \begin{Vmatrix} 0 & 1 & 0 & 0 & 0 & 1 & 0 & \dots \\ 1 & & 0 & 1 & 0 & & 0 & \dots \\ & & 1 & & 0 & & 1 & \dots \\ & & & & 1 & & & \dots \end{Vmatrix} = \begin{Vmatrix} \alpha(2) & \alpha(4) & \dots \\ \alpha(1) & \alpha(2) & \dots \\ \alpha(1) & \alpha(2) & \dots \\ \alpha(1) & . & \dots \end{Vmatrix}.$$

The locations of the blanks in row j are determined by the locations of the ones in earlier rows. The values of the non-blank symbols in row j are independent of one another and of the values in earlier rows. Thus the *non-blank symbols in each of the first $k - 1$ rows are output sequences of $k - 1$ independent sources in A_2.*

An encoding algorithm is as follows. For $j \in \{1, ..., k-1\}$, place a marker j in row j, initially to the left of column 1. Start and choose the leftmost column containing any marker and the smallest (highest) marker in that column, say marker j. Move marker j to the right in row j, passing $r - 1$ zeroes, and halting on the first one to the right of the initial position of marker j in row j, and send the codeword $\rho(r)$. Then return to start.

The received sequence always determines the source sequence up to the column from which the last marker moved, and determines at most $k - 2$ message values lying to the right of that point.

The encoding algorithm has the following formal description.

1. Set $M_1 = \cdots = M_k = 0$.
2. Find the minimal marker, i.e., set

$$m = \min_{1 \le j \le k} M_j, \quad j^* = \arg \min_{1 \le j \le k} M_j.$$

3. Calculate l such that

$$a_{j^*,m+s} \ne 1, \; s = 1, ..., l - 1, \quad a_{j^*,m+l} = 1$$

and

$$r = |\{s = 1, ..., l - 1 Z\, a_{j^*,m+s} = 0\}|.$$

Send $\rho(r + 1)$ and increase M_{j^*} by l, i.e., set $M_{j^*} := M_{j^*} + l$. Go to 2.

The algorithm gives the following order for the codeword for the messages written in the matrix A of the example above

$$\begin{Vmatrix} \alpha(2) \to 1 \; \alpha(4) \to 6 \ldots \\ \alpha(1) \to 2 \; \alpha(2) \to 5 \ldots \\ \alpha(1) \to 3 \; \alpha(2) \to 7 \ldots \\ \alpha(1) \to 4 . \qquad \ldots \end{Vmatrix}.$$

Note that there can be some difficulties if we want to apply the algorithm to specific data.

(i) The searching domain for l in step 3 is unrestricted; if the probability that the integer m corresponding to the marker m is small, the search can require a long time.

(ii) The algorithm moves all k markers, where $P(m) > 0, m = 1, ..., k$. However individual sequences to be encoded have finite length and may contain integers from a restricted alphabet; hence it is needed to examine the sequence in advance, encode k and also form special codewords that inform the decoder when the integer $m \in \{1, ..., k\}$ occurs for the last time.

(iii) We have to classify the source as belonging to the classes $\mathcal{M}_k$, $\mathcal{A}_k$, $\mathcal{M}$, or $\mathcal{A}$ (the algorithms for $\mathcal{M}$ and $\mathcal{A}$ will be described later) since different algorithms are developed for these classes. It is not always possible, especially when the individual sequences, generated by the source, appear in a real time mode.

Let us take some ideas of the approach above and develop a universal algorithm that can be used for individual sequences. The formal description of the encoding and decoding procedures are given below.

4.1.5.4 Encoding

1. *Initialize the encoder.*
   Set $t = 1$ and $k = 0$.
2. *Reconstruct the markers.*
   If $k > 0$, then

   - shift **M** by one position to the left;
   - decrease k by 1;
   - decrease all components of **M** by 1.

3. *Awake a sleeping marker.*
   If $k = 0$ or $M_0 > 0$, then

   - shift **M** by one position to the right;
   - increase k by 1;
   - send $f(L + 2)\rho(u_t)$.

4. *Put the marker to sleep.*
   If $u_{t+s} \neq u_t$ for all $s = 1, ..., L$, then

   - send $f_m(L + 1)$, where $m = u_t$;
   - increase t by 1 and, if $t \leq T$, go to 2.

5. *Move the marker.*

   - find the minimal l such that $u_{t+l} = u_t$ and set $r = |\{s = 1, ..., l - 1 : u_{t+s} > u_t\}|$;
   - find the minimal index $i \in \{1, ..., k - 1\}$ such that $M_i \geq l$ and set $\mathbf{M}' = (M_0, ..., M_{i-1}, l, M_i, ..., M_{k-1})$;
   - increase k by 1;
   - set $(M_0, ..., M_{k-1}) = \mathbf{M}'$;
   - send $f_m(r + 1)$, where $m = u_t$;
   - increase t by 1 and, if $t \leq T$, go to 2.

4.1.5.5 Decoding

1. *Initialize the decoder.*
   Set $t = 0$, $\tau = 1$, and $\hat{k} = 0$.
2. *Reconstruct the markers.*
   Increase t by 1 and, if $\hat{k} > 0$, then

   - shift $\hat{\mathbf{M}}$ by one position to the left;
   - decrease $\hat{k}$ by 1;
   - decrease all components of $\hat{\mathbf{M}}$ by 1.

3. *Awake a sleeping marker.*
   If $(y_\tau, y_{\tau+1}, ...) = (f(L+2), \rho(u), ...)$, then

   - for all $j = \hat{k} - 1, ..., 0$ such that $\hat{u}_{t'} > u$, where $t' = t + \hat{M}_j$, set $\hat{u}_{t'+1} = \hat{u}_{t'}$
     and increase the value of the marker $\hat{M}_j$ by 1;
   - shift $\hat{\mathbf{M}}$ by one position to the right;
   - increase $\hat{k}$ by 1;
   - set $\hat{u}_t = u$;
   - increase τ by $|f(L+2)| + |\rho(u)|$.

4. *Put the marker to sleep.*
   If $y = (f_{\hat{m}}(L+1), ...)$, where $\hat{m} = \hat{u}_t$, then

   - increase τ by $|f(L+1)|$ and, if $\tau \leq T'$, go to 2.

5. *Move the marker.*
   If $y = (f_{\hat{m}}(r+1), ...)$, where $\hat{m} = \hat{u}_t$ and $r < L$, then

   - find l using the following procedure:

     (a) set $i = j = 1$ and $l = 0$;
     (b) increase l by 1;
     (c) if $l = \hat{M}_j$, then

        −   set $t' = t + \hat{M}_j$ and increase j by 1;
        −   if $\hat{u}_{t'} < \hat{u}_t$, then go to (b);

     (d) increase i by 1 and, if $i \leq r+1$, then go to (b);

   - find the minimal index $i \in \{1, ..., \hat{k} - 1\}$ such that $\hat{M}_i \geq l$ and set
     $\hat{\mathbf{M}}' = (\hat{M}_0, ..., \hat{M}_{i-1}, l, \hat{M}_i, ..., \hat{M}_{k-1})$;
   - increase $\hat{k}$ by 1;
   - set $(\hat{M}_0, ..., \hat{M}_{\hat{k}-1}) = \hat{\mathbf{M}}'$;
   - for all $j = i+1, ..., \hat{k} - 1$, if $\hat{u}_{t'} > \hat{u}_t$, where $t' = t + \hat{M}_j$, then set $\hat{u}_{t'+1} = \hat{u}_{t'}$
     and increase $\hat{M}_j$ by 1;
   - set $\hat{u}_{t+l} = \hat{u}_t$;
   - increase τ by $|f(r+1)|$ and, if $\tau \leq T'$, go to 2.

4.1.5.6 Universal Codes for $\mathcal{M}$ and $\mathcal{A}$

When $\mathcal{X}$ is infinite the algorithm used for $\mathcal{A}_k$ will never send the second pair of runs for any symbol and a less symmetric algorithm is needed. To start the new marker-moving algorithm, start a cycle by moving marker 1 to the right side of its first pair and send the encoding of that pair of runs. Return to start a cycle when every column to the left of marker 1 has appeared as the end of the run, so that decoding is complete up to the column occupied by marker 1. If a new cycle is not started, the next move is made by the smallest (highest) marker that lies to the left of marker 1.

The formal description is as follows.

1. Set $m = 0$.
2. Find l_1 such that

$$a_{1,m+s} = 0, \ s = 1, ..., l_1 - 1, \quad a_{1,m+l_1} = 1.$$

   Set $\Delta = 0$, $r = l_1 - 1$ and go to 4.
3. Find j such that $a_{j,m+l_1+\Delta} = 1$. Let

$$a_{j,m+l_1+\Delta-s} \neq 1, \ s = 1, ..., l - 1, \quad a_{j,m+l_1+\Delta-l} = 1$$

and

$$r = |\{s = 1, ..., l - 1 : a_{j,m+l_1+\Delta-s} = 0\}|.$$

4. Send $\rho(r + 1)$ and set $\Delta := \Delta + 1$. If $\Delta < l_1$ go to 3.
5. $m := m + l_1$ and go to 2.

The new algorithm works for $\mathcal{M}$ but not for $\mathcal{A}$ since $P(1) = 0$ is possible for $\mathcal{A}$, in which case marker 1 never moves. To encode $\mathcal{A}$, add labels to the rows of the array A and send the labels to the receiver. The label l_1 on row 1 is that message $l_1 \in \mathbb{N}$ which first completes a pair

$$\{l_1\}^j (\mathbb{N}\backslash\{l_1\})^k \ \text{ or } \ (\mathbb{N}\backslash\{l_1\})\{l_1\}^k$$

of runs, and the label on row $j + 1$ is the message which first completes such a pair in that subsequence of the message sequence that remains when occurrences of $l_1, ..., l_j$ have been deleted. Then $\rho(l_j)$ is sent as a prefix to the code $m(s)\rho(j)\rho(k)$, for the first pair of runs that occur in row j, where $s \in \{0, 1\}$ is a symbol that begins the run and $m(0), m(1)$ are special codewords reserved for 0 and 1.

4.1.5.7 Optimal Sequences of Universal Codes

Universal codes for $\mathcal{M}_2$ and $\mathcal{A}_2$ (and thus $\mathcal{A}_k$ and $\mathcal{A}$) that have better entropy performance than the bound of order 1.6 in [3] are constructed by using an asymptotically optimal representation ρ and an intermediate mapping between the two steps $\alpha(j) \to j$ and $j \to \rho(j)$.

Let $\psi : \mathbb{N} \times \mathbb{N} \to \mathbb{N}$ be the one-to-one mapping of pairs of integers onto integers

$$\psi(j_1, j_2) = \frac{(j_1 + j_2 - 1)(j_1 + j_2 - 2)}{2} + j_1.$$

For $\mathcal{M}_2$ the probability of successive pair (j_1, j_2) of runs is a function of $j_1 + j_2$ alone since

$$P(j_1)P(j_2) = q^{j_1 + j_2}(p/q)^2.$$

therefore the distribution

$$P_2(\psi) = q^j (p/q)^2, \quad \frac{(j-1)(j-2)}{2} + 1 \leq \psi \leq \frac{j(j-1)}{2}$$

of ψ is a non-increasing function. Note that

$$H(P_2) = 2H(P) \geq 4, \quad p \in (0, 1/2].$$

Hence, using the encoding

$$\alpha(j_1)\alpha(j_2) \to j_1 j_2 \to \psi(j_1, j_2) \to \rho(\psi(j_1, j_2))$$

gives an entropy performance

$$\frac{\mathbb{E}_{P_2}(L_\rho)}{H(P_2)} = \frac{\mathbb{E}_P(L_\rho)}{2H(P)}$$

and an n-fold iteration of the ψ-mapping stage in the encoding process maps each 2^n-tuple $j_1, \ldots, j_{2^n}$ into a single integer with decreasing distribution Q_n and then into a single codeword with entropy performance

$$\frac{\mathbb{E}_{P_{2^n}}(L_\rho)}{H(P_2)} = \frac{\mathbb{E}_P(L_\rho)}{2^n H(P)},$$

which approaches 1 uniformly on $\mathcal{M}_2$ as n increases for any asymptotically optimal representation ρ since $H(P) \geq 2$.

The similar considerations can be extended to $\mathcal{A}_2$, $\mathcal{A}_k$, $\mathcal{M}$, and $\mathcal{A}$.

4.2 Lecture on Representation of the Rational Numbers

4.2.1 Upper Bounds on the Encoding Length of Rational Numbers

Every rational number r can be written uniquely as p/q with $q > 0$ and p and q coprime integers; so the space needed to encode r is

$$\langle r \rangle = \langle p \rangle + \langle q \rangle,$$

where we assume that there is a prefix property which allows us to separate p from q. The integer $\langle r \rangle$ will be considered as the *encoding length* of r. Similarly, if x and D are a rational vector and matrix, respectively, then the *encoding length* of x and D are defined as the sum of the encoding lengths of the entries and we will denote these lengths by $\langle x \rangle$ and $\langle D \rangle$.

Lemma 31 (i) *For every rational number* r,

$$|r| \le 2^{\langle r \rangle - 1} - 1. \tag{4.19}$$

(ii) *For every vector* $x \in \mathbb{Q}^n$,

$$\| x \| \le 2^{\langle x \rangle - n} - 1, \tag{4.20}$$

where

$$\| x \| = \left(\sum_{t=1}^{n} x_t^2 \right)^{1/2}$$

denotes the Eucledean norm of x.
(iii) *For every matrix* $D \in \mathbb{Q}^{n \times n}$,

$$\det D \le 2^{\langle D \rangle - n^2} - 1. \tag{4.21}$$

Furthermore,

$$\langle \det D \rangle \le 2\langle D \rangle - n^2 \tag{4.22}$$

and if $D \in \overline{Z}(n, p)^{n \times n}$ *then*

$$\langle \det D \rangle \le \langle D \rangle - n^2 + 1. \tag{4.23}$$

Proof Equation (4.19) follows directly from definition. $x = (x_1, \ldots, x_n)^T$. Using (4.19) we write

$$1 + \| x \| \le 1 + \sum_{i=1}^{n} |x_i| \le \prod_{i=1}^{n} (1 + |x_i|) \le \prod_{i=1}^{n} 2^{\langle x_i \rangle - 1} = 2^{\langle x \rangle - 1}$$

and (4.20) is proved. To prove (4.21), let us denote the rows of D by $d_1, \ldots, d_n$. Then by Hadamard's inequality and (4.20),

$$1 + \det D \le 1 + \prod_{i=1}^{n} \| d_i \| \le \prod_{i=1}^{n} (1 + \| d_i \|) \le \prod_{i=1}^{n} 2^{\langle d_i \rangle - n} = 2^{\langle D \rangle - n^2}.$$

To prove (4.22), let $D = (p_{ij}/q_{ij})$, $i, j = 1, \ldots, n$, be such that p_{ij} and q_{ij} are coprime integers for all $i, j = 1, \ldots, n$. Then the same argument as above shows that

$$|\det D| \leq 2^{\sum_{i,j} \langle p_{ij} \rangle - n^2} - 1.$$

Let $Q = \prod_{i,j} p_{ij}$. Then $\det D = (Q \det D)/Q$, where $Q \det D$ and Q are integers. Hence

$$
\begin{aligned}
\langle \det D \rangle &\leq \langle Q \det D \rangle + \langle Q \rangle \\
&= 1 + \lceil \log_2(|Q \det D| + 1) \rceil + \langle Q \rangle \\
&\leq 1 + \lceil \log_2(1 + Q(2^{\sum_{i,j} \langle p_{ij} \rangle - n^2} - 1)) \rceil + \langle Q \rangle \\
&\leq 1 + \lceil \log_2 Q + \sum_{i,j} \langle p_{ij} \rangle - n^2 \rceil + \langle Q \rangle \\
&\leq 2\langle Q \rangle + \sum_{i,j} \langle p_{ij} \rangle - n^2 \\
&\leq 2\langle D \rangle - n^2.
\end{aligned}
$$

Equation (4.23) directly follows from (4.21). □

4.2.2 Polynomial and Strongly Polynomial Computations with Rational Numbers

We will assume that every integral or rational number is represented in binary notation in the way described above. Then, in the Turing machine model, an arithmetic operation like addition of two integers takes a number of steps (moves of read-write head) which is bounded by a polynomial in the encoding lengths of the two integers.

It is often more natural and realistic to consider the *elementary arithmetic operations: addition, subtraction, multiplication, division, comparison,* rather than a number of moves of a head of some fictitious Turing machine, as one step. If we count elementary arithmetic operations, we say that we analyze an algorithm in the *arithmetic model*. For instance, computing the product of the $n \times n$ matrices in the standard way takes $o(n^3)$ steps in the arithmetic model, while in the Turing machine model its running time depends on the encoding lengths of the entries of the matrix. In the arithmetic model, the *encoding length* of an instance of a problem is the number of numbers occurring in the input (so we do not count the encoding length of the input numbers). If the input has some non-numerical part (a graph or a name) we assume that it is encoded as a $\{0, 1\}$-sequence. Each entry of this sequence is considered a number. Correspondingly, we consider non-numeric steps in the algorithm (like setting or removing a label, deleting an edge from a graph) as arithmetic operations.

The running time of an algorithm may be bounded by a polynomial in the Turing machine model but not in the arithmetic model, and vise versa. For instance, the well-known Euclidean algorithm to compute the greatest common divisor of two integers runs in polynomial time in the Turing machine model but not in the arithmetic model. On the other hand, the algorithm that reads n numbers and computes 2^{2^n} by repeating squaring a polynomial in the arithmetic model but not in the Turing model since the number of digits of the result is exponentially large.

It is easy to see that the elementary arithmetic operations can be executed on a Turing machine in polynomial time. Using this we can further develop the idea of counting arithmetic operations (instead of moves of a read-write head), if we supplement it by verifying that the number of digits of the numbers occurring during the run of the algorithm is bounded by a polynomial in the encoding length. Then, of course, a polynomial number of arithmetic operations will be executed in polynomial time on a Turing machine.

We say that an algorithm runs in *strongly polynomial time* if the algorithm is a polynomial space algorithm and performs a number of elementary arithmetic operations which is bounded by a polynomial in the number of input members. So, a strongly polynomial algorithm is a polynomial space algorithm in the Turing machine model and a polynomial time algorithm in arithmetic model.

Since the definition of strong polynomiality mixes the arithmetic and the Turing machine models in some sense we have to be precise about how numbers and the results of arithmetic operations are encoded. Integers are encoded in the usual binary notation. Adding, subtracting, multiplying, and comparing two integers do not cause any problem. However, we have to specify what dividing two integers means. There are at least four alternatives to define the result of division $a : b$ for $a, b \in \overline{Z}(n, p)$, $b \neq 0$:

 (i) the rational number a/b;
(ii) the rational number a/b and if it is known in advance that a/b is an integer, the integer a/b;
(iii) the rational number a/b and if a/b is an integer, the integer a/b (this operation involves a routine testing whether b divides a);
(iv) the integer $\lfloor a/b \rfloor$.

We reduce the arithmetic for rational numbers to that for integers by setting

$$\frac{a}{b} + \frac{c}{d} = \frac{ad + bc}{bd},$$

$$\frac{a}{b} - \frac{c}{d} = \frac{ad - bc}{bd},$$

$$\frac{a}{b} \cdot \frac{c}{d} = \frac{ac}{bd},$$

$$\frac{a}{b} : \frac{c}{d} = \frac{ad}{bc}.$$

It is important to note that in this model of computation we do not assume that a rational number is represented in a coprime form.

The four versions of division of integers lead to many powerful models of computation. However, it has to be pointed out that a strongly polynomial algorithm—using any of these kinds of division—is always a polynomial time algorithm in the Turing machine model.

4.2.3 Polynomial Time Approximation of Real Numbers

Since there is no encoding scheme with which all irrational numbers can be represented by finite strings of 0's and 1's, it is customary to accept only rational numbers as inputs or outputs of algorithms. It will, however, be convenient to speak about certain irrational numbers as inputs or outputs. This is done by making use of the fact that irrational numbers can be approximated by rationals. To be precise, we have to discuss how errors of approximations can be measured.

We say that a rational number r approximates the real number ρ with *absolute error* $\varepsilon > 0$ if

$$|\rho - r| \le \varepsilon$$

holds, and we say that a rational number r approximates the real number ρ with *relative error* $\varepsilon > 0$ if

$$|\rho - r| \le \varepsilon|\rho|.$$

These two kinds of error measurements are in general not equivalent. In particular, statements about relative errors are very sensitive with respect to addition of constants, while statements about absolute errors are sensitive with respect to multiplication by constants.

Given some input, we shall say that a real number ρ is *polynomially computable with absolute (relative) error* from its input if for every rational $\varepsilon > 0$ a rational number can be computed in time polynomial in the encoding length of the input and in $\langle \varepsilon \rangle$ that approximates ρ with absolute (relative) error ε. An algorithm accomplishing this will be called a *polynomially computation algorithm with absolute (relative) error.*

We now want to show that a polynomial computation algorithm with relative error can be used to obtain a polynomial computation algorithm with absolute error. Suppose we have a polynomial computation algorithm B which guarantees a relative error and we want to find a polynomial computation algorithm A which guarantees an absolute error. Assume $\varepsilon > 0$ is the absolute error we want to achieve. Then we call B with relative error $\delta_1 = 1/2$ and obtain a rational number r_1 satisfying $|\rho - r_1| \le |\rho|/2$, and hence $|\rho| \le 2|r_1|$. If $r_1 = 0$, then $\rho = 0$ and we finish. Since B is polynomial computation algorithm, $\langle r_1 \rangle$ is bounded by a polynomial of the original encoding length. Now we run B again with relative error $\delta_2 = \varepsilon/(2|r_1|)$ which yields a rational number r_2 satisfying $|\rho - r_2| \le \varepsilon|\rho|/(2|r_1|) \le \varepsilon$ and obtain a desired polynomial computation with absolute error ε.

The reverse implication is not true in general: small numbers cause problems. If we assume, however, that ρ is non-zero and that a positive lower bound on $|\rho|$ can be computed in polynomial time $b \in (0, |\rho|]$, then the existence of a polynomial computation algorithm A with absolute error implies the existence of a polynomial computation algorithm B with relative error as follows. We simply call A with absolute error $\delta = \varepsilon b$ and obtain a rational number r satisfying $|\rho - r| \leq \varepsilon b \leq \varepsilon |\rho|$.

We call a real number ρ *polynomially computable from a given input* if it is polynomially computable with relative error. This is equivalent to the existence of a polynomial computation algorithm with absolute error, together with a polynomial auxiliary algorithm that determines whether or not $\rho = 0$ and if not, computes a positive lower bound on $|\rho|$. In fact, it suffices to have a polynomial auxiliary algorithm that computes a positive rational b such that either $\rho = 0$ or $|\rho| \geq b$. Then whether or not $\rho = 0$ can be decided by running the computation algorithm with error $b/3$.

If the number ρ to be computed is known to be an integer, then the existence of a polynomial computation algorithm with absolute error (and hence with relative error) means that ρ can be computed exactly in polynomial time. We only have to run the algorithm with absolute value 1/3 and round the result to the nearest integer. More generally, if ρ is a rational number polynomially computable from a given input, then it can be computed exactly in polynomial time. This can be done using a procedure above, but for the final rounding one must use the technique of continued fractions.

Sometimes, an approximation algorithm with relative error ε can be obtained whose running time is bounded by a polynomial in $1/\varepsilon$ and the encoding length of the original input. Such an algorithm is called a *fully polynomial approximation algorithm*. An even weaker notion is an algorithm that is polynomial in the original encoding length for every fixed $\varepsilon > 0$. Such an algorithm is called a *polynomial approximation algorithm*.

4.3 Lecture on the Efficient Construction of an Unbiased Random Sequence

We present some results of [4] and consider procedures for converting input sequences of symbols generated by a stationary random process into sequences of independent output symbols having equal probabilities. We will restrict our attention to the binary case (a possible generalization is evident) and measure the efficiency of the procedures by the expected value of the ratio of the number of output bits to a fixed number of input bits. First we give an information-theoretic upper bound to the efficiency and then construct specific procedures which asymptotically attain this bound for two classes of processes (the process parameters are not used in these procedures).

4.3.1 Definitions and an Upper Bound to the Efficiency

Definition 23 A random process $X = \{x_n, n \geq 1\}$ is acceptable if

(i) X is a stationary process;
(ii) x_n take values in a countable set $\mathcal{X}$;
(iii) the entropy of X is finite, i.e.,

$$H(X) = -\sum_{x \in \mathcal{X}} P(x) \log P(x) < \infty,$$

where $P(x) = \Pr\{x_n = x\}$ for all $n \geq 1$.

Definition 24 Let $Z = \{0, 1, \Lambda\}$ and, given $N \geq 1$, let B be a function defined for all $x \in X^N$ and taking values in the set of sequences whose components belong to Z. The function is referred to as a randomizing function for X^N iff

$$\Pr\{z_m = 0|z^{m-1}\} = \Pr\{z_m = 1|z^{m-1}\} \tag{4.24}$$

for each $m > 1$ and each $z^{m-1} = (z_1, \ldots, z_{m-1})$. The efficiency of B is defined as

$$\eta_N = \frac{\mathbb{E}[\, t(B(X^N))\,]}{N}, \tag{4.25}$$

where

$$t(z^m) = \sum_{i=1}^{m} \chi\{\, z_i \neq \Lambda \,\} \tag{4.26}$$

denotes the number of occurrences of 0 and 1 in z^m for all $z^m \in Z^m$.

Theorem 20 *If X is an acceptable random process and B is a randomizing function for X^N, then the efficiency η_N is upper-bounded as follows:*

$$\eta_N \leq \frac{H(X^N)}{N}. \tag{4.27}$$

Proof We may write

$$
\begin{aligned}
H(X^N) &\geq H(X^N) - H(X^N|Z) \\
&= H(Z) - H(Z|X^N) \\
&= H(Z) \\
&= \sum_{m \geq 1} H(Z_m|Z^{m-1}) \\
&= \sum_{m \geq 1} \sum_{z_m \in Z} \sum_{z^{m-1} \in Z^{m-1}} \Pr\{z^m\} \log \Pr\{z_m|z^{m-1}\}
\end{aligned}
$$

$$\geq \sum_{m\geq 1} \sum_{z_m \in \{0,1\}} \sum_{z^{m-1} \in Z^{m-1}} \Pr\{z^m\} \log \Pr\{z_m | z^{m-1}\}$$

$$= \sum_{m\geq 1} [\Pr\{z_m = 0\} + \Pr\{z_m = 1\}] ,$$

where the last equation follows from (4.24). Finally, the transformations above lead to the inequality

$$H(X^N) \geq \sum_{m\geq 1} \Pr\{z_m \neq \Lambda\}. \tag{4.28}$$

On the other hand, using (4.25–4.26) we write

$$N\eta_N = \mathbb{E}[t(B(X^N))] \tag{4.29}$$

$$= \sum_{m\geq 1} \Pr\{t(z) \geq m\}$$

$$= \sum_{m\geq 1} \Pr\{z_m \neq \Lambda\}.$$

Combining (4.28) and (4.29) we obtain (4.27).   □

4.3.2 Asymptotically Optimal Procedures for Two Random Processes

4.3.2.1 Bernoulli Process

Let X^N be a Bernoulli random process such that $x_i = 1$ with probability p and $x_i = 0$ with probability $q = 1 - p$ for all $i = 1, ..., N$, where p is some fixed but unknown parameter. Then

$$\frac{H(X^N)}{N} = h(p), \tag{4.30}$$

where $h(p) = -p \log p - q \log q$ is the binary entropy function.

In 1951, von Neumann [7] described a procedure for generating an output sequence of statistically independent bits from an input sequence $x_1, x_2, ...$ using on each pair (x_{2i-1}, x_{2i}), where $t = 1, 2, ...$, the mapping

$$00 \to \Lambda, \quad 01 \to 0, \quad 10 \to 1, \quad 11 \to \Lambda. \tag{4.31}$$

Since, for each input pair, the probability of generating a non-null output digit is $2pq$, and the efficiency is equal to pq. This quantity is rather far from $h(p)$, which is the value of the upper bound stated in Theorem 20 (see (4.27), (4.30)).

We generalize the mapping (4.31) in the following way. Let us divide the set of 2^N input sequences into the $N+1$ composition classes and set that the class S_k contains all binary sequences of the Hamming weight k,

$$|S_k| = \binom{N}{k}, \quad k = 0, ..., N.$$

Given k, let us number the elements of S_k by integers $1, ..., |S_k|$ and construct a mapping of the element numbered by i to a vector $z_k(i)$. If $i = 1$, then we set $z_k(1) = \Lambda$. Otherwise, we find the integer l such that $i + 1 \in [2^l + 1, 2^{l+1}]$ and assign $z_k(i)$ as a standard binary representation of the integer $i - 2^l$ having the length l.

Example Let $N = 6$ and $k = 2$. Then $\binom{N}{k} = 10$ and the mapping described above is shown in the following matrix:

$$\left\| \begin{matrix} 1 & 2 & 3 & 4 & 5 & 6 & 7 & 8 & 9 & 10 \\ \Lambda & 0 & 1 & 00 & 01 & 10 & 11 & 000 & 001 & 010 \end{matrix} \right\| .$$

Let $(x_1, ..., x_N)$ be an input vector. Suppose that it has the Hamming weight k, numbered by i in S_k, and $z_k(i)$ is corresponding output vector. Then we assign $B(x) = z_k(i)$.

Let us denote

$$l_k = \lfloor \log |S_k| \rfloor \tag{4.32}$$

and define $(\alpha_{n_k}, ..., \alpha_0)$ as a binary representation of $|S_k|$, i.e.,

$$|S_k| = \sum_{i=0}^{l_k} \alpha_i 2^i. \tag{4.33}$$

Then, as it is easy to see, the expectation of the length of the output vector is expressed by

$$\mathbb{E}[t(B(X^N))] = \sum_{k=0}^{N} \Pr\{S_k\} \bar{t}_k(B(X^N)), \tag{4.34}$$

where

$$\bar{t}_k(B(X^N)) = \sum_{i=1}^{l_k} i \alpha_i 2^i. \tag{4.35}$$

The estimates

$$|S_k| \geq 2^{l_k},$$

$$\sum_{i=1}^{l_k} (l_k - i)\alpha_i 2^i \leq 2^{l_k+1}$$

are obviously valid. Thus, using (4.33) we can estimate (4.35) and write

$$l_k - 2 \leq \bar{t}_k(B(X^N)) \leq l_k,$$

or, using (4.32),

$$\log |S_k| - 3 \leq \bar{t}_k(B(X^N)) \leq \log |S_k|.$$

Hence (see (4.25) and (4.34)),

$$\frac{1}{N} \sum_{k=0}^{N} |S_k| \cdot p^k q^{N-k} \log |S_k| - \frac{3}{N} \leq \eta_N \leq \frac{1}{N} \sum_{k=0}^{N} |S_k| \cdot p^k q^{N-k} \log |S_k|.$$

Using the Stirling bounds on factorials gives, for fixed $\rho = k/N$,

$$\lim_{N \to \infty} \frac{1}{N} \log \binom{N}{N\rho} = h(\rho). \qquad (4.36)$$

By the weak law for the binomial distribution, given any $\varepsilon > 0$ and $\delta > 0$ there is an N_0 such that for $N > N_0$, all but at most δ of the probability in the binomial occurs in terms for which $|(k/N) - p| < \varepsilon$. This, together with the inequalities for η_N and the fact that $\log |S_k| \leq N$ gives, for $N > N_0$,

$$(1 - \varepsilon) \min_{|\gamma| \leq \varepsilon} h(p + \gamma) - \frac{3}{N} \leq \eta_N \leq \max_{|\gamma| \leq \varepsilon} h(p + \gamma) + \delta.$$

Therefore, using the continuity of the binary entropy function h, we obtain

$$\lim_{N \to \infty} \eta_N = h(p).$$

If the input sequence $x_1, x_2, \ldots$ has a length greater than N, then it can be divided into N-tuples each of which is mapped by B and the non-null results concatenated. It proves the following statement (4.36).

Theorem 21 *For any Bernoulli process X, there is a randomizing procedure which is independent on p, whose limiting efficiency realizes the bound of Theorem 20.*

4.3.2.2 Two-State Markov Process

Let X^N be a Markov random process such that

$$\Pr\{x_i = 1 | x_{i-1} = 0\} = p_0, \quad \Pr\{x_i = 1 | x_{i-1} = 1\} = p_1.$$

Then

$$P = \frac{p_0}{1 + p_0 - p_1} = \Pr\{x_i = 1\}$$

and

$$H(X^N) = h(P) + (N - 1)[(1 - P) \cdot h(p_0) + P \cdot h(p_1)]. \tag{4.37}$$

Given a sequence $(x_1, ..., x_N)$ generated by X^N, we use x_1 only to determine the state of the process when x_2 is generated, and decompose $(x_2, ..., x_N)$ into two sequences, X_0 and X_1. X_0 consists of all x_i for which $x_{i-1} = 0$, concatenated in increasing index order, produced by the mapping

$$x_i \rightarrow \begin{cases} x_i, & \text{if } x_{i-1} = 0, \\ \lambda, & \text{if } x_{i-1} = 1 \end{cases}$$

and X_1 consists of all x_i for which $x_{i-1} = 1$, similarly concatenated.

Note that X_0 and X_1 are pocesses that generate independent bits and can be viewed as Bernoulli processes with probabilities p_0 and p_1, respectively. The initial process is ergodic, so there exists an ε_M which tends to zero with increasing (M) such that, with probability $> (1 - \varepsilon_M)$, X_0 generates at least $M(1 - P - \varepsilon_M)$ bits and X_1 generates at least $M(P - \varepsilon_M)$ bits when the process generates M bits. Applying the procedure described above independently to X_0 and X_1, and concatenating the output sequences, the expected number of total output symbols for an input M-tuple can be lower-bounded as

$$(1 - \varepsilon_M)M(1 - P - \varepsilon_M)\eta_{M(1-P-\varepsilon_M)} + (1 - \varepsilon_M)M(P - \varepsilon_M)\eta_{M(P-\varepsilon_M)}.$$

Dividing by M, taking the limit as $M \rightarrow \infty$, using (4.37) and the result of Theorem 21 we obtain the following statement.

Theorem 22 *Given a two-state Markov process, X, there is a randomizing procedure which is independent on p_0 and p_1 and has a limiting efficiency which attains the bound of Theorem 20.*

References

1. R. Ahlswede, T.S. Han, K. Kobayashi, Universal coding of integers and unbounded search trees. IEEE Trans. Inf. Theory **43**(2), 669–682 (1997)
2. L.D. Davisson, Universal noiseless coding. IEEE Trans. Inform. Theory **19**, 783–795 (1973)
3. P. Elias, Predictive coding. IRE Trans. Inform. Theory **1**, 16–33 (1955)
4. P. Elias, The efficient construction of an unbiased random sequence. Ann. Math. Statist. **43**, 865–870 (1972)
5. P. Elias, Universal codeword sets and representations of the integers. IEEE Trans. Inform. Theory **21**(2), 194–203 (1975)
6. S. Golomb, Run-length encodings. IEEE Trans. Inform. Theory **12**(4), 399–401 (1966)
7. J. von Neumann, *The general and logical theory of automata, cerebral mechanisms in behavior* (The Hixon Symposium, London, 1951)
8. A.D. Wyner, An upper bound on the entropy series. Inform. Contr. **20**, 176–181 (1972)

Part II
Transmitting Data

Chapter 5
Coding Theorems and Converses for the DMC

5.1 Lecture on Introduction

In [13] 1948 C.E. Shannon initiated research in Information Theory with his paper 'A Mathematical Theory of Communication', in which he investigated several problems concerning the transmission of information from a source to a destination (see Chap. 1). In this lecture we shall discuss the three fundamental results of this pioneering paper: (1) the Source Coding Theorem for the Discrete Memoryless Source (DMS), (2) the Coding Theorem for the Discrete Memoryless Channel (DMC), (3) the Rate Distortion Theorem. In each of these fundamental theorems the entropy arises as a measure for data compression.

(1) The Source Coding Theorem for the DMS
The Source Coding Theorem was intensively discussed in Chap. 2, where two proofs were given. Let us reformulate the theorem in terms appropriate for the following paragraphs. An (n, ε)-compression can be represented by a pair of functions (f_n, g_n), where $f_n : \mathcal{X}^n \to \mathbb{N}$ and $g_n : \mathbb{N} \to \mathcal{X}^n$ are encoding and decoding function of the source, respectively, with $\mathrm{Prob}\big(g_n\big(f_n(X^n)\big) = X^n\big) \geq 1 - \varepsilon$ for a random variable X^n on a finite set $\mathcal{X}^n$.

($\mathbb{N}$ may obviously be replaced by $\{0, 1\}^*$). If $\mathrm{rate}(f_n) = \frac{1}{n} \log \|f_n\| = R$ ($\|f_n\| \triangleq$ cardinality of the range of f_n) we have an (n, R, ε)-compression. R is said to be an achievable rate, if for all n large enough there exists an (n, R, ε)-compression. Denoting by $R_X(\varepsilon)$ the set of all ε-achievable rates and by $R_X \triangleq \bigcap_{\varepsilon > 0} R_X(\varepsilon)$ we have the

Theorem 23 (Source Coding Theorem for the DMS)

$$R_X = H(X) = - \sum_{x \in \mathcal{X}} P(x) \cdot \log P(x)$$

(2) The Coding Theorem for the DMC
Remember that a (n, N, λ)-code is a set of pairs $\big\{(u_i, D_i), 1 \leq i \leq N, u_i \in \mathcal{X}^n,$ $D_i \subset \mathcal{Y}^n, D_i \cap D_j = \varnothing$ for $i \neq j, w^n(D_i|u_i) \geq 1 - \lambda$ for all $i\big\}$, where

A. Ahlswede et al. (eds.), *Storing and Transmitting Data*, 113
Foundations in Signal Processing, Communications and Networking 10,
DOI: 10.1007/978-3-319-05479-7_5, © Springer International Publishing Switzerland 2014

$W = \left(w(y|x)\right)_{x \in \mathcal{X}, y \in \mathcal{Y}}$ is a stochastic matrix characterizing the channel over which the data are transmitted, and n, N and λ are the block length, cardinality and maximum probability of error of the code, respectively. Denoting by $N(n, \lambda)$ the maximum cardinality of a code with given parameters n and λ we have the

Theorem 24 (Coding Theorem for the DMC)

$$\lim_{n \to \infty} \frac{1}{n} N(n, \lambda) = C \text{ for every } \lambda \in (0, 1)$$

where $C \triangleq \max\limits_{X} I(X \wedge Y)$ is the capacity of the channel.

This fundamental theorem was intensively discussed. We present a proof for the direct part $\left(\underline{\lim}_{n \to \infty} \frac{1}{n} \log N(n, \lambda) \geq C\right)$ using a random choice argument in order to achieve list size 1 in a list code for the DMC with feedback. Further the weak converse $\left(\inf_{\lambda > 0} \overline{\lim}_{n \to \infty} \frac{1}{n} \log N(n, \lambda) \leq C\right)$ will be proved using Fano's Lemma. In Lecture 5.2 we shall present Feinstein's proof of the direct part, which makes use of a "greedy" construction. Further, several proofs of the strong converse $\left(\overline{\lim}_{n \to \infty} \frac{1}{n} \log N(n, \lambda) \leq C\right)$ are introduced.

(3) The Rate Distortion Theorem

In Lecture 3.3 of Chap. 2 we briefly discussed Rate Distortion Theory. Again a discrete memoryless source $(X_t)_{t=1}^{\infty}$ has to be compressed. In Source Coding Theory the encoding function f_n was chosen bijective on a support (typical or entropy-typical sequences). Hence all the typical sequences were correctly reproduced, whereas all the other sequences (a set A with $P^n(A) \leq \varepsilon$) cannot be reproduced.

In Rate Distortion Theory there is another criterion for the quality of the reproduction than the error probability ε. Here our model is as follows. We are given an encoding function $f_n : \mathcal{X}^n \to \mathbb{N}$ and a decoding function $g_n : \mathbb{N} \to \hat{\mathcal{X}}^n$. Hence words of length n over an alphabet $\mathcal{X}$ are reproduced over an alphabet $\hat{\mathcal{X}}$. A function $d : \mathcal{X} \times \hat{\mathcal{X}} \to \mathbb{R}_+$ is introduced as a measure for the quality of the reproduction for each letter. Often d is a distance function, e.g., the Hamming distance or the Lee distance. To obtain a measure for the quality of the reproduction of a word $x^n \in \mathcal{X}^n$, d is summed up over all letters $t = 1, \ldots, n$. This way we can define our measure of distortion $d : \mathcal{X}^n \times \hat{\mathcal{X}}^n \to \mathbb{R}_+$ by $d(x^n, \hat{x}^n) = \sum_{t=1}^{n} d(x_t, \hat{x}_t)$.

If the average distortion per letter

$$\frac{1}{n} \mathbb{E} d\left(x^n, g_n\left(f_n(x^n)\right)\right) \leq D$$

and $\frac{1}{n} \log \|f_n\| \triangleq R$, we say that (f_n, g_n) is a (n, R, D)-reproduction. The problem now is to determine the rate for a specified D and large n.

Theorem 25 (Rate Distortion Theorem) *For all $D \geq 0$ and all distortion measures d*

$$R(D) = \min_{\substack{W: \mathcal{X} \to \hat{\mathcal{X}} \\ \mathbb{E}d(X,\hat{X}) \leq D}} I(X \wedge \hat{X})$$

where X is the generic variable of the source and $\hat{X}$ is the output variable of a channel W.

5.2 Lecture on the Coding Theorem for the DMC

5.2.1 Shannon's Coding Theorem: A Preview

In the channel coding theory we investigate the question of the values N, n and λ for which (N, n, λ) codes exist. Recall that a code is a set of pairs

$$\left\{ (u_i, D_i) : 1 \leq i \leq N, u_i \in \mathcal{X}^n, D_i \subset \mathcal{Y}^n, D_i \cap D_j = \varnothing \text{ for } i \neq j \right\}.$$

For a given block length n, we would like to transmit as many messages N as possible with as small an error probability λ as possible over the channel specified by the stochastic matrix $W = \big(w(y|x) \big)_{x \in \mathcal{X}, y \in \mathcal{Y}}$. These two goals, of course, conflict, and we therefore limit ourselves to two questions:

1. Given n and $\lambda \in (0, 1)$ how large can we choose N?
2. Given n and N, how small can λ become?

We denote the extreme values by $N(n, \lambda)$ and by $\lambda(n, R)$, $R = \frac{1}{n} \log N$. Both questions have been studied in detail in the literature. Shannon, using heuristic reasoning, arrived at the relation

$$\lim_{n \to \infty} \frac{1}{n} \log N(n, \lambda) = C \text{ for } \lambda \in (0, 1),$$

and for the constant C he presented a formula which includes the information function I, which he also introduced. This is defined as follows for a pair of discrete random variables (U, V):

$$I(U \wedge V) \triangleq H(U) + H(V) - H(U, V).$$

Elementary transformations yield

$$I(U \wedge V) = H(U) - H(U|V) = H(V) - H(V|U),$$

A random variable X with values in $\mathcal{X}$ is called the input variable of the channel. The corresponding output variable Y is defined by the specification

$$P(Y = y, X = x) = w(y|x)P(X = x) \text{ for } y \in \mathcal{Y}, x \in \mathcal{X}.$$

Hence $P(Y = y) = \sum_{x \in \mathcal{X}} w(y|x)P(X = x)$ for all $y \in \mathcal{Y}$ and the distribution of the output variable Y is induced by the distribution P of the input variable and the transmission matrix W. For this reason the notation $I(P, W)$ instead of $I(X \wedge Y)$ is also convenient.

These auxiliary quantities serve to describe the behaviour of codes.

Theorem 26 (Shannon's Coding Theorem) *For the DMC, for all $\lambda \in (0, 1)$*

$$\lim_{n \to \infty} \frac{1}{n} \log N(n, \lambda) = \max_X I(X \wedge Y).$$

Feinstein and Shannon proved

$$\lim_{n \to \infty} \frac{1}{n} \log N(n, \lambda) \geq \max_X I(X \wedge Y), \tag{5.1}$$

and Fano proved

$$\inf_{\lambda > 0} \overline{\lim_{n \to \infty}} \frac{1}{n} \log N(n, \lambda) \leq \max_X I(X \wedge Y). \tag{5.2}$$

Wolfowitz showed that for all $n \in \mathbb{N}$

$$|\log N(n, \lambda) - n \max_X I(X \wedge Y)| \leq c_1(\lambda, w)\sqrt{n}. \tag{5.3}$$

This implies for all $\lambda \in (0, 1)$,

$$\overline{\lim_{n \to \infty}} \frac{1}{n} \log N(n, \lambda) \leq \max_X I(X \wedge Y). \tag{5.4}$$

In order to emphasize the differences, Wolfowitz called inequalities 5.1, 5.2 and 5.4 the coding theorem, the weak converse of the coding theorem and the strong converse of the coding theorem. Shannon's Coding Theorem follows from 5.1 and 5.4. The largest rate which can be achieved asymptotically in the block length for an arbitrarily small error probability is called the capacity C of a channel. It therefore already follows from 5.1 and 5.2 that $\max_X I(X \wedge Y)$ is the capacity of the DMC.

Engineers are more often interested in the second question. We preset rate R and estimate $\lambda(n, R)$. The reliability function

$$E(R) \triangleq \overline{\lim_{n \to \infty}} -\frac{1}{n} \log \lambda(n, R)$$

describes the exponential decrease of the error. The 'sphere packing' bound gives an upper estimate for $E(R)$ and the random coding bound gives a lower estimate for $E(R)$. Above a critical rate $R_{\text{crit}} < C$, the two bounds are identical.

Even for the binary symmetric channel (BSC), $E(R)$ is not known for all rates (Shannon et al.).

The coding theorem is an existential statement. Explicit code constructions are known only for 'regular' channels. We shall see in the next lecture that feedback permits the explicit statement of asymptotically optimal codes.

5.2.2 Information as List Reduction

In this lecture we shall introduce a block coding procedure for the DMC with feedback with rate asymptotically $\frac{1}{\ell} \log N(\ell, \lambda) \geq \max_x I(X \wedge Y) = C$ the capacity of the DMC. In the next lecture we shall show that this rate is optimal, hence, feedback does not increase the capacity of the discrete memoryless channel.

Observe that the block length was denoted ℓ instead of the usual n. We shall keep this notation throughout this lecture, since we are working with a new code concept now, which was first introduced by Elias.

Definition 25 An (N, ℓ, λ, L) **list code** is a set of pairs $\{(u_i, D_i) : 1 \leq i \leq N\}$ with

(a) $u_i \in \mathcal{X}^\ell, D_i \subset \mathcal{Y}^\ell$,
(b) $w(D_i^c | u_i) \leq \lambda$ for all $i = 1, \ldots, N$
(c) $\sum_{i=1}^{N} 1_{D_i(y^\ell)} \leq L$ for all $y^\ell \in \mathcal{Y}^\ell$.

Remember that usually a code consists of code words u_i and decoding sets D_i which are disjoint for $i \neq j$. In a list code the disjointness is no longer required (only for $L = 1$). Now we demand that a word $y^\ell \in \mathcal{Y}^\ell$ does not allow too many ($\leq L$) interpretations. So for each $y^\ell \in \mathcal{Y}^\ell$ the list of possible code words u_i with $y^\ell \subset D_i$ does not contain more than L elements.

A list code induces the hypergraph $(y^\ell, \{D_i : i = 1, \ldots, N\})$. A hypergraph is a pair $(\mathcal{V}, \xi)$ consisting of a set $\mathcal{V}$ and system ξ of subsets of $\mathcal{V}$. In the special case that all subsets in ξ are of size 2, we have a usual graph. Analogously to Graph Theory the elements of $\mathcal{V}$ are denoted as vertices, the elements of ξ as hyper-edges.

Since the sets D_i associated with the code words $u_i, i = 1, \ldots, N$ may overlap, it makes also sense to define the dual hypergraph $(\mathcal{U}, \{\mathcal{U}(y^\ell) : y^\ell \in \mathcal{Y}^\ell\})$, where $\mathcal{U} \triangleq \{u_1, \ldots, u_N\}$ and for every $y^\ell \in \mathcal{Y}^\ell \ \mathcal{U}(y^\ell) \triangleq \{u_i : 1 \leq i \leq N, y^\ell \in D_i\}$ denotes the list of ($\leq L$) code words associated with y^ℓ.

5.2.3 Typical and Generated Sequences

Remember that a δ_ℓ-typical sequence $x^\ell \in \mathcal{X}^\ell$ is characterized by the property

$$|\ell P(x) - <x^\ell|x>| \leq \delta_\ell \cdot \ell \text{ for all } x \in \mathcal{X},$$

where $<x^\ell|x>$ denotes the frequency of the letter x in the sequence x^ℓ. For the special choice $\delta_\ell = 0$ the typical sequences arise. When Chebyshev's Inequality and the weak law of large numbers are involved, usually $\delta_\ell = \frac{c}{\sqrt{\ell}}$ for some constant c is chosen. Throughout this paragraph we choose $\delta_\ell \triangleq \delta$ for some constant $\delta > 0$.

The set of all δ-typical sequences is denoted by $T^\ell_{P,\delta}$ or $T^\ell_{X,\delta}$ if X is a random variable with probability distribution P.

Observe that $T^\ell_{P,\delta} = \bigcup_{Q:|P(x)-Q(x)|\leq\delta} T^\ell_Q$ is the union of type classes.

Definition 26 According to the notation for typical sequences we define for pairs of sequences $x^\ell \in \mathcal{X}^\ell$, $y^\ell \in \mathcal{Y}^\ell$ and $x \in \mathcal{X}, y \in \mathcal{Y}$

$$<x^\ell, y^\ell|x, y> \triangleq |\{i : 1 \leq i \leq \ell, x_i = x \text{ and } y_i = y\}|.$$

In words, $<x^\ell, y^\ell|x, y>$ is the number of components in which x^ℓ assumes the value x and y^ℓ assumes the value y.

Definition 27 Let $\mathcal{X}$ and $\mathcal{Y}$ be the input and output alphabet of a DMC specified by the transmission matrix $W = (w(y|x))_{x\in\mathcal{X},y\in\mathcal{Y}}$. A sequence $y^\ell \in \mathcal{Y}^\ell$ is called $(\mathbf{x}^\ell, \delta)$-**generated** or **conditionally typical**, if for all $x \in \mathcal{X}, y \in \mathcal{Y}$

$$\left| <x^\ell, y^\ell|x, y> - <x^\ell|x> \cdot w(y|x) \right| \leq \delta \cdot \ell.$$

The set of all $(\mathbf{x}^\ell, \delta)$-**generated** sequences is denoted as

$$T^\ell_{Y|X,\delta}(x^\ell).$$

Let us give a short interpretation of the previous definitions. Assume that a sequence $x^\ell \in \mathcal{X}^\ell$ of absolute type $(<x^\ell|x>)_{x\in\mathcal{X}}$ is transmitted over the channel W. Then the received sequence y^ℓ will "typically" contain $<x^\ell|x> \cdot w(y|x)$ y's in those positions where the original sequence x^ℓ had the letter x. So with high probability a (x^ℓ, δ)-generated sequence will be received, when x^ℓ is transmitted over the channel.

Lemma 32 *For all $\ell \in \mathbb{N}$*

(a) $\exp\{\ell \cdot H(P) - g_1(P, \delta) \cdot \ell\} \leq |T^\ell_{P,\delta}| \leq \exp\{\ell \cdot H(P) + g_1(P, \delta) \cdot \ell\}$ *with*
$g_1(P, \delta) \to 0$ *for* $\delta \to 0$
(b) $Prob(x^\ell \in T^\ell_{P,\delta}) \geq 1 - \exp\{-g_2(P, \delta) \cdot \ell\}$ *with* $g_2(P, \delta) > 0$ *for* $\delta > 0$.

Proof We shall prove two slightly weaker statements, namely
(a') For all $\delta > 0$ there exists a positive integer $\ell(\delta)$, such that for all $\ell > \ell(\delta)$

$$\exp\{\ell \cdot H(P) - g_1(P, \delta) \cdot \ell\} \leq |T_{P,\delta}^\ell| \leq \exp\{\ell \cdot H(P) + g_1(P, \delta) \cdot \ell\}$$

with $g_1(P, \delta) \to 0$ for $\delta \to 0$.
(b') For all $\ell \in \mathbb{N}$

$$\mathrm{Prob}\big(x^\ell \in T_{P,\delta}^\ell\big) \geq 1 - 2a \cdot \exp\{-g_2(P, \delta) \cdot \ell\}$$

with $g_2(P, \delta) > 0$ for $\delta > 0$

(a') Recall that for the type class T_Q^ℓ

$$(\ell + 1)^{-|\mathcal{X}|}\exp\{\ell \cdot H(Q)\} \leq |T_Q^\ell| \leq \exp\{\ell \cdot H(Q)\} \tag{5.5}$$

and that for the set $\mathcal{P}(\ell, \mathcal{X})$ of all types for a given glock length ℓ over the alphabet $\mathcal{X}$

$$|\mathcal{P}(\ell, \mathcal{X})| \leq (\ell + 1)^{|\mathcal{X}|}. \tag{5.6}$$

Since $T_{P,\delta}^\ell = \bigcup_{Q:||P-Q||_1 \leq \delta} T_Q^\ell$ and the type classes are obviously disjoint, we have $|T_{P,\delta}|^\ell = \sum_{Q:||P-Q||_1 \leq \delta} |T_Q^\ell|$. With 5.5 we can conclude

$$(\ell + 1)^{-|\mathcal{X}|} \sum_{Q:||P-Q||_1 \leq \delta} \exp\{\ell \cdot H(Q)\} \leq |T_{P,\delta}^\ell|$$

$$\leq \sum_{Q:||P-Q||_1 \leq \delta} \exp\{\ell \cdot H(Q)\}.$$

By Lemma 82 (Chap. 8) for $||P - Q||_1 \leq \delta < \frac{1}{2}$ we have $|H(P) - H(Q)| \leq -\delta \cdot \log \frac{\delta}{|\mathcal{X}|}$. Hence we can estimate

$$(\ell + 1)^{-|\mathcal{X}|} \sum_{Q:||P-Q||_1 \leq \delta} \exp\{\ell(H(P) + \delta \cdot \log \tfrac{\delta}{|\mathcal{X}|})\} \leq |T_{P,\delta}^\ell|$$

$$\leq \sum_{Q:||P-Q||_1 \leq \delta} \exp\{\ell(H(P) - \delta \cdot \log \tfrac{\delta}{|\mathcal{X}|})\}$$

and further

$$(\ell + 1)^{-|\mathcal{X}|}\exp\{\ell(H(P) + \delta \cdot \log \frac{\delta}{|\mathcal{X}|})\} \leq |T_{P,\delta}^\ell|$$

$$\leq \sum_{Q \in \mathcal{P}(\ell, \mathcal{X})} \exp\{\ell(H(P) - \delta \cdot \log \frac{\delta}{|\mathcal{X}|})\}.$$

By 5.6 $|\mathcal{P}(\ell, \mathcal{X})| \leq (\ell + 1)^{|\mathcal{X}|}$. Hence

$$(\ell + 1)^{-|\mathcal{X}|} \exp\{\ell(H(P) + \delta \cdot \log \frac{\delta}{|\mathcal{X}|})\} \leq |T_{P,\delta}^\ell|$$

$$\leq (\ell + 1)^{|\mathcal{X}|} \exp\{\ell(H(P) - \delta \cdot \log \frac{\delta}{|\mathcal{X}|})\}$$

or equivalently

$$\exp\{\ell(H(P) + \delta \cdot \log \frac{\delta}{|\mathcal{X}|} - \frac{1}{\ell}|\mathcal{X}|\log(\ell + 1))\} \leq |T_{P,\delta}^\ell|$$

$$\leq \exp\{\ell(H(P) - \delta \cdot \log \frac{\delta}{|\mathcal{X}|} + \frac{1}{\ell}|\mathcal{X}|\log(\ell + 1))\}.$$

Now for every $\ell \geq \ell(\delta)$ it is $\frac{1}{\ell}|\mathcal{X}|\log(\ell + 1) \leq -\delta \cdot \log \frac{\delta}{|\mathcal{X}|}$ and with $g_1(\delta, P) = -2\delta \cdot \log \frac{\delta}{|\mathcal{X}|}$ (a') is proved.

(b') $\text{Prob}(x^\ell \in T_{P,\delta}^\ell)$

$$= \text{Prob}(|< x^\ell|x > -\ell \cdot P(x)| \leq \delta \cdot \ell \text{ for all } x \in \mathcal{X})$$
$$= 1 - \text{Prob}(\exists x \in \mathcal{X} : |< x^\ell|x > -\ell \cdot P(x)| > \delta \cdot \ell)$$
$$\geq 1 - \sum_{x \in \mathcal{X}} \text{Prob}(|< x^\ell|x > -\ell \cdot P(x)| > \delta \cdot \ell)$$
$$\geq 1 - a \cdot \max_{x \in \mathcal{X}} \text{Prob}(|< x^\ell|x > -\ell \cdot P(x)| > \delta \cdot \ell)$$
$$\geq 1 - 2a \cdot \max_{x \in \mathcal{X}} \cdot \max\{\text{Prob}(< x^\ell|x > -\ell \cdot P(x) > \delta \cdot \ell),$$
$$\text{Prob}(< x^\ell|x > -\ell \cdot P(x) < -\delta \cdot \ell)\}$$

For all $x \in \mathcal{X}$ we now define the random variables

$$T_i^x \triangleq \begin{cases} 1, & x_i = x \\ 0, & x_i \neq x \end{cases}, \text{ where } x^\ell = (x_1, \ldots, x_\ell).$$

For every $x \in \mathcal{X}$ the random variables $T_i^x, i = 1, \ldots, \ell$ are independent and identically distributed with $\mathbb{E} T_i^x = P(x)$. Now

$$\text{Prob}(< x^\ell|x > -\ell \cdot P(x) > \delta \cdot \ell) = \text{Prob}\left(\sum_{i=1}^\ell T_i^x > \ell(P(x) + \delta)\right)$$

$$\leq \exp\{-\ell \cdot D(P(x) + \delta \| P(x))\}$$

by application of Bernstein's trick (Inequalities, Lemma 7) with $\lambda = P(x) + \delta$ and $\mu = P(x)$. Accordingly it can be shown that

$$\text{Prob}\big(< x^\ell | x > -\ell \cdot P(x) < -\delta \cdot \ell\big) \leq \exp\{-\ell \cdot D\big(P(x) - \delta || P(x)\big)\}.$$

Combining these two results we have

$$\text{Prob}\big(x^\ell \in T^\ell_{P,\delta}\big) \geq 1 - 2a \cdot \max_{x \in \mathcal{X}} \cdot \max\{\exp\{-\ell \cdot D(P(x) + \delta || P(x))\},$$

$$\exp\{-\ell \cdot D(P(x) - \delta || P(x))\}\}$$

$$= 1 - 2a \cdot \exp\{-\ell \cdot g_2(P, \delta)\},$$

where $g_2(P, \delta) = \min\{\min_{x \in \mathcal{X}} D\big(P(x) + \delta || P(x)\big), \min_{x \in \mathcal{X}} D\big(P(x) - \delta || P(x)\big)\}$ $\qquad \square$

Since a generated sequence is a special case of a typical sequence (under the condition that $x^\ell \in \mathcal{X}^\ell$ is given) the proof of the next lemma follows the same lines as the proof of Lemma 32.

Lemma 33 *For all $\ell \in \mathbb{N}$ and $x^\ell \in T^\ell_{X,\delta}$*

(a) $\exp\{\ell \; H(Y|X) - g_3(P, W, \varepsilon, \delta) \cdot \ell\} \leq \left| T^\ell_{Y|X,\varepsilon}(x^\ell) \right| \leq \exp\{\ell \; H(Y|X) +$
  $g_3(P, W, \varepsilon, \delta) \cdot \ell\}$ *with* $g_3(P, W, \varepsilon, \delta) \to 0$ *for* $\varepsilon, \delta \to 0$
(b) $\text{Prob}\big(y^\ell \in T^\ell_{Y|X,\varepsilon} | X^\ell = x^\ell\big) \geq 1 - \exp\{-g_4(P, W, \varepsilon, \delta) \cdot \ell\}$ *with* $g_4(P, W, \varepsilon, \delta)$
  > 0 *for* $\varepsilon > 0$.

For a channel $W = \big(w(y|x)\big)_{x \in \mathcal{X}, y \in \mathcal{Y}}$ with input alphabet $\mathcal{X}$ and output alphabet $\mathcal{Y}$ the dual channel W^* is defined by the probabilities $w^*(x|y) = P_{X|Y}(x|y)$, where X and Y are random variables on $\mathcal{X}$ and $\mathcal{Y}$, respectively, with distributions P_X and P_Y. By the Bayes formula, $P_{XY}(x, y) = P_X(x) \cdot P_{Y|X}(y|x) = P_Y(y) \cdot P_{X|Y}(x|y)$. Applied to the channels W and W^* this yields

$$P_X(x) \cdot w(y|x) = P_Y(y) \cdot w^*(x|y).$$

Lemma 34 (Duality Lemma) *For $u \in T^\ell_{X,\delta}$ and $v \in T^\ell_{Y|X,\delta}(u)$ we have*

(a) $v \in T^\ell_{Y,\delta_1}, \; \delta_1 = 2\delta |\mathcal{X}|$
(b) $u \in T^\ell_{X|Y,\delta_2}(v), \; \delta_2 = 2\delta(|\mathcal{X}| + 1)$.

Proof By the assumptions $| < u|x > -P(x) \cdot \ell | \leq \delta \cdot \ell$ and $| < u, v|x, y > -w(y|x)$
$< u|x > | \leq \delta \cdot \ell$ for $x \in \mathcal{X}, y \in \mathcal{Y}$.
  From this follows $| < u, v|x, y > -w(y|x) \cdot P(x) \cdot \ell | \leq 2\delta \cdot \ell$ and since
$< v|y > = \sum_{x \in \mathcal{X}} < u, v|x, y >$ we can further conclude

$$| < v|y > -P_Y(y) \cdot \ell| = \left| \sum_{x \in \mathcal{X}} (< u, v|x, y > -w(y|x) \cdot P(x) \cdot \ell \right|$$

$$\leq \sum_{x \in \mathcal{X}} |< u, v|x, y > -w(y|x)P(x)\ell| \leq 2\delta\ell|\mathcal{X}|$$

and (a) is proved.

In order to prove (b) we apply the Bayes formula and replace $w(y|x)P(x)$ by $w^*(x|y) \cdot P_Y(y)$ $(P(x) = P_X(x))$ to obtain $| < u, v|x, y > -w^*(x|y) \cdot P_Y(y) \cdot \ell| \leq 2\delta \cdot \ell$ and further by (a)

$$| < u, v|x, y > -w^*(x|y) < v|y > | \leq 2\delta\ell(|\mathcal{X}| + 1). \qquad \Box$$

Now we have all the preliminaries for the list reduction lemma that will be central in the proof of the direct part of the coding theorem for the DMC with feedback.

Lemma 35 (List Reduction Lemma) *The canonical (N, ℓ, λ, L) list code $\{T_{Y|X,\delta}^\ell(x^\ell) : x^\ell \in T_{X,\varepsilon}^\ell\}$ has the following properties:*

(a) $\exp\{H(X) \cdot \ell - g_1(P, \delta) \cdot \ell\} \leq |T_{X,\delta}^\ell| \leq \exp\{\ell \cdot H(X) + g_1(P, \delta)\}$ *with $g_1(P, \delta) \to 0$ for $\delta \to 0$*
(b) $\lambda \leq \exp\{-g_4(\delta, P, W)\ell\}$ *with $g_4(\delta, P, W) > 0$*
(c) $L \leq \exp\{H(X|Y) \cdot \ell + g_5(\delta, P, W) \cdot \ell\}$ *with $g_5(\delta, \varepsilon, P, W) \to 0$ for $\delta, \varepsilon \to 0$.*

Proof (a) is just Part (a) of Lemma 32 and (b) is Part (b) in Lemma 33. (c) follows from the duality lemma and Part (a) in Lemma 33. $\qquad \Box$

5.2.4 Ahlswede's Block Coding Procedure

We shall now present Ahlswede's block coding procedure [1] which allows to interpret information as list reduction. This procedure yields a sequential block code for the Discrete Memoryless Channel with Feedback (DMCF), i.e. the sender knows the received sequence, that asymptotically has rate C, hence we shall finally give a constructive proof for the direct part of the coding theorem for the DMCF.

The idea of the procedure is an iterative application of the list reduction lemma. In the first step the sender chooses the canonical list code with block length $\ell_1 \triangleq \min\{\ell : |T_{X\delta}^\ell| \geq N\}$, which has the typical sequences as code words and the generated sequences as decoding sets. So the sender transmits as code word u the typical sequence corresponding to his message $z \in \{1, \ldots, N\}$. With high probability $(\geq 1 - \exp\{-g_4(\delta, P, W)\ell_1\})$ the received sequence y^{ℓ_1} will be generated by u and hence u is on the list $\mathcal{U}(y^{\ell_1})$ which has size about $H(X|Y) \cdot \ell_1$.

Because of the feedback the sender knows the received sequence y^{ℓ_1} and hence the list $\mathcal{U}(y^{\ell_1})$ of code words which are decoded as y^{ℓ_1}. If his code word is not on this list, the procedure will fail. With high probability, however, this will not take place and after the first step sender and receiver know that the message $z \in \mathcal{U}(y^{\ell_1})$. Hence the list of possible messages is reduced from $\{1, \ldots, N\}$ to $\mathcal{U}(y^{\ell_1})$ after the transmission of the first code word. The procedure is now iterated with $\mathcal{U}(y^{\ell_1})$ taking the role of $\{1, \ldots, N\}$. Hence, in the second step the sender has to encode a typical sequence of block length $\ell_2 \triangleq \min\{\ell : |T_{X,\delta}^\ell| \geq |\mathcal{U}(y^{\ell_1})|\}$, etc..

Since $|T_{X,\delta}^{\ell_2}| \sim |\mathcal{U}(y^{\ell_1})|$, we know by the list reduction lemma that $\exp\{H(X) \cdot \ell_2\} \sim \exp\{H(X|Y) \cdot \ell_1\}$ and hence

$$\ell_2 \sim \frac{H(X|Y)}{H(X)}\ell_1.$$

The procedure is iterated until ℓ_i is small enough. In each step the block length is reduced by a factor about $\frac{H(X|Y)}{H(X)}$.

Hence, at the end of the procedure we have a total block length of

$$n = \ell_1 + \ell_2 + \ell_3 + \cdots \sim \ell_1\left(1 + \frac{H(X|Y)}{H(X)} + \left(\frac{H(X|Y)}{H(X)}\right)^2 + \cdots\right)$$

$$= \ell_1\left(\frac{H(X)}{H(X) - H(X|Y)}\right).$$

Since $\ell_1 \triangleq \min\{\ell : |T_{X\delta}^\ell| \geq N\}$, it holds $\ell_1 \cdot H(X) \sim N$ and hence for the rate $R = \frac{N}{n}$ of the code we have

$$R \sim \frac{\ell_1 H(X)}{n} = H(X) - H(X|Y) = I(X \wedge Y).$$

If we now choose the input variable X with probability distribution such that $I(X \wedge Y) = \max_X I(X \wedge Y) = C$, we have an explicit coding procedure with rate assuming (asymptotically) the capacity.

The probability of error can be controlled, since there are only about $\log \ell_1$ iterations and in each step the contribution to the error probability is exponentially small.

An exact analysis of the procedure can be found in the proof of the direct part of the coding theorem for the DMCF in the next paragraph.

Observe that there are similarities to Shannon's error-free encoding procedure for the DMCF since, because of the feedback, sender and receiver have the same information after each transmission and hence the list of possible messages is iteratively reduced. However, we now have to control the probability of error and the list reduction now is blockwise (with decreasing block length), whereas in Shannon's procedure each letter reduces the list of possible messages.

5.2.5 The Coding Theorem for the DMCF

Theorem 27 *For the DMCF we have*

$$\lim_{n \to \infty} \frac{1}{n} \log N_f(n, \lambda) \geq C.$$

With given conditional probability $P(Y = y|X = x) = w(y|x)$, the list reduction is the greatest if the distribution of X is chosen in such a way that $I(X \wedge Y)$ is maximal, i.e., equal to the capacity C. We imagine this choice to have been made. We can assume that W has at least two distinct row vectors, let us say, $w(\cdot|1)$ and $w(\cdot|2)$, since otherwise $C = 0$. Let ε be chosen such that

$$H(X) - g_1(\varepsilon) > H(X|Y) + g_5(\varepsilon).$$

Feedback now enables us to apply Lemma 35 iteratively. Before we do this, we first present two technical results which are needed to describe the method.

Lemma 36 *Let $(\ell_i)_{i=1}^\infty$ be a sequence of natural numbers with $q\ell_i \leq \ell_{i+1} \leq q\ell_i + 1$ $(i \in \mathbb{N})$ for a $q \in (0, 1)$. Then it holds that for $I = 1 + \lceil(-\log \ell_1 + \log \log \ell_1)(\log q)^{-1}\rceil$*

(a) $q \log \ell_1 \leq \ell_I \leq \log \ell_1 + (1 - q)^{-1}$
(b) $\sum_{i=1}^I \ell_i \leq (1 - q)^{-1}(\ell_1 + I)$.

Proof It follows from the hypothesis that $\ell_i \leq \ell_1 q^{i-1} + q^{i-2} + \cdots + 1$ and therefore $\ell_1 q^{I-1} \leq \ell_I \leq \ell_1 q^{I-1} + (1 - q)^{-1}$. Now by selection of I, $q \log \ell_1 \leq \ell_1 q^{I-1} \leq \log \ell_1$, and therefore (a) holds. Since

$$\sum_{i=1}^I \ell_i \leq \sum_{i=1}^I \left(\ell_1 q^{i-1} + \sum_{j=0}^{i-2} q^j \right) \leq \ell_1 (1 - q)^{-1} + I(1 - q)^{-1},$$

(b) is also true. $\square$

Lemma 37 *Let $R^s = \prod_1^s R$, $\tilde{R}^s = \prod_1^s \tilde{R}$ be different distributions on $\mathcal{X}^s$. Then for the disjoint sets*

$$A \triangleq \left\{ x^s \,|\, \log \frac{R^s(x^s)}{\tilde{R}^s(x^s)} > 0 \right\}, \; \tilde{A} \triangleq \left\{ x^s \,|\, \log \frac{\tilde{R}^s(x^s)}{R^s(x^s)} > 0 \right\}$$

$$R^s(A) \geq 1 - \beta(R, \tilde{R})^s, \; \tilde{R}^s(\tilde{A}) \geq 1 - \beta(R, \tilde{R})^s,$$

where $\beta(R, \tilde{R}) \in (0, 1)$.

Proof It follows from Chebychev's inequality for $\alpha > 0$ that

$$R^s(A^c) = R^s \left(\left\{x^s \mid \log \tfrac{\tilde{R}^s(x^s)}{R^s(x^s)} \geq 0\right\}\right) = R^s \left(\left\{x^s \mid \exp\left(\alpha \log \tfrac{\tilde{R}^s(x^s)}{R^s(x^s)}\right) \geq 1\right\}\right)$$

$$\leq \mathbb{E} \exp\left(\alpha \log \tfrac{\tilde{x}^s}{x^s}\right) = \left(\sum_{x \in \mathcal{X}} R(x)^{1-\alpha} \tilde{R}(x)^\alpha\right)^s.$$

If we now set $\beta(R, \tilde{R}) = \sum_{x \in \mathcal{X}} R(x)^{\frac{1}{2}} \tilde{R}(x)^{\frac{1}{2}}$, it follows, since $R \neq \tilde{R}$, from the property of the scalar product that

$$\beta(R, \tilde{R}) < \left(\sum_{x \in \mathcal{X}} R(x)\right)^{\frac{1}{2}} \left(\sum_{x \in \mathcal{X}} \tilde{R}(x)\right)^{\frac{1}{2}} = 1. \qquad \square$$

Likewise, $\tilde{R}(\tilde{A}^c) \leq \beta(R, \tilde{R})^s$.

Proof of the theorem Now we have all the means we need for the exact analysis of the coding method. For $g_i(\delta, \varepsilon, P, W)$ we write g_i. We shall use the notation $f(n) = 0(g(n))$ which means that $f(n) \leq c \cdot g(n)$ for all $n \in \mathbb{N}$ and some constant $c \in \mathbb{R}$.

Let $\mathcal{Z} \triangleq \{1, \ldots, N\}$ be the set of the possible messages and let $\ell_1 \in \mathbb{N}$ be the smallest number with $N \leq \exp((H(X) - g_1)\ell_1)$. We define the sequence (ℓ_i) inductively. Then $\ell_{i+1} \in \mathbb{N}$ is the smallest number satisfying

$$\exp((H(X|Y) + g_5)\ell_i) \leq \exp((H(X) - g_1)\ell_{i+1}).$$

If we set $q \triangleq (H(X|Y) + g_5)(H(X) - g_1)^{-1} \in (0, 1)$, then we obtain $q\ell_i \leq \ell_{i+1} \leq q\ell_i + 1$. With this, the condition of Lemma 5 is fulfilled.

Now we consider the elements in $T_{X,\delta}^{\ell_i}(1 \leq i \leq I)$ to be ordered lexicographically. Let $\Phi_1 : \mathcal{Z} \to T_{X,\delta}^{\ell_1}$ map $i \in \mathcal{Z}$ into the ith element in $T_{X,\delta}^{\ell_1}$. Let such an injective mapping between two ordered sets be called canonical.

If $z \in \mathcal{Z}$ is to be transmitted, then we first transmit $\Phi_1(z)$. The receiver receives a sequence $v(1) = (v_1, \ldots, v_{\ell_1})$, which is also known to the sender due to feedback. Therefore, the list

$$\mathcal{U}(v(1)) = \{x^{\ell_1} \mid v(1) \in T_{Y|X,\varepsilon}^{\ell_1}(x^{\ell_1})\}$$

is known to both. By Lemma 35

$$|\mathcal{U}(v(1))| \leq \exp((H(X|Y) + g_5)\ell_1), \quad \lambda_1 \leq \exp(-g_4\ell_1).$$

Now let Φ_2 map $\mathcal{U}(v(1))$ canonically into $T_{X,\delta}^{\ell_1}$. If $\Phi_1(x)$ is in $\mathcal{U}(v(1))$, i.e., no error was made, then $\Phi_2(\Phi_1(x))$ is transmitted now. Otherwise, an arbitrary element from $\Phi_2(\mathcal{U}(v(1)))$ is transmitted, let us say the first. Since there is already one error, it doesn't matter which element is transmitted.

If $u(1), u(2), \ldots, u(i)$ have already been transmitted and $v(1), v(2), \ldots, v(i)$ have been received, then the next element transmitted will be $\Phi_{i+1}(u(i))$ if $u(i) \in \mathcal{U}(v(i))$. Otherwise the first element in $\Phi_i\big(\mathcal{U}(v(i))\big)$ will be transmitted.

We carry out a total of I iterations. At the time of the ith iteration, we have a contribution of $\lambda_i \leq \exp\{-g_4 \ell_i\}$ to the error probability, and because $\ell_1 \geq \ell_2 \geq \cdots \geq \ell_I, \exp\{-g_4 \ell_i\} \leq \exp\{-g_4 \ell_I\}$ holds. Therefore, it follows from Lemma 36 that

$$\sum_{i=1}^{I} \lambda_i \leq I \exp\{-g_5 \ell_I\} \leq I \exp\{-g_5 q \log \ell_1\}.$$

Since $I = O(\log \ell_1)$, this error will become arbitrarily small as ℓ_1 (and therefore N) increases.

By Lemma 36(b), the previously necessary block length is

$$m_I = \sum_{i=1}^{I} \ell_i \leq (1-q)^{-1}(\ell_1 + I).$$

Since $\ell_1 \leq \big(H(X) - g_1\big)^{-1} \log N + 1, q = \big(H(X|Y) + g_5\big)\big(H(X) - g_1\big)^{-1}$ and $I = O(\log \ell_1)$ it follows that

$$m_I \leq \big(H(X) - g_1 - H(X|Y) - g_5\big)^{-1} \log N + O(\log \log N) \text{ and therefore}$$

$$\log N + O(\log \log N) \geq m_I \big(I(X \wedge Y) - g_2 - g_5\big).$$

Therefore, with this rate we approach the capacity C arbitrary closely. Of course, after the Ith step we still have $L_I \leq \exp\big((H(X|Y) + g_5)\ell_I\big)$ candidates on the list. However, by Lemma 5, these are only $\exp(O(\log \ell_1))$. These messages can be separated with a block code which we explicitly construct with the help of Lemma 37, using relatively little additional block length.

Let $R \triangleq w(\cdot|1), \tilde{R} \triangleq w(\cdot|2)$, let d_1 be a sequence of s ones and let d_2 be a sequence of s twos. We set $D_{d_1} \triangleq A$ and $D_{d_2} \triangleq \tilde{A}$.

Let $\{b_1 \ldots b_t : b_i \in \{d_1, d_2\}, 1 \leq i \leq t\}$ be the set of code words and let $D_{b_1 \ldots b_t} = \prod_{i=1}^{t} D_{b_i}$ be the decoding set for code word $b_1 \ldots b_t$. Then by Lemma 6 and Bernoulli's inequality $\big((1+x)^t \geq 1 + tx \text{ for all } x \geq -1, t \in \mathbb{N}\big)$,

$$W(D_{b_1 \ldots b_t} | b_1 \ldots b_t) \geq 1 - t\beta(w(\cdot|1), w(\cdot|2))^s. \tag{5.7}$$

If we select $t = s$ and s as small as possible in such a way that $2^s \geq L_I = \exp(O(\log \ell_1))$, thus $s = O(\log \ell_1)$, then the L_I messages can be assigned in a one-to-one correspondence to code words and because of 5.7 the error probability can be made arbitrarily small. The total block length of the method satisfies $n \leq m_I + O(\log^2 m_I)$. And so the method has all the claimed properties.

5.2.6 The Coding Theorem for the DMC

If there is no feedback, an explicit coding scheme as for the DMC with feedback is not known. However, the list reduction lemma can also be applied to demonstrate the existence of an (N, n, λ)-code with rate $\frac{\log N}{n} \geq C$.

Theorem 28 *For the discrete memoryless channel we have for all $\lambda \in (0, 1)$*

$$\lim_{n \to \infty} \frac{1}{n} \log N(n, \lambda) \geq C = \max_X I(X \wedge Y)$$

Proof The idea of the proof is to choose the code words $u_1, \ldots, u_N$ at random from the set $T^\ell_{X,\delta}$ of δ-typical sequences of length ℓ (again $\ell \triangleq n$, to keep the notation of the list reduction lemma). So the code words are realizations of random variables $U_1, \ldots, U_N$ which we assume to be independent and uniformly distributed on $T^\ell_{X,\delta}$. If the channel output was $v \in \mathcal{Y}^\ell$ the receiver voted for the list $\mathcal{U}_v \triangleq \mathcal{U}(v) = \{x^\ell : v \in T^\ell_{Y|X,\delta'}(x^\ell)\}$. Since now there is no feedback, this list is not known to the sender and we cannot proceed as in the proof of Theorem 27. Now we restrict the lists to $\mathcal{U}_v(U_1, \ldots, U_n) \triangleq \mathcal{U}_v \cap \{U_1, \ldots, U_n\}$. We are done if these lists consist of only one element. Loosely spoken, we want to achieve list length 1 in one step.

If, e.g., $U_1 = u_1$ has been sent, then with probability at least $1 - \lambda$, say, u_1 is contained in the list of the received word v. The receiver correctly votes for u_1, if
$$\mathcal{U}_v \cap \{U_2, \ldots, U_N\} = \varnothing.$$
Now remember that by the list reduction lemma and by Lemma 32

$$|\mathcal{U}_v| \leq \exp\{H(X|Y) \cdot \ell + g_5 \ell\} \text{ and } |T^\ell_{X,\delta}| \geq \exp\{H(X)\ell - g_1 \ell\},$$

where g_5 and g_1 tend to 0 for $\delta \to 0$.

The probability that other code words than u_1 are contained in the list $\mathcal{U}_v$ can hence be upper bounded by

$$P\left(\mathcal{U}_v \cap \{U_2, \ldots, U_N\} \neq \varnothing\right) \leq (N-1)\frac{\exp\{H(X|Y) \cdot \ell + g_5 \ell\}}{\exp\{H(X) - g_1 \ell\}}$$
$$= (N-1) \cdot \exp\{(H(X|Y) - H(X) + g_5 + g_1) \cdot \ell\}$$
$$= (N-1) \cdot \exp\{-I(X \wedge Y) \cdot \ell + (g_5 + g_1) \cdot \ell\}$$

For every rate below capacity, i.e., $\frac{\log N}{\ell} < C$ or equivalently $N < \exp\{C \cdot \ell\}$, and δ sufficiently small it is possible to make this probability of error arbitrarily small.

If, e.g., we want to assure that $P\left(\mathcal{U}_v \cap \{U_2, \ldots, U_N\} \neq \varnothing\right) < \bar{\lambda}$, we have for N the condition

$$N < \bar{\lambda} \cdot \exp\{I(X \wedge Y) \cdot \ell - (g_5 + g_1) \cdot \ell\}.$$

If we denote by $\lambda_i(U_1, \ldots, U_N)$ the probability of error for the code word U_i, it can hence be assured that the average error $\mathbb{E}\frac{1}{N}\sum_{i=1}^{N}\lambda_i(U_1, \ldots, U_N)$ is smaller than λ'', say. Hence, there exists a realization $(u_1, \ldots, u_N)$ of the random variable $(U_1, \ldots, U_N)$ with average error $\frac{1}{N}\sum_{i=1}^{N}\lambda_i(u_1, \ldots, u_N) \leq \lambda''$.

Now we choose the $\frac{N}{2}$ code words indexed by i_j, $j = 1, \ldots, \frac{N}{2}$, say, with the smallest error probabilities λ_{i_j}. For these code words the maximum error is surely smaller than $2 \cdot \lambda''$.   □

Remark In the proof of the coding theorem for the DMC we used the method of random choice. This probabilistic method is often applied in Combinatorial Theory to show the existence of a configuration with certain properties. A random experiment according to a probability distribution on the set of all possible configurations is performed of which the expected value can often be estimated. Hence there exists a realization assuming a value as least as "good" as the expected value. The problem is to choose the random experiment appropriately. In the proof of the coding theorem we chose the code words successively with the same random experiment (uniform distribution on the set of typical sequences). Observe that it is possible that the same code word occurs twice in the random code obtained this way. However, the probability of such an event is negligibly small.

Observe that the existence of a good code can also be expressed in terms of hypergraphs. The problem is to find a maximum number of vertices in the dual hypergraph $(\mathcal{U}, \{\mathcal{U}(y^\ell) : y^\ell \in \mathcal{Y}^\ell\})$, such that with high probability each hyper-edge contains only one vertex. This approach has further been investigated by Ahlswede in his paper "Coloring Hypergraphs".

5.2.7 The Proofs of Wolfowitz

5.2.7.1 Typical Sequences

For the proofs of the following two theorems we need some facts concerning typical sequences.

$$T_{P,\delta}^n (= T_{X,\delta}^n) \triangleq \left\{x^n \in \mathcal{X}^n : |< x^n|x > -nP(x)| \leq \delta\sqrt{n} \text{ for all } x \in \mathcal{X}\right\}$$

and generated sequences (conditionally typical given $x^n \in \mathcal{X}^n$)

$$T_{W,\delta}^n(x^n) \quad \left(= T_{Y|X,\delta}^n(x^n)\right)$$
$$\triangleq \left\{y^n \in \mathcal{Y}^n : |< x^n, y^n|x, y > -w(y|x) < x^n|x >| \leq \delta\sqrt{n} \right.$$
$$\left. \text{for all } x \in \mathcal{X}, y \in \mathcal{Y}\right\}.$$

Observe that now $\delta_n \triangleq \frac{\delta}{\sqrt{n}}$ is chosen in the definition of typical sequences. The idea is to make use of Chebyshev's Inequality instead of Bernstein's trick in the proofs of the following lemmas.

Lemma 38 $P^n(T_{P,\delta}^n) \geq 1 - \frac{|\mathcal{X}|}{4\delta^2}$.

Proof

$$P^n(T_{P,\delta}^n) = P^n\left(\{x^n \in \mathcal{X}^n : |< x^n|x > -nP(x)| \leq 2\delta\sqrt{\frac{n}{4}} \; \forall x \in \mathcal{X}\}\right)$$

$$\geq P^n\left(\{x^n \in \mathcal{X}^n : |< x^n|x > -nP(x)| \leq 2\delta\sqrt{nP(x)(1-P(x))} \; \forall x \in \mathcal{X}\}\right)$$

$$\geq 1 - \frac{|\mathcal{X}|}{4\delta^2}$$

by Chebyshev's Inequality. $\qquad\square$

Analogously, the following lemma is proven.

Lemma 39 $w^n\left(T_{W,\delta}^n(x^n)|x^n\right) \geq 1 - \frac{|\mathcal{X}|\cdot|\mathcal{Y}|}{4\delta^2}$.

The elements in $T_{W,\delta_2}^n(x^n)$ for some $x^n \in T_{P,\delta_1}^n$ are also typical sequences for the probability distribution $Q = P \cdot W$. The proof of the next lemma is the same as for Part (a) of the Duality Lemma.

Lemma 40 *For* $x^n \in T_{P,\delta_1}^n$ *and* $y^n \in T_{W,\delta_2}^n(x^n)$ *it is (with* $Q = P \cdot W$)

$$y^n \in T_{Q,\delta'}^n, \; \delta' = (\delta_1 + \delta_2)|\mathcal{X}|.$$

Lemma 41 *For all* $y^n \in T_{W,\delta_2}^n(x^n)$, $x^n \in T_{P,\delta_1}^n$, $Q = P \cdot W$

$$\exp\{-nH(Q) - k_1\sqrt{n}\} < Q^n(y^n) < \exp\{-nH(Q) + k_1\sqrt{n}\}$$

for some constant $k_1 \triangleq (\delta_1 + \delta_2) \cdot |\mathcal{X}| \cdot \left(-\sum_{y \in \mathcal{Y}} \log Q(y)\right)$.

Proof $Q^n(y^n) = \prod_{y \in \mathcal{Y}} Q(y)^{<y^n|y>}$. By the previous lemma y^n is Q-typical, hence with $\delta' = (\delta_1 + \delta_2) \cdot |\mathcal{X}|$

$$\prod_{y \in \mathcal{Y}} Q(y)^{nQ(y) - \delta'\sqrt{n}} < Q^n(y^n) < \prod_{y \in \mathcal{Y}} Q(y)^{nQ(y) + \delta'\sqrt{n}}$$

from which the statement is immediate. $\qquad\square$

Lemma 42 *Let* $T_{W,\delta_2}^n = \bigcup_{x^n \in T_{P,\delta_1}^n} T_{W,\delta_2}^n(x^n)$. *Then*

$$\exp\{nH(Q) - k_2\sqrt{n}\} < |\; T_{W,\delta_2}^n \;| < \exp\{nH(Q) + k_2\sqrt{n}\}$$

with some constant k_2.

Proof By Lemma 41 obviously, $|\; T_{W,\delta_2}^n \;| < \exp\{nH(Q) + k_1\sqrt{n}\}$, since $1 = \sum_{y^n \in \mathcal{Y}^n} Q^n(y^n) \geq \sum_{y^n \in T_{W,\delta_2}^n} Q^n(y^n)$.
By the Lemmas 38 and 39

$$
\begin{aligned}
Q^n(T_{W,\delta_2}^n) &= \sum_{x^n \in \mathcal{X}^n} P^n(x^n) w^n(T_{W,\delta_2}^n | x^n) \\
&\geq \sum_{x^n \in T_{P,\delta_1}^n} P^n(x^n) w^n(T_{W,\delta_2}^n | x^n) \\
&\geq \left(1 - \frac{|\mathcal{X}|}{4\delta_1^2}\right)\left(1 - \frac{|\mathcal{X}| \cdot |\mathcal{Y}|}{4\delta_2^2}\right).
\end{aligned}
$$

Hence, with Lemma 41

$$|\; T_{W,\delta_2}^n \;| \geq \left(1 - \frac{|\mathcal{X}|}{4\delta_1^2}\right)\cdot\left(1 - \frac{|\mathcal{X}| \cdot |\mathcal{Y}|}{4\delta_2^2}\right)\cdot\exp\{nH(Q) - k_1\sqrt{n}\} \qquad \square$$

Lemma 43 *For $x^n \in T_{P,\delta_1}^n$ and $y^n \in T_{W,\delta_2}^n(x^n)$ it is*

$$\exp\left\{-n\sum_{x \in \mathcal{X}} P(x)H\big(w(\cdot|x)\big) - k_3\sqrt{n}\right\} < w^n(y^n|x^n)$$

$$< \exp\left\{-n\sum_{x \in \mathcal{X}} P(x)H\big(w(\cdot|x)\big) + k_3\sqrt{n}\right\}$$

for some constant $k_3 = (\delta_1 + \delta_2)\sum_{x \in \mathcal{X}}\sum_{y \in \mathcal{Y}} \log w(y|x)$.

Proof

$$
\begin{aligned}
w^n(y^n|x^n) &= \prod_{\substack{x \in \mathcal{X} \\ y \in \mathcal{Y}}} w(y|x)^{<x^n, y^n|x,y>} \\
&= \exp\left\{\sum_{x \in \mathcal{X}}\sum_{y \in \mathcal{Y}} < x^n, y^n|x, y > \cdot \log w(y|x)\right\}.
\end{aligned}
$$

Since x^n and y^n are typical sequences, for all $x \in \mathcal{X}, y \in \mathcal{Y}$ $n \cdot w(y|x) \cdot P(x) - \sqrt{n}(\delta_1 + \delta_2) \leq < x^n, y^n|x, y > \leq n \cdot w(y|x)P(x) + \sqrt{n}(\delta_1 + \delta_2)$.

$\square$

Analogously to Lemma 42 it can be derived.

Lemma 44 *For $x^n \in T_{P,\delta_1}^n$ and a constant k_4*

$$\exp\left\{n \sum_{x \in \mathcal{X}} P(x)H\big(w(\cdot|x)\big) - k_4\sqrt{n}\right\} < T_{W,\delta_2}^n(x^n)$$

$$< \exp\left\{n \sum_{x \in \mathcal{X}} P(x)H\big(w(\cdot|x)\big) + k_4\sqrt{n}\right\}.$$

5.2.7.2 Proof of the Direct Part

In Lecture 5.2.2 we already presented a proof for the direct part of the Coding Theorem for the DMC ($\underline{\lim}_{n\to\infty} N(n, \lambda) \geq C$).

It followed the lines of the proof of the analogous statement for the DMC with feedback, which made use of the list reduction lemma. It was observed that the list reduction lemma could also be applied to codes for the DMC, when list size 1 is achieved in one iteration step. This condition could be guaranteed for a code of size $\exp\{n \cdot C\}$, when the code words were chosen at random as typical sequences. Shannon's original proof also made use of a random choice argument.

Feinstein [9] gave a proof of the direct part of the coding theorem, Theorem 2 below, based on his idea of maximal coding, in which a non-extendable code is constructed by the Greedy Algorithm. Whereas he used entropy-typical sequences we follow here Wolfowitz's version of the maximal code method in terms of (letterwise) typical sequences (in the sense of 2 in Lecture 2). Abstract forms of Feinstein's method were formulated by Blackwell, Breiman, and Thomasian [7].

Theorem 29 (Feinstein 1954 [9]) For all $\lambda \in (0, 1)$ there exists some constant $k = k(\lambda)$, such that for all $n \in \mathbb{N}$ a $\big(n, \exp\{nC - k\sqrt{n}\}, \lambda\big)$ code exists.

Proof Let $\{(u_i, D_i), 1 \leq i \leq N\}$ be a (N, n, λ)-code, which has the following properties:

(1) $u_i \in T_{P,\delta_1}^n$,
(2) $D_i \triangleq T_{W,\delta_2}^n(u_i) \smallsetminus \left(\bigcup_{j=1}^{i-1} D_j\right)$,
(3) The code is maximal, i.e., it is impossible to add some pair (u_{N+1}, D_{N+1}) fulfilling (1) and (2), such that a $(n, N + 1, \lambda)$-code is obtained.

Here δ_2 is chosen such that in Lemma 39 $w^n\big(T_{W,\delta_2}^n(u)|u\big) \leq \frac{\lambda}{2}$. Observe that the possible codes obtained by this construction do not necessarily have the same size. However, it will turn out that the approach is quite useful, since the size is determined asymptotically. Such a maximal code obviously exists, since it is constructible for $N = 1$ (by Lemma 39).

A code constructed as above has the following properties:

(i) For $u \in T^n_{P,\delta_1}$ with $u \notin \{u_1, \dots, u_N\}$ we have

$$w^n\big(T^n_{W,\delta_2}(u) \cap \big(\bigcup_{i=1}^{N} D_i\big)|u\big) > \frac{\lambda}{2} \qquad (5.8)$$

(otherwise the code would be extendable by $\big(u, T^n_{W,\delta_2}(u) \setminus \bigcup_{i=1}^{N} D_i\big)$).

(ii) For $u \in \{u_1, \dots, u_N\}$ it is $T^n_{W,\delta_2}(u) \subset \bigcup_{i=1}^{N} D_i$ (by construction)

Since $\lambda < 1$, obviously $1 - \frac{\lambda}{2} > \frac{\lambda}{2}$ and 5.8 also holds for the code words. Now

$$Q^n\big(\bigcup_{i=1}^{N} D_i\big) = \sum_{x^n \in \mathcal{X}^n} P^n(x^n) w^n\big(\bigcup_{i=1}^{N} D_i|x^n\big)$$

$$\geq \sum_{x^n \in T^n_{P,\delta_1}} P^n(x^n) w^n\big(\bigcup_{i=1}^{N} D_i|x^n\big)$$

$$\geq P^n\big(T^n_{P,\delta_1}\big) \cdot \frac{\lambda}{2}$$

$$\geq (1 - \varepsilon) \cdot \frac{\lambda}{2} \text{ for some appropriate } \varepsilon > 0.$$

By the pigeonhole principle we now can conclude that

$$\big|\bigcup_{i=1}^{N} D_i\big| \geq (1 - \varepsilon)\frac{\lambda}{2}\exp\{n\, H(Q) - k_1\sqrt{n}\}$$

for a constant k_1 as in Lemma 41.

Further, since $D_i \subset T^n_{W,\delta_2}(u_i)$, it follows that for each $i = 1, \dots, N : |D_i| \leq |T^n_{W,\delta_2}(u_i)| \leq \exp\{n\, H(W|P) + k_4\sqrt{n}\}$ for a constant k_4. Therefore by Lemma 44

$$\big|\bigcup_{i=1}^{N} D_i\big| \leq N\exp\{n\, H(W|P) + k_4\sqrt{n}\}.$$

We thus estimated $|\bigcup_{i=1}^{N} D_i|$ in two different ways. This yields a condition for the size N of the code.

$$N \geq (1 - \varepsilon)\frac{\lambda}{2}\exp\{n\big(H(Q) - H(W|P)\big) - (k_1 + k_4)\sqrt{n}\}.$$

Since $C = \max_{P}\{H(Q) - H(W|P)\}$ and the estimations are correct for all probability distributions P on $\mathcal{X}$, the theorem is immediate. $\qquad\square$

5.2.7.3 Wolfowitz' Proof of the Strong Converse

Wolfowitz [14] proved the somehow stronger statement.

Theorem 30 (Wolfowitz 1957) *Let $\lambda \in (0, 1)$, then there is a constant $k(\lambda)$ such that for all $n \in \mathbb{N}$ a $\left(n, \exp\{nC + k(\lambda)\sqrt{n}\}, \lambda\right)$-code does not exist.*

Proof Let $\mathcal{U} \triangleq \{(u_i, D_i) : 1 \le i \le N\}$ be a (n, N, λ)-code, hence $w^n(D_i|u_i) \ge 1 - \lambda$ for all $i = 1, \ldots, N$.

From this code $\mathcal{U}$ a subcode $\mathcal{U}(P)$ is obtained by allowing only those pairs (u_i, D_i) in which u_i is a typical sequence for the distribution P on $\mathcal{X}$. Denoting by P^* the distribution for which $|\mathcal{U}(P^*)|$ is maximal it suffices to estimate $|\mathcal{U}(P^*)|$, since there are only fewer than $(n + 1)^{-|\mathcal{X}|}$ types P and hence $|\mathcal{U}(P^*)| \ge |\mathcal{U}| \cdot (n + 1)^{-|\mathcal{X}|}$. Observe that $(n + 1)^{-|\mathcal{X}|}$ is polynomial in n, whereas $|\mathcal{U}|$ is usually exponential. Hence we have reduced $\mathcal{U}$ to a subcode $\mathcal{U}(P^*) = \{(u_i, D_i) : 1 \le i \le N'\}$ by concentration on the typical sequences.

In a next step the decoding sets in $\mathcal{U}(P^*)$ are reduced via

$$D_i' \triangleq D_i \smallsetminus T_{W,\delta}^n(u_i)^c$$

Here δ is chosen such that in Lemma 39 $w^n\left(T_{W,\delta}^n(u)|u\right) \ge 1 - \frac{1-\lambda}{2}$.

So, we only consider the generated sequences in D_i. Obviously,

$$w^n(D_i'|u_i) \ge w^n(D_i|u_i) - w^n\left(T_{W,\delta}^n(u_i)^c\right)$$
$$\ge 1 - \lambda - \frac{1-\lambda}{2} = \frac{1-\lambda}{2} = 1 - \lambda',$$

where $\lambda' \triangleq \frac{1+\lambda}{2}$.

As in the proof of the direct part, we estimate $| \bigcup_{i=1}^{N'} D_i' |$ in two ways.

(i) $| \bigcup_{i=1}^{N'} D_i' | \ge N'(1 - \lambda') \cdot \exp\{nH(W|P) - k_3\sqrt{n}\}$ (since by Lemma 43 and the above computation

$$N'(1 - \lambda') \le \sum_{i=1}^{N} w^n(D_i'|u_i)$$

$$= \sum_{i=1}^{N'} \sum_{x^n \in D_i'} w^n(x^n|u_i)$$

$$\leq \mid \bigcup_{i=1}^{N'} D_i' \mid \cdot \exp\{-nH(W|P) + k_3\sqrt{n}\})$$

(ii) Since the elements in $\bigcup_{i=1}^{N'} D_i'$ are Q-typical, $\bigcup_{i=1}^{N'} D_i' \subset T_{Q,\delta'}^n$ for $\delta' = \delta \cdot |\mathcal{X}|$ and by Lemma 42

$$\mid \bigcup_{i=1}^{N'} D_i' \mid \leq \mid T_{Q,\delta'}^n \mid \leq \exp\{nH(Q) + k_2\sqrt{n}\}$$

Combining these two estimations, again a condition for N' is obtained.

$$N' \leq \exp\{n(H(Q) - H(W|P) + (k_2 + k_3)\sqrt{n}) - \log(1 - \lambda')\}$$

$$= \exp\{n\,I(P, W) + (k_2 + k_3)\sqrt{n} - \log(1 - \lambda')\}$$

$$\leq \exp\{n\,C + (k_2 + k_3)\sqrt{n} - \log(1 - \lambda')\},$$

since the estimations hold for all P on $\mathcal{X}$ and $C = \max_P I(P, W)$.   □

5.2.7.4 Further Proofs of the Strong Converse: An Outline

In the following paragraphs we shall present several proofs of the strong converse:

$$\overline{\lim_{n \to \infty}} \frac{1}{n} \log N(n, \lambda) \leq C \text{ for all } \lambda \in (0, 1).$$

In his pioneering paper in 1948 Shannon already mentioned the statement but he only gave a sketch of a proof. Wolfowitz (1957) [14] gave the first rigorous proof, hereby using typical sequences. Later Kemperman et al. analyzed Wolfowitz' proof and found out that the typical sequences have an important property, specified in the Packing Lemma. The proofs of Kemperman (I, II) [10, 11], Augustin [5] and Ahlswede/Dueck [2] presented in Lecture 5.2 make use of this Packing Lemma. A new idea is introduced in Arimoto's proof [4]. Here the components in the code words are permuted and symmetries in the total system are exploited.

Quite another idea is the background of the last proofs presented. The weak converse will be made "strong". Ahlswede and Dueck [2] show that every code with larger probability of error λ contains a subcode with small λ and similar rate. In [3] the decoding sets are "blown up" to obtain a strong converse from a weak converse. The theory behind can be found in Lecture 5.3.3.

The reason why we present such a variety of proofs for the strong converse is a methodical one. Later on, we shall see that the proofs allow generalizations in several directions. For instance, Kemperman's second proof can also be applied to channels with feedback, Augustin's proof is appropriate for channels with infinite alphabets and the method of Ahlswede/Dueck [2] yields a proof of the converse for channels with input constraints.

5.3 Lecture on Strong Converses

5.3.1 Proofs of the Strong Converse Using the Packing Lemma

5.3.1.1 A Packing Lemma

Let us analyze Wolfowitz' proof of the strong converse. The code words u_i and the words in the decoding sets D_i, $i = 1, \ldots, N$ were chosen as typical and generated sequences, respectively. The total size of the decoding sets $| \bigcup_{i=1}^{N} D_i |$ could then be estimated from above and below using some well-known combinatorial properties of the generated sequences, namely

1. $w^n(y^n|x^n) \approx \exp\{-nH(W|P)\}$ for all $y^n \in T^n_{W,\delta_2}$, $x^n \in T^n_{P,\delta_1}$,
2. $Q^n(y^n) \approx \exp\{-nH(Q)\}$, for $y^n \in T^n_{W,\delta_2}$, $Q = P \cdot W$,
3. $w^n\left(T^n_{W,\delta_2}(u_i)^c|u_i\right) < \frac{1-\lambda}{2} \triangleq \gamma$ for all $u_i \in T^n_{P,\delta_1}$.

From the upper and lower bound obtained this way a condition on N could be derived, in which the mutual information is the central expression.

The properties 1, 2, and 3 are abstracted in the following Packing Lemma (the decoding sets form a packing of $\mathcal{Y}^n$).

As Fano's Lemma it does not make use of the time structure, i.e., the block length n is not involved and can hence be chosen w.l.o.g. as $n = 1$.

Let $\mathcal{X}$ and $\mathcal{Y}$ denote finite sets, $W = \left(w(y|x)\right)_{x \in \mathcal{X}, y \in \mathcal{Y}}$ a stochastic matrix and let Q be an arbitrary probability distribution on $\mathcal{Y}$. Further, for all $x \in \mathcal{X}$ we have some $\Theta_x > 0$ and define

$$B_x(\Theta_x, Q) \triangleq \left\{ y \in \mathcal{Y} : \frac{w(y|x)}{Q(y)} \geq 2^{\Theta_x} \right\}.$$

Lemma 45 (Packing Lemma) *If for a (N, λ)-code $\{(u_i, D_i) : i = 1, \ldots, N\}$ ($u_i \in \mathcal{X}$, $D_i \subset \mathcal{Y}$ for all i, $D_i \cap D_j = \emptyset$ for all $i \neq j$) there exists a probability distribution Q on $\mathcal{Y}$ with $\max_{u_i} \sum_{y \in B_{u_i}(\Theta_{u_i}, Q)} w(y|u_i) < \gamma$ and $\lambda + \gamma < 1$, then*

$$N < (1 - \lambda - \gamma)^{-1} 2^{\frac{1}{N} \sum_{i=1}^{N} \Theta_{u_i}}.$$

Proof We shortly write Θ_i instead of Θ_{u_i} and set for $i = 1, \ldots, N$

$$A_i \triangleq \left\{ y \in D_i : \frac{w(y|u_i)}{Q(y)} < 2^{\Theta_i} \right\} = D_i \cap \left(B_{u_i}(\Theta_i, Q)\right)^c.$$

For $y \in A_i$ then by definition $2^{\Theta_i} Q(y) > w(y|u_i)$ and hence for all i

$$
2^{\Theta_i} Q(D_i) \geq 2^{\Theta_i} Q(A_i) > w(A_i|u_i) = w(D_i|u_i) - w(D_i \setminus A_i|u_i)
$$
$$
\geq w(D_i|u_i) - B_{u_i}(\Theta_i, Q) \geq 1 - \lambda - \gamma
$$

by the assumptions. Division by $Q(D_i)$ and taking logarithm on both sides yields for all $i = 1, \ldots, n$ $\Theta_i \geq \log\left(\frac{1-\lambda-\gamma}{Q(D_i)}\right)$ and hence

$$
\frac{1}{N}\sum_{i=1}^{N}\Theta_i \geq \frac{1}{N}\sum_{i=1}^{N}\log\left(\frac{1-\lambda-\gamma}{Q(D_i)}\right) = -\sum_{i=1}^{N}\frac{1}{N}\log Q(D_i) + \log(1-\lambda-\gamma)
$$

$$
> -\sum_{i=1}^{N}\frac{1}{N}\log\frac{1}{N} + \log(1-\lambda-\gamma) = \log N + \log(1-\lambda-\gamma),
$$

since the relative entropy $\sum_{i=1}^{N}\frac{1}{N}\log\left(\frac{\frac{1}{N}}{Q(D_i)}\right) > 0$.

Exponentiation on both sides yields the statement of the lemma. $\qquad\square$

5.3.1.2 Kemperman's First Proof of the Strong Converse

With the notation of the preceding lemma we now choose $\mathcal{X} \triangleq \mathcal{X}^n$, $\mathcal{Y} \triangleq \mathcal{Y}^n$ and $W \triangleq W^n$ and let $\{(u_i, D_i) : i = 1, \ldots, N\}$ be an (n, N, λ)-code. We partition $\mathcal{U} \triangleq \{u_1, \ldots, u_N\}$ into subcodes $\mathcal{U}(P)$, where each code word in $\mathcal{U}(P)$ is of type $P \in \mathcal{P}$.

Let the type P^* be such that $\mathcal{U}(P^*)$ is a subcode of maximum length $M \triangleq |\mathcal{U}(P^*)|$. Then $N \leq (n+1)^{|\mathcal{X}|} \cdot M$.

Further let $Q \triangleq P^* \cdot W$. In each code word $u = (u_{(1)}, \ldots, u_{(n)}) \in \mathcal{U}(P^*)$ by definition every $x \in \mathcal{X}$ has frequency exactly $< u|x > = P^*(x) \cdot n$. We denote by $\mathbb{E}_P$ and Var_P the expected value and the variance of a random variable on a set $\mathcal{X}$ with probability distribution P.

$$
\mathbb{E}_{w^n(\cdot|u)}\log\frac{w^n(\cdot|u)}{Q^n(\cdot)} = \sum_{t=1}^{n}\mathbb{E}_{w(\cdot|u_{(t)})}\log\frac{w(\cdot|u_{(t)})}{Q(\cdot)}
$$

$$
= \sum_{t=1}^{n}\sum_{y\in\mathcal{Y}}w(y|u_{(t)})\log\frac{w(y|u_{(t)})}{Q(y)}
$$

$$
= \sum_{x\in\mathcal{X}}\sum_{y\in\mathcal{Y}}nP^*(x)w(y|x)\cdot\log\frac{w(y|x)}{Q(y)}
$$

$$
= nI(P^*, W).
$$

Next we shall show that $\mathrm{Var}_{w^n(\cdot|u)} \log \frac{w^n(\cdot|u)}{Q^n(\cdot)}$ is bounded.

$$\mathrm{Var}_{w^n(\cdot|u)} \log \frac{w^n(\cdot|u)}{Q^n(\cdot)} = \sum_{t=1}^{n} \mathrm{Var}_{w(\cdot|u_{(t)})} \log \frac{w(\cdot|u_{(t)})}{Q(\cdot)}$$

$$\leq \sum_{t=1}^{n} \mathbb{E}_{w(\cdot|u_{(t)})} \left(\log \frac{w(\cdot|u_{(t)})}{Q(\cdot)} + \log P^*(u_{(t)}) \right)^2$$

$$\text{(since } \mathrm{Var} X \leq \mathbb{E}(X + a)^2 \text{ for all } a \in \mathbb{R})$$

$$= \sum_{t=1}^{n} \sum_{y \in \mathcal{Y}} w(y|u_{(t)}) \cdot \left(\log \frac{P^*(u_{(t)})w(y|u_{(t)})}{Q(y)} \right)^2$$

$$= \sum_{y \in \mathcal{Y}} n \, P^*(u_{(t)})w(y|u_{(t)}) \cdot \left(\log \frac{P^*(u_{(t)})w(y|u_{(t)})}{Q(y)} \right)^2$$

$$= n \sum_{y \in \mathcal{Y}} Q(y) \frac{P^*(u_{(t)}) \cdot w(y|u_{(t)})}{Q(y)} \left(\log \frac{P^*(u_{(t)})w(y|u_{(t)})}{Q(y)} \right)^2.$$

Now observe that the function $x \cdot \log^2 x$ is uniformly continuous and therefore bounded by some constant. Hence the function $\sum_{i=1}^{b} x_i \log^2 x_i$ is also bounded for every fixed b. It can be shown that $\sum_{i=1}^{b} x_i \log^2 x_i \leq \max\{\log^2(3), \log^2(b)\}$, when $\sum_{i=1}^{b} x_i = 1$. Hence

$$\mathrm{Var}_{w^n(\cdot|u)} \log \frac{w^n(\cdot|u)}{Q^n(u)} \leq n \cdot c, \quad \text{where } c \triangleq \max\{\log^2 3, \log^2 |\mathcal{Y}|\}.$$

Now choose for all $i = 1, \ldots, N \ \Theta_i = \Theta \triangleq nI(P^*, W) + \sqrt{\frac{2n}{1-\lambda}c}$ and $Q \triangleq Q^n$. Then by Chebyshev's Inequality

$$w^n\big(B_u(\Theta, Q)|u\big) = \mathrm{Pr}\left(\frac{w^n(\cdot|u)}{Q^n(\cdot)} \geq 2^{\Theta}|u \right)$$

$$= \mathrm{Pr}\left(\log \frac{w^n(\cdot|u)}{Q^n(\cdot)} - \mathbb{E}_{w(\cdot|u)} \log \frac{w^n(\cdot|u)}{Q^n(\cdot)} \geq \sqrt{\frac{2nc}{1-\lambda}} \Big| u \right)$$

$$\leq \frac{nc}{\left(\sqrt{\frac{2nc}{1-\lambda}}\right)^2} = \frac{1-\lambda}{2}.$$

If we use the preceding lemma with $\gamma \triangleq \frac{1-\lambda}{2}$ we obtain

$$N \leq \big(1 - \lambda - \frac{1-\lambda}{2}\big)^{-1} 2^{nI(P^*, W) + \sqrt{\frac{2nc}{1-\lambda}}}$$

$$= \frac{2}{1-\lambda} \exp\left\{ nI(P^*, W) + \sqrt{\frac{2nc}{1-\lambda}} \right\}.$$

From this we can conclude that

$$N \leq \exp\left\{ Cn + \sqrt{\frac{2nc}{1-\lambda}} + a\log(n+1) + \log \frac{2}{1-\lambda} \right\}.$$

This proves the strong converse.   □

5.3.1.3 Kemperman's Second Proof of the Strong Converse

Let $\{(u_i, D_i) : 1 \leq i \leq N\}$ be a (n, N, λ)-code. From Shannon's Lemma (Chapter on Inequalities) we know that for the maximal output distribution $\overline{Q}$

$$\sum_{y\in\mathcal{Y}} w(y|x) \log \frac{w(y|x)}{\overline{Q}(y)} \leq C \text{ for all } x \in \mathcal{X} \tag{5.9}$$

Since $\log \frac{w^n(y^n|u_i)}{\overline{Q}^n(y^n)} = \sum_{t=1}^n \log \frac{w(y_t|u_{i(t)})}{\overline{Q}(y_t)}$, it follows that

$$\mathbb{E}_{w^n(\cdot|u_i)} \log \frac{w^n(\cdot|u_i)}{\overline{Q}^n(\cdot)} \leq n\,C \text{ for all } u_i$$

From (5.9) also follows that

$$\overline{Q}(y) = 0, \text{ if and only if } w(y|x) = 0 \text{ for all } x \in \mathcal{X}.$$

In order to bound the variance we now define

$$q' \triangleq \min_{\overline{Q}(y)>0} \overline{Q}(y)$$

and obtain

$$\text{Var}_{w^n(\cdot|u)} \log \frac{w^n(\cdot|u)}{\overline{Q}^n(\cdot)} = \sum_{t=1}^n \text{Var}_{w(\cdot|u_{(t)})} \log \frac{w(\cdot|u_{(t)})}{\overline{Q}(\cdot)}$$

$$\leq \sum_{t=1}^n \mathbb{E}_{w(\cdot|u_{(t)})} \log^2 \frac{w(\cdot|u_{(t)})}{\overline{Q}(\cdot)}$$

$$= \sum_{t=1}^n \sum_{\substack{y\in\mathcal{Y}: \\ w(y|u_{(t)})>0}} \overline{Q}(y)\frac{w(y|u_{(t)})}{\overline{Q}(y)} \log^2 \frac{w(y|u_{(t)})}{\overline{Q}(y)}.$$

Since the function $x \cdot \log^2 x$ is continuous in the interval $[0, \frac{1}{q'}]$, it assumes its maximum, M say, and the variance is bounded by

$$\text{Var}_{w^n(\cdot|u)} \log \frac{w^n(\cdot|u)}{\overline{Q}^n(\cdot)} \leq n \cdot \sum_{w(y|u_{(t)})>0} \overline{Q}(y) \cdot M \leq n \cdot M$$

Now we proceed as in Kemperman's first proof and set

$$\Theta_x = \Theta \triangleq nC + \sqrt{\frac{2nd}{1-\lambda}} \text{ for all } x \in \mathcal{X} \text{ and } Q \triangleq \overline{Q}^n$$

with Chebyshev's Inequality $w^n(B_u(\Theta, Q)|u) \leq \frac{1-\lambda}{2}$ for all code words u, such that the Packing Lemma finally yields:

$$N \leq \left(1 - \lambda - \frac{1-\lambda}{2}\right)^{-1} \exp\left\{nC + \sqrt{\frac{2nd}{1-\lambda}}\right\} \qquad \square$$

Remark In the estimation of the variance the upper bound M now depends on the distribution Q (not on the alphabet size as in the previous proof). Therefore the bound is not as sharp as the previous one.

5.3.1.4 Augustin's Proof of the Strong Converse

Again let $\{(u_i, D_i), 1 \leq i \leq N\}$ be a (n, N, λ)-code. In the previous proofs we obtained our probability distribution by choosing code words with the same composition. This procedure makes use of the rows in the matrix $(u_{i(j)})_{i=1,\ldots,N, j=1,\ldots,n}$ obtained by writing the code words one beneath another. Now we shall obtain a probability distribution suitable for the application of the Packing Lemma from the columns of this matrix.

The canonical idea would be to choose the distribution defined by

$$\overline{Q}(y^n) \triangleq \frac{1}{N} \sum_{i=1}^{N} w(y^n|u_i),$$

which already occurred in Fano's Lemma and hence is usually denoted as Fano Source. Indeed, with $\overline{\Theta}_{u_i} \triangleq c_i \sum_{y^n \in \mathcal{Y}^n} w^n(y^n|u_i) \log \frac{w^n(y^n|u_i)}{\overline{Q}(y^n)}$ with some appropriate constant c_i, it turns out that $\Theta \triangleq \frac{1}{N} \sum_{i=1}^{N} \overline{\Theta}_{u_i}$ is a mutual information up to a constant factor. However, following the lines of the previous proofs we only obtain a weak converse of the Coding Theorem, not a strong one. The crucial point is that, contrasting to Kemperman's proofs, the denominator now is not of product structure. Hence, expected value and variance are no longer sums of independent random variables, and the variances may become very large.

In order to get rid of this difficulty, Augustin chose as probability distribution Q the product of the one-dimensional marginal distributions of the *Fano Source*, defined by

$$Q^n(y^n) = \prod_{t=1}^n Q_t(y_t),$$

where $Q_t(y_t) \triangleq \frac{1}{N} \sum_{t=1}^N w(y_t|u_{i(t)})$ for $t = 1, \dots, n$.

Counting letter by letter the relative frequencies $P_t(x) \triangleq \frac{1}{N} \mid \{u \in \mathcal{U} : u_{i(t)} = x\} \mid$, $t = 1, \dots, n$, we can also write $Q_t(y_t) = \sum_{x \in \mathcal{X}} P_t(x) w(y_t|x)$. We are now able to estimate the average variance and expected value

$$\frac{1}{N} \sum_{i=1}^N \mathbb{E}_{w^n(\cdot|u_i)} \log \frac{w^n(\cdot|u_i)}{Q^n(\cdot)} = \frac{1}{N} \sum_{i=1}^N \sum_{y^n \in \mathcal{Y}^n} w^n(y^n|u_i) \log \frac{w^n(y^n|u_i)}{Q^n(y^n)}$$

$$= \frac{1}{N} \sum_{i=1}^N \sum_{y^n \in \mathcal{Y}^n} \prod_{t=1}^n w(y_t|u_{i(t)}) \log \prod_{t=1}^n \frac{w(y_t|u_{i(t)})}{Q(y_t)}$$

$$= \frac{1}{N} \sum_{i=1}^N \sum_{y^n \in \mathcal{Y}^n} \sum_{t=1}^n \log \frac{w(y_t|u_{i(t)})}{Q(y_t)} \cdot \prod_{s=1}^n w(y_s|u_{i(s)})$$

$$= \sum_{t=1}^n \frac{1}{N} \sum_{i=1}^N \sum_{y_t \in \mathcal{Y}} w(y_t|u_{i(t)}) \log \frac{w(y_t|u_{i(t)})}{Q(y_t)} \prod_{\substack{s=1 \\ s \neq t}}^n \sum_{y_s \in \mathcal{Y}} w(y_s|u_{i(s)})$$

$$= \sum_{t=1}^n \frac{1}{N} \sum_{i=1}^N \sum_{y_t} w(y_t|u_{i(t)}) \log \frac{w(y_t|u_{i(t)})}{Q_t(y_t)} \quad (\text{since } \sum_{y_s} w(y_s|u_{i(s)}) = 1$$

$$= \sum_{t=1}^n \sum_{x \in \mathcal{X}} \sum_{y \in \mathcal{Y}} P_t(x) w(y|x) \log \frac{w(y|x)}{Q_t(y)}$$

$$= \sum_{t=1}^n I(P_t, W) \leq n\, C.$$

For the average variance we have (with the same computation as above)

$$\frac{1}{N} \sum_{i=1}^N \mathrm{Var}_{w^n(\cdot|u_i)} \log \frac{w^n(\cdot|u_i)}{Q^n(\cdot)} = \frac{1}{N} \sum_{i=1}^N \sum_{t=1}^n \mathrm{Var}_{w^n(\cdot|u_i)} \log \frac{w^n(\cdot|u_i)}{Q_t(\cdot)}$$

$$= \sum_{t=1}^n \sum_{x \in \mathcal{X}} \sum_{y \in \mathcal{Y}} P_t(x) w(y|x) \left(\log \frac{w(y|x)}{Q_t(y)} - \mathbb{E}_{w(\cdot|x)} \log \frac{w(\cdot|x)}{Q(\cdot)} \right)^2$$

$$\leq n \cdot c$$

for an appropriate constant c, as in Kemperman's first proof.

Now again the Packing Lemma can be applied with

$$\Theta_{u_i} \triangleq \mathbb{E}_{w^n(\cdot|u_i)} \log \frac{w^n(\cdot|u_i)}{Q^n(\cdot)} + \sqrt{\frac{1}{\gamma} \mathrm{Var}_{w^n(\cdot|u_i)} \log \frac{w^n(\cdot|u_i)}{Q^n(\cdot)}} \text{ for all } i = 1, \ldots, N.$$

By Chebyshev's Inequality for $i = 1, \ldots, N$ $w^n\big(B_{u_i}(\Theta_{u_i}, Q^n)|u_i\big) \le \gamma$ and with the Packing Lemma we can conclude that

$$N \le (1 - \lambda - \gamma)^{-1} \exp\{\frac{1}{N} \sum_{i=1}^{N} \Theta_{u_i}\}.$$

Now

$$\frac{1}{N} \sum_{i=1}^{N} \Theta_{u_i} = \frac{1}{N} \sum_{i=1}^{N} \mathbb{E}_{w^n(\cdot|u_i)} \log \frac{w^n(\cdot|u_i)}{Q^n(\cdot)} + \frac{1}{N} \sum_{i=1}^{N} \sqrt{\frac{1}{\gamma} \mathrm{Var}_{w^n(\cdot|u_i)} \log \frac{w^n(\cdot|u_i)}{Q^n(\cdot)}}$$

$$\le nC + \sqrt{\frac{1}{\gamma} \frac{1}{N} \sum_{i=1}^{N} \mathrm{Var}_{w^n(\cdot|u_i)} \log \frac{w^n(\cdot|u_i)}{Q^n(\cdot)}}$$

(by the concavity of the square root)

$$\le nC + \sqrt{\frac{1}{\gamma} c \cdot n}. \qquad \square$$

5.3.1.5 Proof of Ahlswede and Dueck

In Augustin's proof the probability distribution Q was obtained by averaging in the single components. Now we average over all components simultaneously and obtain the distribution

$$\overline{Q}(y) \triangleq \frac{1}{nN} \sum_{i=1}^{N} \sum_{t=1}^{n} w(y|u_{i(t)})$$

With $\overline{P}(x) \triangleq \frac{1}{nN} | \{(i, t) : u_{i(t)} = x; i = 1, \ldots, N, t = 1, \ldots, n\} |$ it is easy to see that $\overline{Q} = \overline{P} \cdot W$. We have

$$\sum_{i=1}^{N} \mathbb{E}_{w^n(\cdot|u_i)} \log \frac{w^n(\cdot|u_i)}{\overline{Q}^n(u_i)} = \sum_{t=1}^{n} \frac{1}{N} \sum_{i=1}^{N} \sum_{y \in \mathcal{Y}} w(y|u_{i(t)}) \log \frac{w(y|u_{i(t)})}{\overline{Q}(y)}$$

$$= n \sum_{x \in \mathcal{X}} \sum_{y \in \mathcal{Y}} \overline{P}(x) \cdot w(y|x) \log \frac{w(y|x)}{\overline{Q}(x)} = nI(\overline{P}, W)$$

and

$$\sum_{i=1}^{N} \mathrm{Var}_{w^n(\cdot|u_i)} \log \frac{w^n(\cdot|u_i)}{\overline{Q}^n(\cdot)} = \sum_{t=1}^{n} \sum_{x \in \mathcal{X}} \sum_{y \in \mathcal{Y}} \overline{P}(x) w(y|x) \left(\log \frac{w(y|x)}{\overline{Q}(y)} - \mathbb{E}_{w(\cdot|x)} \log \frac{w(\cdot|x)}{\overline{Q}(\cdot)} \right)^2.$$

Again, we can proceed as in the previous proofs.

Application of the Packing Lemma yields the desired result. $\square$

5.3.1.6 The Strong Converse for the DMC with Feedback

Theorem 31 *Every (n, N, λ, f) code fullfills the condition*

$$N < \exp\{c \cdot n + k(\lambda)\sqrt{n}\}$$

with a constant $k(\lambda)$ not depending on $n \in \mathbb{N}$.

Proof The proof follows the lines of Kemperman's second proof of the strong converse for the DMC. As in Kemperman's second proof let $\overline{P}$ be a distribution on $\mathcal{X}$ such that $\overline{Q} = \overline{P} \cdot W$ is the maximal output distribution, hence $I(\overline{P}, W) = C$. For all code words $u \in \{1, \ldots, N\}$ let

$$\Theta_u \triangleq C \cdot n + \alpha\sqrt{n},$$

where $\alpha = \alpha(\lambda)$ depends on the probability of error λ.

The theorem follows from the Packing Lemma, if we are able to show that for all u_i

$$w^n \left(\left\{ y^n \in \mathcal{Y}^n : \log \frac{w^n(y^n|u_i)}{\overline{Q}^n(y^n)} > C \cdot n + \alpha\sqrt{n} \right\} \mid u_i \right) < \frac{1 - \lambda}{2}.$$

According to the preceding proofs we only have to estimate

$$\mathbb{E}_{w^n(\cdot|u_i)} \log \frac{w^n(\cdot|u_i)}{\overline{Q}^n(\cdot)} \mathrm{and} \mathrm{Var}_{w^n(\cdot|u_i)} \log \frac{w^n(\cdot|u_i)}{\overline{Q}^n(\cdot)}.$$

In this case this is not as easy as before, since $\log \frac{w^n(y^n|u_i)}{\overline{Q}^n(y^n)}$ is no longer a sum of independent random variables. However, it can be shown that it is a sum of uncorrelated random variables, and hence the variance of the sum is the sum of the variances.

Therefore we define random variables $Z_t, t = 1, \ldots, n$ by

$$\log \frac{w^n(y^n|u_i)}{\overline{Q}(y^n)} = \sum_{t=1}^{n} \log \frac{w(y_t|u_{i(t)}(y_1, \ldots, y_{t-1}))}{\overline{Q}(y_t)} \triangleq \sum_{t=1}^{n} Z_t.$$

(single summands are denoted as $Z_t, t = 1, \ldots, n$).

From Shannon's Lemma we know that

$$\mathbb{E}_{w(\cdot|u_{i(t)})}(Z_t \mid Y_1, \ldots, Y_{t-1}) \leq C \qquad (5.10)$$

We further define random variables V_t by

$$V_t \triangleq Z_t - \mathbb{E}_{w(\cdot|u_{i(t)})}(Z_t \mid Y_1, \ldots, Y_{t-1}),$$

such that $\mathbb{E}_{w(\cdot|u_{i(t)})}(V_t \mid Y_1, \ldots, Y_{t-1}) = 0$.

Since V_ℓ is a function of $Y_1, \ldots, Y_\ell$ for $\ell = 1, \ldots, t-1$ we also have

$$\mathbb{E}_{w(\cdot|u_{i(t)})}(V_t \mid V_1, \ldots, V_{t-1}) = 0,$$

from which follows that

$$\mathbb{E}_{w(\cdot|u_{i(t)})} V_\ell V_k = 0.$$

This shows that the V_t's and therefore the Z_t's are uncorrelated.

Now we continue as in the preceding proofs.

For all $t = 1, \ldots, n$ $\mathbb{E}_{w(\cdot|u_{i(t)})} V_t^2 < \beta$ for some $\beta \in \mathbb{R}$ and with Chebyshev's Inequality

$$\Pr\left(\sum_{t=1}^{n} V_t > \alpha\sqrt{n} \mid u_i\right) \leq \frac{1}{\alpha^2}$$

From 5.10 and the definition of V_t we further conclude that

$$\Pr\left(\sum_{t=1}^{n} Z_t > nC + \alpha\sqrt{n} \mid u_i\right) \leq \frac{\beta}{\alpha^2}$$

If now $\alpha = \alpha(\lambda)$ is chosen such that $\frac{\beta}{\alpha^2} < \frac{1-\lambda}{2}$, the Packing Lemma can be applied to obtain the desired result. $\square$

5.3.2 Arimoto's Proof of the Strong Converse

Let $\mathcal{C} = \{(u_i, D_i) : i = 1, \ldots, N\}$ be a (n, N, λ)-code. W.l.o.g. we further assume that maximum-likelihood decoding is applied, i.e.

$$y^n \in D_i \text{ exactly if } w^n(y^n|u_i) \geq w^n(y^n|u_j) \text{ for } j \neq i.$$

Then it is possible to express the maximum error λ of the code $\mathcal{C}$ in the following form

$$\lambda(\mathcal{C}) = 1 - \sum_{y^n \in \mathcal{Y}^n} \frac{1}{N} \max_{i=1,\dots,N} w^n(y^n|u_i)$$

Since we can estimate (with $\beta > 0$)

$$\max_{i=1,\dots,N} w^n(y^n|u_i) = \left(\left(\max_{i=1,\dots,N} w^n(y^n|u_i) \right)^{\frac{1}{\beta}} \right)^{\beta}$$

$$\leq \left(\sum_{i=1}^{N} w^n(y^n|u_i)^{\frac{1}{\beta}} \right)^{\beta},$$

we can conclude that

$$\lambda(\mathcal{C}) \geq 1 - \frac{1}{N} \sum_{y^n \in \mathcal{Y}^n} \left(\sum_{i=1}^{N} w^n(y^n|u_i)^{\frac{1}{\beta}} \right)^{\beta}. \tag{5.11}$$

This inequality already yields a bound on N. However, it is too complicated and gives no further information on the asymptotic behaviour for $n \to \infty$. Hence it is necessary to simplify the above inequality 5.11. The right hand side will be averaged over all codes. We denote by $\mathcal{C}^o \triangleq \{u_1^o, \dots, u_N^o\}$ some code for which the minimal maximum-error probability, further on denoted by λ, is assumed.

Further, we denote by Q a probability distribution on the set of all codes of length N. We are interested in distributions Q such that

$$\mathbb{E}_Q \lambda(\mathcal{C}) = \lambda \tag{5.12}$$

Obviously there exists at least one distribution fulfilling 5.12, namely

$$Q(u_1, \dots, u_N) = \begin{cases} \frac{1}{N!}, & \text{if } (u_1, \dots, u_N) \text{ is a permutation of } (u_1^o, \dots, u_N^o) \\ 0 & \text{else} \end{cases}.$$

Since the function $-x^{\beta}$ is convex for every $x \geq 0$ and $1 \geq \beta > 0$, we obtain by averaging on both sides of (1) for each $1 \geq \beta > 0$

$$\mathbb{E}_Q \lambda(\mathcal{C}) \geq 1 - \frac{1}{N} \sum_{y^n \in \mathcal{Y}^n} \left(\sum_{i=1}^{N} \mathbb{E}_Q[w^n(y^n|u_i)^{\frac{1}{\beta}}] \right)^{\beta}. \tag{5.13}$$

From now on, we further assume that the probability distribution Q defined as above is invariant under permutations of its arguments. Then it is clear that all the marginal distributions

$$Q_i(u_i) = \sum_{u_1 \in \mathcal{X}^n} \cdots \sum_{u_{i-1} \in \mathcal{X}^n} \sum_{u_{i+1} \in \mathcal{X}^n} \cdots \sum_{u_N \in \mathcal{X}^n} Q(u_1, \ldots, u_N)$$

are equal. Therefore, for this special type of probability distributions 5.13 is reduced to

$$\mathbb{E}_Q \lambda(\mathcal{C}) \geq 1 - N^{\beta-1} \sum_{y^n \in \mathcal{Y}^n} \left(\sum_{u \in \mathcal{X}^n} Q(u) w^n(y^n|u)^{\frac{1}{\beta}} \right)^\beta$$

(where $Q = Q_1 = \cdots = Q_N$ defined as above).

As shown above, there exists a Q^o invariant under permutations of its arguments fulfilling 5.12, hence we can also write

$$\lambda \geq \inf_{Q \ PD \ \mathrm{on} \ \mathcal{X}^n} \left[1 - N^{\beta-1} \sum_{y^n \in \mathcal{Y}^n} \left(\sum_{u \in \mathcal{X}^n} Q(u) w^n(y^n|u)^{\frac{1}{\beta}} \right)^\beta \right]. \tag{5.14}$$

Up to now we did not require that the channel should be memoryless. However, we now need the following lemma.

Lemma 46 *For the DMC $W = \big(w(y|x)\big)_{x \in \mathcal{X}, y \in \mathcal{Y}}$ with input distribution P on $\mathcal{X}$ and output distribution Q on $\mathcal{Y}$ we have for $0 < \beta \leq 1$*

$$\max_{Q(u)} \sum_{y^n \in \mathcal{Y}^n} \left(\sum_{u \in \mathcal{X}^n} Q(u) w^n(y^n|u)^{\frac{1}{\beta}} \right)^\beta = \left(\max_P \left\{ \sum_{y \in \mathcal{Y}} \left(\sum_{x \in \mathcal{X}} P(x) w(y|x)^{\frac{1}{\beta}} \right)^\beta \right\} \right)^n$$

Application of Lemma 46 to 5.14 yields

$$\lambda \geq 1 - N^{\beta-1} \cdot \left(\max_{P \ PD \ \mathrm{on} \ \mathcal{X}} \left\{ \sum_{y \in \mathcal{Y}} \left(\sum_{x \in \mathcal{X}} P(x) w(y|x)^{\frac{1}{\beta}} \right)^\beta \right\} \right)^n.$$

We now define

$$E(\sigma, P) \triangleq -\log \left(\sum_{y \in \mathcal{Y}} \left(\sum_{x \in \mathcal{X}} P(x) w(y|x)^{\frac{1}{1+\sigma}} \right)^{1+\sigma} \right).$$

If we further set $N \triangleq \exp\{nR\}$ for some $R \in \mathbb{R}$ and $\beta \triangleq 1 + \sigma$

$$\lambda \geq 1 - \exp\left\{-n\big(-\sigma R + \min_{P \ PD \ \mathrm{on} \ \mathcal{X}} E(\sigma, P)\big)\right\} \text{ for } -1 \leq \sigma < 0 \tag{5.15}$$

The following lemma is stated without proof

Lemma 47

$$\lim_{\sigma \nearrow 0} \frac{1}{\sigma} \min_{P \ PD \ on \ \mathcal{X}} E(\sigma, P) = \lim_{\sigma \searrow 0} \frac{1}{\sigma} \max_{P \ PD \ on \ \mathcal{X}} E(\sigma, P) = 0.$$

As a consequence of Lemma 47 we have

Lemma 48 *For $R > C$, the channel capacity,*

$$\max_{-1 \le \sigma < 0} \left\{ -\sigma R + \min_{P} E(\sigma, P) \right\} > 0$$

From 5.15 $\exp\{nR\} = N$ can be calculated to obtain the strong converse.

5.3.3 Blowing Up Lemma: A Combinatorial Digression

Given a finite set $\mathcal{X}$, the set $\mathcal{X}^n$ of all n-length sequences of elements from $\mathcal{X}$ is sometimes considered as a metric space with the Hamming metric. Recall that the Hamming distance of two n-length sequences is the number of positions in which these two sequences differ. The Hamming metric can be extended to measure the distance of subsets of $\mathcal{X}^n$, setting

$$d_H(B, C) \triangleq \min_{x^n \in B, y^n \in C} d_H(x^n, y^n).$$

Some classical problems in geometry have exciting combinatorial analogues in this setup. One of them is the *isoperimetric problem*, which will turn out to be relevant for information theory.

Given a set $B \subset \mathcal{X}^n$, the *Hamming ℓ-neighbour-hood* of B is defined as the set

$$\Gamma^\ell B \triangleq \left\{ x^n : x^n \in \mathcal{X}^n, d_H(\{x^n\}, B) \le 1 \right\}.$$

We shall write Γ for Γ^1. The *Hamming boundary* ∂B of $B \subset \mathcal{X}^n$ is defined by $\partial B \triangleq B \cap \Gamma \overline{B}$.

Considering the boundary ∂B as a discrete analogue of the surface, one can ask how small the "size" $|\partial B|$ of the "surface" of a set $B \subset \mathcal{X}^n$ can be if the "volume" $|B|$ is fixed. Theorem 1 below answers (a generalized form of) this question in an asymptotic sense. Afterwards, the result will be used to see how the probability of a set is changed by adding or deleting relatively few sequences close to its boundary.

One easily sees that if $\frac{\ell_n}{n} \to 0$ then the cardinality of B and of its ℓ_n − neighbour-hood have the same exponential order of magnitude, and the same holds also for their P^n-probabilities, for every distribution P on Y. More precisely, one has

Lemma 49 *Given a sequence of positive integers $\{\ell_n\}_{n=1}^\infty, \frac{\ell_n}{n} \to 0$ and a distribution P on $\mathcal{X}$ with positive probabilities, there exists a sequence $\{\varepsilon_n\}_{n=1}^\infty$ with $\lim_{n\to\infty} \varepsilon_n = 0$ depending only on $\{\ell_n\}_{n=1}^\infty, |\mathcal{X}|$ and $m_p \triangleq \min_{x\in\mathcal{C}} P(x)$ such that for every $B \subset \mathcal{X}^n$*

$$0 \le \frac{1}{n} \log |\Gamma^{\ell_n} B| - \frac{1}{n} \log |B| \le \varepsilon_n,$$

$$0 \le \frac{1}{n} \log P^n(\Gamma^{\ell_n} B) - \frac{1}{n} \log P^n(B) \le \varepsilon_n.$$

Proof Since $B \subset \Gamma^\ell B = \bigcup_{x^n \in B} \Gamma^\ell \{x^n\}$, it suffices to prove both assertions for one-point sets. Clearly

$$|\Gamma^{\ell_n}\{x^n\}| \le \binom{n}{\ell_n} |\mathcal{X}|^{\ell_n}.$$

As $P^n(z^n) \le m_p^{-\ell_n} P^n(x^n)$ for every $z^n \in \Gamma^{\ell_n}\{x^n\}$, this implies

$$P^n\left(\Gamma^{\ell_n}\{x^n\}\right) \le \binom{n}{\ell_n} |\mathcal{X}|^{\ell_n} \cdot m_p^{-\ell_n} P^n(x^n).$$

Since $\binom{n}{\ell_n}$ equals the number of binary sequences of length n and type $\left(\frac{\ell_n}{n}, 1 - \frac{\ell_n}{n}\right)$, we have

$$\binom{n}{\ell_n} \le \exp\left\{nH\left(\frac{\ell_n}{n}, 1 - \frac{\ell_n}{n}\right)\right\} = \exp\left\{nh\left(\frac{\ell_n}{n}\right)\right\}.$$

Thus the assertions follows with

$$\varepsilon_n \triangleq h\left(\frac{\ell_n}{n}\right) + \frac{\ell_n}{n}(\log |\mathcal{X}| - \log m_p). \qquad \square$$

Knowing that the probability of the ℓ_n-neighborhood of a set has the same exponential order of magnitude as the probability of the set itself (if $\frac{\ell_n}{n} \to 0$), we would like to have a deeper insight into the question how passing from a set to its ℓ_n-neighborhood increases probability. The answer will involve the function

$$f(s) \triangleq \begin{cases} \varphi(\Phi^{-1}(s)) & \text{if } s \in (0, 1) \\ 0 & \text{if } s = 0 \text{ or } s = 1 \end{cases}$$

where $\varphi(t) \triangleq (2\pi)^{-\frac{1}{2}} \cdot e^{-\frac{t^2}{2}}$ and $\Phi(t) \triangleq \int_{-\infty}^t \varphi(u)du$ are the density resp. distribution function of the standard normal distribution, and $\Phi^{-1}(s)$ denotes the inverse function of $\Phi(t)$. Some properties of $f(s)$ are summarized in

Lemma 50 *The function $f(s)$, defined on $[0, 1]$, is symmetric around the point $s = \frac{1}{2}$, it is non-negative, concave, and satisfies*

(i) $\lim\limits_{s \to 0} \dfrac{f(s)}{s\sqrt{-2 \, \ell n \, s}} = \lim\limits_{s \to 0} \dfrac{-\Phi^{-1}(s)}{\sqrt{-2 \, \ell n \, s}} = 1,$

(ii) $f'(s) = -\Phi^{-1}(s) \ \big(s \in (0, 1)\big),$

(iii) $f''(s) = -\dfrac{1}{f(s)} \big(s \in (0, 1)\big).$

Corollary 2 *There exists a positive constant K_0 such that*
   $f(s) \le K_0 \cdot s \cdot \sqrt{-\ell n \, s} \ \text{for all } s \in \big[0, \frac{1}{2}\big].$

Proof   The obvious relation (ii) implies (iii), establishing the concavity of the non-negative function $f(s)$. The symmetry is also clear, so that it remains to prove (i). Observe first that because of (ii)

$$\lim_{s \to 0} \frac{f(s)}{s\sqrt{-2 \, \ell n \, s}} = \lim_{s \to 0} \frac{f'(s)}{\sqrt{-2 \, \ell n \, s}} \frac{1}{1 + \frac{1}{2 \, \ell n \, s}} = \lim_{s \to 0} \frac{-\Phi^{-1}(s)}{\sqrt{-2 \, \ell n \, s}}.$$

Hence, applying the substitution $s \triangleq \Phi(t)$ and using the well-known fact

$$\lim_{t \to -\infty} \frac{-t \Phi(t)}{\varphi(t)} = 1$$

(cf. Feller (1968), p. 175), it follows that the above limit further equals

$$\lim_{t \to -\infty} \frac{-t}{\sqrt{-2 \, \ell n \Phi(t)}} = \lim_{t \to -\infty} \frac{-t}{\sqrt{-2 \, \ell n \frac{\varphi(t)}{-t}}} = \lim_{t \to \infty} \frac{t}{\sqrt{-2 \, \ell n \varphi(t) + 2 \, \ell n \, t}} = 1. \ \ \square$$

Now we are ready to give an asymptotically rather sharp lower bound on the probability of the boundary of an arbitrary set in terms of its own probability. This and the following results will be stated somewhat more generally than Lemma 49, for conditional probabilities, because this is the form needed in the subsequent parts of the script.

Theorem 32 *For every stochastic matrix $W = \big(w(y|x)\big)_{x \in \mathcal{X}, y \in \mathcal{Y}}$, integer n, set $B \subset \mathcal{Y}^n$ and $x^n \in \mathcal{X}^n$ one has*

$$w^n(\partial B|x) \ge \frac{a}{\sqrt{n}} f\big(w^n(B|x)\big).$$

Here $a = a_W \triangleq K \frac{m_W}{\sqrt{-\ell n \ m_W}}$, m_W *is the smallest positive entry of W, and K is an absolute constant.*

Proof The statement is trivial if all the positive entries of W equal 1. In the remaining cases the smallest positive entry m_W of W does not exceed $\frac{1}{2}$.

The proof goes by induction. The case $n = 1$ is simple. In fact, then $\partial B = B$ for every $B \subset Y$. Hence one has to prove that for some absolute constant K

$$t \geq K \frac{m_W}{\sqrt{-\ell n \ m_W}} f(t) \text{ for } t \geq m_W.$$

As $m_W \leq \frac{1}{2}$ and by Lemma 50 $f(t) \leq f\left(\frac{1}{2}\right) = \frac{1}{\sqrt{2\pi}}$, this inequality obviously holds if

$$K \leq \sqrt{2\pi \ \ell n \ 2}. \tag{5.16}$$

Suppose now that the statement of the theorem is true for $n - 1$. Fix some set $B \subset Y^n$ and sequence $x^n = (x_1, \ldots, x_{n-1}, x_n) \in X^n$. Write $x^{n-1} \triangleq (x_1, \ldots, x_{n-1})$; further, for every $y \in Y$ denote by B_y the set of those sequences $(y_1, \ldots, y_{n-1}) \in Y^{n-1}$ for which $(y_1, \ldots, y_{n-1}) \in B$. Then, obviously,

$$w^n(B|x^n) = \sum_{y \in Y} w(y|x_n) \cdot w^{n-1}(B_y|x^{n-1}), \tag{5.17}$$

and since

$$\partial B \subset \bigcup_{y \in Y} [\partial B_y \times \{y\}],$$

also

$$w^n(\partial B|x^n) \geq \sum_{y \in Y} w(y|x_n) \cdot w^{n-1}(\partial B_y|x^{n-1}). \tag{5.18}$$

Put $S \triangleq \{y : w(y|x_n) > 0\}$ and

$$d \triangleq \max_{y \in S} w^{n-1}(B_y|x^{n-1}) - \min_{y \in S} w^{n-1}(B_y|x^{n-1}).$$

Since $\partial B \supset (B_{y'} - B_{y''}) \times \{y'\}$ for any y', y'' in Y, one gets

$$w^n(\partial B|x^n) \geq m_W \cdot d. \tag{5.19}$$

If

$$d \geq \frac{a}{m_W \sqrt{n}} f(w^n(B|x^n)),$$

the statement of the theorem for n immediately follows from 5.19. Let us turn therefore to the contrary case of

$$d < \frac{a}{m_W \sqrt{n}} f\left(w^n(B|x^n)\right). \tag{5.20}$$

Combining 5.18 and the induction hypothesis we see that

$$w^n(\partial B|x^n) \geq \sum_{y \in \mathcal{Y}} w(y|x_n) \cdot w^{n-1}(\partial B_y|x^{n-1})$$

$$\geq \frac{a}{\sqrt{n-1}} \sum_{y \in \mathcal{Y}} w(y|x_n) \cdot f\left(w^{n-1}(B_y|x^{n-1})\right). \tag{5.21}$$

Write

$$s \triangleq w^n(B|x^n), \quad s_y \triangleq w^{n-1}(B_y|x^{n-1}),$$

and consider the interval of length d

$$\Delta \triangleq \left[\min_{y \in S} s_y, \max_{y \in S} s_y\right].$$

Since by (5.17) $\sum_{y \in \mathcal{Y}} w(y|x_n)s_y = s$, it follows from Taylor's formula

$$f(s_y) = f(s) + (s_y - s)f'(s) + \frac{1}{2}(s_y - s)^2 f''(\sigma_y)$$

(where $\sigma_y \in \Delta$ if $y \in S$) that if $\sigma \in \Delta$ satisfies

$$|f''(\sigma)| = \max_{s \in \Delta} |f''(s)| \tag{5.22}$$

then

$$\sum_{y \in \mathcal{Y}} w(y|x_n) \cdot f(s_y) \geq f(s) - \frac{1}{2}d^2 |f''(\sigma)|.$$

This, 5.20 and Lemma 50 yield, by substitution into 5.21, the estimate

$$w^n(\partial B|x^n) \geq \frac{a}{\sqrt{n-1}}\left[f(s) - \frac{a^2}{2m_W^2 \cdot n}\frac{f^2(s)}{f(\sigma)}\right].$$

Rearranging this we get

$$w^n(\partial B|x^n) \geq \frac{a}{\sqrt{n}} \cdot f(s)\left[\sqrt{\frac{n}{n-1}} - \frac{a^2 f(s)}{2m_W^2 \sqrt{n(n-1)}f(\sigma)}\right].$$

To complete the proof, we show that

$$\sqrt{\frac{n}{n-1}} - \frac{a^2 f(s)}{2m_W^2 \sqrt{n(n-1)} \cdot f(\sigma)} \geq 1,$$

or, equivalently, that

$$\frac{f(\sigma)}{f(s)} \geq \frac{a^2}{m_W^2} \cdot \frac{\sqrt{n} + \sqrt{n-1}}{2\sqrt{n}}.$$

It is sufficient to prove

$$\frac{f(\sigma)}{f(s)} \geq \frac{a^2}{m_W^2}. \tag{5.23}$$

Notice that on account of Lemmas 50 and 15, σ is an endpoint of the interval Δ. Thus, with the notation $\bar{r} \triangleq \min(r, 1-r)$, one sees that

$$\bar{s} - d \leq \bar{\sigma} \leq \bar{s},$$

and therefore, by the symmetry and the concavity of $f(s)$,

$$\frac{f(\sigma)}{f(s)} = \frac{f(\bar{\sigma})}{f(\bar{s})} \geq \frac{\bar{\sigma}}{\bar{s}} \geq 1 - \frac{d}{\bar{s}}.$$

Thus, using 5.20 and the Corollary of Lemma 50,

$$\frac{f(\sigma)}{f(s)} \geq 1 - \frac{a}{m_W \sqrt{n}} \frac{f(\bar{s})}{\bar{s}} \geq 1 - \frac{a}{m_W \sqrt{n}} K_0 \sqrt{-\ell n \bar{s}}.$$

Hence, substituting a and using the fact that $\bar{s} \geq m_W^n$, we get

$$\frac{f(\sigma)}{f(s)} \geq 1 - \frac{K K_0}{\sqrt{-n \ell n \, m_W}} \sqrt{-\ell n \, m_W^n} = 1 - K K_0.$$

Choosing a K satisfying 5.16 and $1 - K K_0 \geq \frac{K^2}{\ell n \, 2}$ 5.23 will follow, since

$$\frac{a}{m_W K} = \frac{1}{\sqrt{\ell n \frac{1}{m_W}}} < \frac{1}{\sqrt{\ell n \, 2}}. \qquad \square$$

For our purpose, the importance of this theorem lies in the following corollary establishing a lower bound on the probability of the ℓ-neighborhood of a set in terms of its own probability.

Corollary 3 *For every n, ℓ, $B \subset \mathcal{Y}^n$ and $x^n \in \mathcal{X}^n$*

$$w^n(\Gamma^\ell B | x^n) \geq \Phi\left[\Phi^{-1}(w^n(B|x^n)) + \frac{\ell - 1}{\sqrt{n}} \cdot a_W\right].$$

Proof We shall use the following two obvious relations giving rise to estimates of the probability of 1-neighborhoods by that of boundaries:

$$\Gamma B \smallsetminus B \supset \partial(\Gamma B), \quad \Gamma B \smallsetminus B = \partial \overline{B}.$$

Denoting $t_k \triangleq \Phi^{-1}(w^n(\Gamma^k B | x^n))$, one has

$$\Phi(t_{k+1}) - \Phi(t_k) = w^n(\Gamma^{k+1} B - \Gamma^k B | x^n),$$

and hence the above relations yield by Theorem 32 that

$$\Phi(t_{k+1}) - \Phi(t_k) \geq \max\{w^n(\partial(\Gamma^{k+1} B)|x^n), w^n(\partial(\overline{\Gamma^k B})|x^n)\}$$

$$\geq \frac{a}{\sqrt{n}} \max\{\varphi(t_{k+1}), \varphi(t_k)\}. \tag{5.24}$$

However, φ is monotone on both $(-\infty, 0)$ and $(0, \infty)$, and therefore, unless $t_k < 0 < t_{k+1}$,

$$\max_{t_k \leq u \leq t_{k+1}} \varphi(u) = \max\{\varphi(t_k), \varphi(t_{k+1})\}.$$

This, substituted into 5.24, yields by Lagrange's theorem

$$t_{k+1} - t_k \geq \left[\Phi(t_{k+1}) - \Phi(t_k)\right] \cdot \left(\max_{t_k \leq u \leq t_{k+1}} \varphi(u)\right)^{-1} \geq \frac{a}{\sqrt{n}}$$

unless $t_k < 0 < t_{k+1}$. Hence

$$t_\ell - t_0 = \sum_{k=0}^{\ell-1} (t_{k+1} - t_k) \geq \frac{\ell - 1}{\sqrt{n}} a. \qquad \square$$

We conclude this series of estimates by a counterpart of Lemma 49. In fact, Lemma 49 and the next Lemma 51 are those results of this lecture which will be often used in the sequel.

Lemma 51 *(Blowing Up): To any finite sets $\mathcal{X}$ and $\mathcal{Y}$ and sequence $(\varepsilon_n)_{n=1}^\infty$ with $\varepsilon_n \to 0$ there exists a sequence of positive integers $\{\ell_n\}_{n=1}^\infty$ with $\frac{\ell_n}{n} \to 0$ and a sequence $(\eta_n)_{n=1}^\infty$ with $\eta_n \to 1$ such that for every stochastic matrix $W = (w(y|x))_{x \in \mathcal{X}, y \in \mathcal{Y}}$ $\mathcal{X} \to \mathcal{Y}$ and every n, $x^n \in \mathcal{X}^n$, $B \subset \mathcal{Y}^n$*

$$w^n(B|x^n) \geq \exp\{-n\varepsilon_n\} \text{ implies } w^n(\Gamma^{\ell_n} B|x^n) \geq \eta_n. \tag{5.25}$$

Proof For a fixed W, the existence of sequences $\{\ell_n\}_{n=1}^\infty$ and $\{\eta_n\}_{n=1}^\infty$ satisfying (5.25) is an easy consequence of the preceding Corollary. The bound of the Corollary depends on W through m_W, as $a_W = K\frac{m_W}{\sqrt{-\ell n\, m_W}}$. Thus, in order to get such sequences which are good for every W, for matrices with small m_W an approximation argument is needed.

Let $\mathcal{X}, \mathcal{Y}$ and the sequence $\varepsilon_n \to 0$ be given. We first claim that for a suitable sequence of positive integers $\{k_n\}_{n=1}^\infty$ with $k_n/n \to 0$ the following statement is true:
Setting

$$\delta_n \triangleq \frac{k_n}{2n|\mathcal{X}||\mathcal{Y}|}, \tag{5.26}$$

for every pair of stochastic matrices

$$W = \big(w(y|x)\big)_{x\in\mathcal{X}, y\in\mathcal{Y}}, \quad \widetilde{W} = \big(\widetilde{w}(y|x)\big)_{x\in\mathcal{X}, y\in\mathcal{Y}}$$

such that

$$|w(y|x) - \widetilde{w}(y|x)| \leq \delta_n \text{ for every } x \in \mathcal{X}, y \in \mathcal{Y} \tag{5.27}$$

and for every $x^n \in \mathcal{X}^n$, $B \subset \mathcal{Y}^n$, the inequality

$$\overset{n}{\widetilde{w}}(\Gamma^{k_n} B|x^n) \geq w^n(B|x^n) - \frac{1}{2}\exp(-n\varepsilon_n) \tag{5.28}$$

holds. To prove this, notice that 5.27 implies the existence of a stochastic matrix $\hat{W} = \big(\hat{w}(y, \widetilde{y}\,|x)\big)_{x\in\mathcal{X}, y, \widetilde{y}\in\mathcal{Y}}$ having W and $\widetilde{W}$ as marginals, i.e.,

$$\sum_{\widetilde{y}\in\mathcal{Y}} \hat{w}(y, \widetilde{y}\,|x) = w(y|x), \quad \sum_{y\in\mathcal{Y}} \hat{w}(y, \widetilde{y}\,|x) = \widetilde{w}(\widetilde{y}\,|x),$$

such that

$$\sum_{y\in\mathcal{Y}} \hat{w}(y, \widetilde{y}\,|x) \geq 1 - \delta_n|\mathcal{Y}| \text{ for every } x \in \mathcal{X}.$$

By the last property of $\hat{W}$ we have for every $(y^n, \overset{n}{\widetilde{y}}) \in T^n_{\hat{W}, \delta_n}(x^n)$

$$d_H(y^n \overset{n}{\tilde{y}}) = n - \sum_{x \in \mathcal{X}} \sum_{y \in \mathcal{Y}} < x^n, y^n, \overset{n}{\tilde{y}} | x, y, y > \leq 2n\delta_n |X||Y| = k_n,$$

so that

$$(B \times Y^n) \cap T^n_{\hat{W}, \delta_n}(x^n) \subset B \times \Gamma^{k_n} B.$$

Hence

$$\overset{n}{\tilde{w}}(\Gamma^{k_n} B | x) \geq \hat{w}^n \big((B \times Y^n) \cap T^n_{\hat{W}, \delta_n}(x^n) | x^n \big)$$
$$\geq w^n(B|x^n) - \big(1 - \hat{w}^n(T^n_{\hat{W}, \delta_n}(x^n)|x^n) \big).$$

Thus our claim will be established if we show that for $\mathcal{Z} \triangleq \mathcal{Y}^2$ there exists a sequence $\delta_n \to 0$ (depending on $\{\varepsilon_n\}_{n=1}^{\infty}, |\mathcal{X}|, |\mathcal{Z}|$) such that for every $\hat{W} : \mathcal{X} \to \mathcal{Z}$ and $x^n \in \mathcal{X}^n$

$$\hat{w}\big(T^n_{\hat{W}, \delta_n}(x^n)|x^n\big) \geq 1 - \frac{1}{2}\exp(-n\varepsilon_n). \tag{5.29}$$

To verify 5.29, denote by $c(\delta)$ the minimum of $D(V \| W | P)$ for all stochastic matrices $V = (v(z|x))_{x \in \mathcal{X}, z \in \mathcal{Z}}$, $W = (w(z|x))_{x \in \mathcal{X}, z \in \mathcal{Z}}$ and distributions P on $\mathcal{X}$ such that $|P(x)v(z|x) - P(x)W(z|x) \geq \delta$ for at least one pair $(x, z) \in \mathcal{X} \times \mathcal{Z}$. Then by the properties of typical sequences

$$1 - \hat{w}^n\big(T^n_{\hat{W}, \delta}(x^n)x^n\big) \leq (n+1)^{|\mathcal{X}||\mathcal{Z}|}\exp\{-nc(\delta).$$

Choosing $\delta_n \to 0$ suitably, this establishes 5.29 and thereby our claim that 5.27 implies 5.28.

Notice that we are free to choose δ_n to converge to 0 as slowly as desired. We shall henceforth assume that

$$\lim_{n \to \infty} \frac{\delta_n}{\sqrt{-\varepsilon_n \ell n \, \delta_n}} = +\infty. \tag{5.30}$$

Consider now an arbitrary $W = (w(y|x))_{x \in \mathcal{X}, y \in \mathcal{Y}}$ and $x \in \mathcal{X}^n$, $B \subset \mathcal{Y}^n$ for which

$$w^n(B|x^n) \geq \exp\{-n\varepsilon_n\}.$$

Approximate the matrix W in the sense of 5.27 by a $\tilde{W}$ satisfying

$$m_{\tilde{W}} \geq \frac{\delta_n}{|\mathcal{Y}|},$$

and apply the Corollary of Theorem 1 to the matrix $\widetilde{W}$ and the set $\widetilde{B} \triangleq \Gamma^{k_n} B$. Since by 5.28 we have

$$\underset{W}{\overset{n}{\sim}}(\Gamma^{k_n} B | x^n) \geq \exp\{-n\varepsilon_n - 1\},$$

it follows, using also Lemma 50(i), that for every positive integer ℓ

$$\underset{W}{\overset{n}{\sim}}(\Gamma^{k_n + \ell} B | x^n) \geq \Phi\left[\Phi^{-1}\left(\exp(-n\varepsilon_n - 1)\right) + \frac{\ell - 1}{\sqrt{n}} K \frac{m \underset{\sim}{\widetilde{W}}}{\sqrt{-\ell n m \underset{\sim}{\widetilde{W}}}}\right]$$

$$\geq \Phi\left(-b\sqrt{n\varepsilon_n + 1} + \frac{\ell - 1}{\sqrt{n}} K \frac{\delta_n}{|\mathcal{Y}|\sqrt{\ell n \frac{|\mathcal{Y}|}{\delta_n}}}\right). \tag{5.31}$$

Here K is the constant from Theorem 32 and $b > 0$ is another absolute constant. By 5.30, there exists a sequence of positive integers $\{\widetilde{\ell}\}_{n=1}^{\infty}$ such that $\frac{\widetilde{\ell}}{n} \to 0$ and

$$-b\sqrt{n\varepsilon_n + 1} + \frac{\frac{\widetilde{\ell}}{n} - 1}{\sqrt{n}} K \frac{\delta_n}{|\mathcal{Y}|\sqrt{\ell n \frac{|\mathcal{Y}|}{\delta_n}}} \to \infty.$$

Denoting by $\underset{n}{\widetilde{\eta}}$ the lower bound resulting from 5.31 for $\ell \triangleq \underset{n}{\widetilde{\ell}}$, we have $\underset{n}{\widetilde{\eta}} \to 1$ as $n \to \infty$. Applying 5.28 once more (interchanging the roles of W and $\widetilde{W}$), the assertion 5.25 follows with $\ell_n \triangleq 2k_n + \underset{n}{\widetilde{\ell}}$, $\eta_n \triangleq \underset{n}{\widetilde{\eta}} - \frac{1}{2}\exp\{-n\varepsilon_n\}$. $\qquad\square$

5.3.3.1 Every Bad Code has a Good Subcode

Recall that the strong converse (for the DMC) states that the asymptotic growth of the maximal codelengths $N(n, \lambda)$ does not depend on the probability of error $\lambda \in (0, 1)$, hence for any error probabilities λ and λ' with $0 < \lambda < \lambda' < 1$ the asymptotic growth of $N(n, \lambda)$ and $N(n, \lambda')$ is the same.

With the results of the preceding section Ahlswede and Dueck [2] obtained an even stronger statement, which they called the local converse.

Theorem 33 (Local Converse) *Let W be a DMC and let $\varepsilon, \lambda, \lambda'$ and R be real numbers such that $\varepsilon > 0$, $R > 0$, $0 < \lambda' < \lambda < 1$. Then one can find (also explicitly) an $n_0(\lambda, \lambda', \varepsilon)$ such that for all $n \geq n_0(\lambda, \lambda', \varepsilon)$ the following is true:*
Every $(n, \exp\{nR\}, \lambda)$-code $\{(u_i, D_i) : i = 1, \ldots, N = \exp\{nR\}\}$ contains a subset of codewords $\{u_{i_k} \mid k = 1, \ldots, N' = \exp\{n(R - \varepsilon)\}\}$ with suitable decoding

sets $F_{ik}(i = 1, \ldots, N')$ such that $\{(u_{ik}, F_{ik}) \mid k = 1, \ldots, N'\}$ is an $(n, exp\{n(R - \varepsilon)\}, \lambda')$-code.

The result gives a new geometric insight into the coding problem. Out of a set of codewords with a certain minimal "distance" one can select a rather big subset of a prescribed larger minimal "distance". In the case of a binary symmetric channel the word "distance" as used here can be replaced by the Hamming metric, a true distance. The result may be of interest for the actual construction of small error codes.

The theorem was stated here for the DMC, the simplest and most familiar channel, even though the phenomenon "bad codes contain good codes" is of a rather general nature and occurs for much more general one-way channels as well as for multi-way channels.

Also, this theorem together with the weak converse implies the strong converse. A general method to prove strong converses was presented by Ahlswede, Gács and Körner. It applies to many multi-user coding problems for which all classical approaches fail. The idea is as follows: One enlarges the decoding sets of a given code in order to decrease the error probability. The new decoding sets are no longer disjoint, that is, one has a list code to which one applies Fano's Lemma. Surprisingly enough one can decrease the error probability significantly with a "small" increase in list size and therefore the idea works. The novelty of the present method of proof for the Theorem lies in the observation that one can select at random a subcode of list size 1 out of a list code with small list size without losing too much in error probability or rate.

Proof of the theorem   Suppose we are given the (n, N, λ)-code $\{(u_i, D_i) : i = 1, \ldots, N\}$. Define

$$E_i \triangleq \Gamma^{\ell_n} D_i \text{ for } i = 1, \ldots, N$$

with $\ell_n \triangleq n^{\frac{1}{2}} \log n$ (actually every ℓ_n with $n^{\frac{1}{2}} \ell_n^{-1} = o(1)$ and $n^{-1} \ell_n = o(1)$ would work). By the corollary of Theorem 1

$$w^n(E_i | u_i) \geq \Phi\left[\Phi^{-1}(1 - \lambda) + n^{-\frac{1}{2}}(\ell_n - 1)a_W\right].$$

Since $\ell_n n^{-\frac{1}{2}} \to \infty$ for $n \to \infty$ and since $\Phi(t) \to 1$ for $t \to \infty$, the right side converges to 1, and therefore certainly exceeds $1 - \lambda'/4$ for $n \geq n_0(\lambda, \lambda')$, suitable. On the other hand, the decoding list $\mathcal{G}(y^n) \triangleq \{u_i \mid 1 \leq i \leq N, y^n \in E_i\}$ satisfies:

$$\mid \mathcal{G}(y^n) \mid \leq \mid \Gamma^{\ell_n}(y^n) \mid \leq \binom{n}{\ell_n} |\mathcal{Y}|^{\ell_n} \leq (n|\mathcal{Y}|)^{\ell_n}$$

and hence

$$\mid \mathcal{G}(y^n) \mid \leq exp(n\delta_n) \text{ for all } y^n \in \mathcal{Y}^n,$$

where $\delta_n \triangleq |\mathcal{Y}| n^{-\frac{1}{2}} \log^2 n \to 0$ as $n \to \infty$.

We complete now the proof by a random coding argument. Let $U_i (i = 1, \ldots, M)$ be independent, identically distributed random variables with distribution

$$\Pr(U_i = u_k) = 1/N \text{ for } k = 1, \ldots, N.$$

With every outcome $(u_{i_1}, \ldots, u_{i_M})$ of $(U_1, \ldots, U_M)$ we associate decoding sets $(F_{i_1}, \ldots, F_{i_M})$, where

$$F_{i_k} = \big\{ y^n \in E_{i_k} : |\mathcal{G}(y^n) \cap \{u_{i_1}, \ldots, u_{i_M}\}| = 1 \big\}.$$

Equivalently: $F_{i_k} = E_{i_k} \setminus \bigcup_{j \neq k} E_{i_j}$.

For reasons of symmetry the expected average error probability for this decoding rule

$$\mathbb{E}\lambda(U_1, \ldots, U_M) = \frac{1}{M} \sum_{k=1}^{M} \mathbb{E} \sum_{y^n \in F_{i_k}^c} w^n(y^n | U_k)$$

equals $\mathbb{E} \sum_{y^n \in F_{i_1}^c} w^n(y^n | U_1)$, and this expression is upper bounded by

$$\mathbb{E} \sum_{y^n \in E_{i_1}^c} w^n(y^n | U_1) + \mathbb{E} \sum_{y^n \in E_{i_1} \cap F_{i_1}^c} w^n(y^n | U_1).$$

The first sum is smaller than $\lambda'/4$. Assume therefore that $y^n \in E_{i_1}$ and also that $U_1 = u_{i_1}$. Since $|\mathcal{G}(y^n)| \leq \exp(n\delta_n)$, the probability for

$$\{U_2, \ldots, U_M\} \cap \mathcal{G}(y^n) \neq \varnothing$$

is smaller than

$$1 - \big(1 - \exp(n\delta_n)/N\big)^{M-1} \leq M\exp(n\delta_n)/N$$

and hence

$$\mathbb{E} \sum_{y^n \in E_{i_1} \cap F_{i_1}^c} w^n(y^n | U_1) \leq \mathbb{E} \sum_{y^n \in E_{i_1}} M\exp(n\delta_n) w^n(y^n | U_1)/N \leq M\exp(n\delta_n)/N.$$

With $M \triangleq 2\exp\big(n(R - \varepsilon)\big)$ we obtain for n large enough

$$\mathbb{E} \sum_{y^n \in F_{i_1}^c} w^n(y^n | U_1) \leq \lambda'/4 + M\exp(n\delta_n)/N \leq \lambda'/2.$$

From a subcode of length M and with average error $\lambda'/2$ we can pass to a further subcode of length $N' = \exp(n(R - \varepsilon))$ and maximal error λ'.

5.4 Lecture on Fano's Lemma and the Weak Converse of the Coding Theorem

5.4.1 Fano's Lemma

Central in the proof of weak converses is the following inequality discovered by Fano [8].

Lemma 52 (Fano's Lemma) *Let* $\{(u_i, D_i) : 1 \leq i \leq N\}$ *be a block code with average error* $\lambda_Q \triangleq \sum_{i=1}^{N} Q(i) w(D_i^c | u_i)$. *Further, let* U *be a random variable with* $P(U = u_i) = Q(i)$ *and let* V *be a random variable induced by the channel, i.e.,* $P(V = y | U = u_i) = w(y | u_i)$ *for all* $i \in \{1, \ldots, N\}$ *and* $y \in \mathcal{Y}$ *and* $P(V = y) = \sum_{i=1}^{N} Q(i) \cdot w(y | u_i)$. *Then*

$$H(U|V) \leq 1 + \lambda_Q \log N.$$

Fano's Lemma states that the conditional entropy is smaller (by a factor λ_Q) than $\log N$, the logarithm of the code length. If, e.g., we would have chosen $Q(i) = \frac{1}{N}$ for $i = 1, \ldots, N$, the uniform distribution, then the uncertainty $H(U) = \log N$ is reduced by a factor at least λ_Q, when we already know the realization of V.

Observe that Fano's Lemma does not make use of the time structure, i.e., the block length n is not important and can be chosen as $n = 1$.

Proof of Lemma 52 Let the decoding function d be given by $d(y) = u_i$ exactly if $y \in D_i$ (we can assume w.l.o.g. that $\bigcup_{i=1}^{N} D_i = Y$, otherwise the "rest" $\mathcal{Y} \setminus \bigcup_{i=1}^{N} D_i$ is added to some D_i). Then

$$\lambda_Q = P(U \neq d(V)) = \sum_{y \in \mathcal{Y}} P(U \neq d(y) | V = y) \cdot P(V = y).$$

Now for $y \in \mathcal{Y}$ let $\lambda(y) \triangleq P(U \neq d(v) | V = y)$ and think of the random experiment "U given $V = y$" divided into "$U \neq d(y)$" and "$U = d(y)$". "$U \neq d(y)$" will take place with probability $\lambda(y)$ by definition and hence "$U = d(y)$" has probability $1 - \lambda(y)$. So, by the grouping axiom for the entropy function

$$H(U|V = y) \leq h(\lambda(y)) + (1 - \lambda(y)) \cdot 0 + \lambda(y) \cdot \log(N - 1),$$

where $h(p) \triangleq H(p, 1 - p)$ for $p \in [0, 1]$.

Multiplication by $P(V = y)$ yields

$$H(U|V) = \sum_{y \in \mathcal{Y}} H(U|V = y) \cdot P(V = y)$$

$$\leq \sum_{y \in \mathcal{Y}} P(V = y) \cdot h(\lambda(y)) + \sum_{y \in \mathcal{Y}} P(V = y) \cdot \lambda(y) \cdot \log(N - 1).$$

Now observe that the second term on the right hand side is just $\lambda_Q \log(N-1)$. Since the entropy function is concave and $h(p)$ can be at most 1 (for $p = \frac{1}{2}$), we can further conclude that

$$H(U|V) \leq h\left(\sum_{y \in \mathcal{Y}} P(V = y) \cdot \lambda(y) \right) + \lambda_Q \cdot \log(N - 1)$$

$$\leq 1 + \lambda_Q \cdot \log(N - 1) \leq 1 + \lambda_Q \cdot \log N. \qquad \square$$

5.4.2 A Fundamental Inequality

A further important tool for the proof of the weak converse is the following inequality for the mutual information.

Lemma 53 *For random variables* $X^n = (X_1, \ldots, X_n)$ *and* $Y^n = (Y_1, \ldots, Y_n)$ *taking values in* $\mathcal{X}^n$ *and* $\mathcal{Y}^n$, *respectively, with*

$$P(Y^n = y^n | X^n = x^n) = \prod_{t=1}^{n} w(y_t | x_t) \text{ we have}$$

$$I(X^n \wedge Y^n) \quad \leq \sum_{t=1}^{n} I(X_t \wedge Y_t).$$

Proof $I(X^n \wedge Y^n) = H(Y^n) - H(Y^n | X^n)$

$$= H(Y^n) - \sum_{x^n \in \mathcal{X}^n} P(X^n = x^n) \cdot H(Y^n | X^n = x^n)$$

$$= H(Y^n) + \sum_{x^n \in \mathcal{X}^n} P(X^n = x^n) \cdot$$

$$\sum_{y^n \in \mathcal{Y}^n} P(Y^n = y^n | X^n = x^n) \log P(Y^n = y^n | X^n = x^n)$$

$$= H(Y^n) + \sum_{x^n \in \mathcal{X}^n} P(X^n = x^n) \sum_{y^n \in \mathcal{Y}^n} \prod_{t=1}^{n} w(y_t | x_t) \cdot \log\left(\prod_{t=1}^{n} w(y_t | x_t) \right)$$

$$= H(Y^n) + \sum_{x^n \in \mathcal{X}^n} P(X^n = x^n) \sum_{y^n \in \mathcal{Y}^n} \sum_{s=1}^n \prod_{t=1}^n w(y_t|x_t) \log w(y_s|x_s)$$

$$= H(Y^n) + \sum_{x^n \in \mathcal{X}^n} P(X^n = x^n) \sum_{t=1}^n \sum_{y_t \in \mathcal{Y}} w(y_t|x_t) \log w(y_t|x_t)$$

$$= H(Y^n) + \sum_{t=1}^n \sum_{x_t \in \mathcal{X}} P(X_t = x_t) \sum_{y_t \in \mathcal{Y}} w(y_t|x_t) \cdot \log w(y_t|x_t)$$

$$= H(Y^n) - \sum_{t=1}^n H(Y_t|X_t).$$

Now recall that $H(Y_1, \ldots, Y_n) \le \sum_{t=1}^n H(Y_t)$, hence

$$I(X^n \wedge Y^n) \le \sum_{t=1}^n \big(H(Y_t) - H(Y_t|X_t)\big) = \sum_{t=1}^n I(X_t \wedge Y_t). \qquad \square$$

5.4.3 Proof of the Weak Converse for the DMC

We now have all the preliminaries to prove the weak converse for the discrete memoryless channel.

Theorem 34 *For the discrete memoryless channel we have*

$$\inf_{\lambda > 0} \overline{\lim_{n \to \infty}} \frac{1}{n} \log N(n, \lambda) \le C.$$

Proof Let $\{(u_i, D_i), i = 1, \ldots, N\}$ be a (N, n, λ)–code and let $Q = (\frac{1}{N}, \ldots, \frac{1}{N})$ the uniform distribution on the code words. Obviously, the average error $\lambda_Q \le \lambda$. Now we define random variables $U = U^n$ and $V = V^n$ as in Fano's Lemma $(V = U \cdot W)$ and we have

$$I(U|V) = H(U) - H(U|V)$$

$$= \log N - H(U|V)$$

$$\ge \log N - 1 - \lambda_Q \log N$$

by application of Fano's Lemma and the fact that Q is the uniform distribution (and hence $H(U) = \log N$).

Since the U_t's and the V_t's have range $\mathcal{X}$ and $\mathcal{Y}$, respectively, we have by the fundamental inequality and by $1 - \lambda_Q \geq 1 - \lambda$

$$n \cdot C = n \cdot \max_X I(X \wedge Y) \geq \sum_{t=1}^{n} I(U_t \wedge V_t) \geq I(U \wedge V) \geq (1 - \lambda) \log N - 1,$$

hence $n \, C \geq (1 - \lambda) \log N - 1$ or, equivalently, $\log N \leq \frac{n\,C-1}{1-\lambda}$ for every (N, n, λ)-Code. For the rate this yields

$$\frac{\log N}{n} \leq \frac{C}{1 - \lambda} + \frac{1}{n(1 - \lambda)} \quad \text{and it follows}$$

$$\inf_{\lambda>0} \overline{\lim_{n\to\infty}} \frac{1}{n} \log N(n, \lambda) \leq C. \qquad\qquad \square$$

Remark Observe that using Fano's Lemma we cannot obtain the strong converse $\overline{\lim}_{n\to\infty} \frac{1}{n} \log N(n, \lambda) \leq C$. However, since only few assumptions have to be fulfilled, this lemma is widely applicable in Information Theory.

5.4.4 On the Existence of the Weak Capacity for the DMC

We call C the weak capacity of a channel, if for the maximal code size $N(n, \lambda)$

$$\underline{\lim_{n\to\infty}} \frac{1}{n} \log N(n, \lambda) \geq C \quad \text{for all } \lambda \in (0, 1) \tag{5.32}$$

and

$$\inf_{\lambda>0} \overline{\lim_{n\to\infty}} \frac{1}{n} \log N(n, \lambda) \leq C. \tag{5.33}$$

The second inequality we know already as weak converse for the DMC.
We can also use the following terminology. Define

$$\overline{C}(\lambda) = \overline{\lim_{n\to\infty}} \frac{1}{n} \log N(n, \lambda)$$

$$\underline{C}(\lambda) = \overline{\lim_{n\to\infty}} \frac{1}{n} \log N(n, \lambda)$$

Then $C = \inf_\lambda \underline{C}(\lambda)$ was called (pessimistic) capacity by Csiszár and Körner and others.

The weak converse can be defined by the statement

$$\inf_\lambda \overline{C}(\lambda) = C.$$

5.4.4.1 A New Proof for the Weak Converse for the DMC

Assume $\inf_\lambda \overline{C}(\lambda) \geq C + \delta$. Then for every $\lambda > 0$ there exists an $n(\lambda)$ such that there is an $(n(\lambda), N(n(\lambda), \lambda), \lambda)$ code $\{(u_i, D_i) : 1 \leq i \leq N(n(\lambda), \lambda)\}$ with

$$\frac{1}{n(\lambda)} \log N(n(\lambda), \lambda) \geq C + \delta \tag{5.34}$$

$$W(D_i|u_i) = 1 - \lambda_i \geq 1 - \lambda \quad \text{for } 1 \leq i \leq N(n(\lambda), \lambda), \tag{5.35}$$

and

$$\lambda \log N(n(\lambda), \lambda) \geq -\lambda \log \lambda. \tag{5.36}$$

Consider now the new DMC with input alphabet $\mathcal{U} = \{u_i : 1 \leq i \leq N(n(\lambda), \lambda)\}$ and output alphabet $\{D_i : 1 \leq i \leq N(n(\lambda), \lambda)\}$.

Let X be uniformly distributed with values in $\mathcal{U}$ and let Y be the corresponding output distribution. Then

$$I(X \wedge Y) = H(X) - H(X|Y) = H(Y) - H(Y|X)$$
$$\geq (1 - 3\lambda) \log N(n(\lambda), \lambda). \tag{5.37}$$

Indeed, with the abbreviation $N = N(n(\lambda), \lambda)$, now $\Pr(Y = y) \geq \frac{1}{N}(1 - \lambda)$ and since $-\delta \log \delta$ is increasing in δ

$$H(Y) \geq -N \frac{1 - \lambda}{N} \log \frac{1 - \lambda}{N} = (1 - \lambda) \log N - (1 - \lambda) \log(1 - \lambda).$$

Furthermore, for $\lambda \leq \frac{1}{2}$

$$H(Y|X) \leq H(1 - \lambda, \frac{\lambda}{N - 1}, \ldots, \frac{\lambda}{N - 1})$$
$$\leq 1 - \lambda \log \frac{1}{1 - \lambda} + \lambda \log \frac{1}{\lambda} + \lambda \log(N - 1),$$

thus

$$I(X \wedge Y) \geq (1 - 2\lambda) \log N + \lambda \log \lambda$$

and by (5.36) we get (5.37) as claimed.

From the coding theorem of the DMC

$$\varlimsup_{m \to \infty} \frac{1}{m} \log N(n(\lambda)m, \varepsilon) \geq (1 - 3\lambda) \log N(n(\lambda), \lambda)$$

$$\geq (1 - 3\lambda)n(\lambda)(C + \delta)$$

and

$$\varlimsup_{m \to \infty} \frac{1}{n(\lambda)m} \log N(n(\lambda)m, \varepsilon) \geq (1 - 3\lambda)(C + \delta).$$

Therefore also, by omitting at most $n(\lambda)$ letters, we get

$$\varlimsup_{n \to \infty} \frac{1}{n} \log N(n, \varepsilon) \geq (1 - 3\lambda)(C + \delta). \tag{5.38}$$

By choosing $3\lambda(C + \delta) < \frac{\delta}{2}$ or $\lambda \leq \frac{\delta}{6(C+\delta)}$ we get

$$\varlimsup_{n \to \infty} \frac{1}{n} \log N(n, \varepsilon) > C + \frac{\delta}{2}$$

and therefore

$$\inf_{\varepsilon} \underline{C}(\varepsilon) \geq C + \frac{\delta}{2}, \tag{5.39}$$

which contradicts the definition of C.

5.4.5 The Existence of the Weak Capacity for the AVC with Maximal Probability of Error

We follow the same reasoning as for the DMC for the arbitrarily varying channel (AVC) $\mathcal{W}$. The only difference is that (5.35) has to be replaced by

$$W(D_i|u_i, s^n) \geq 1 - \lambda \quad \text{for } 1 \leq i \leq N(n, \lambda) \text{ and all } s^n \in \mathcal{S}^n. \tag{5.40}$$

With $N = N(n(\lambda), \lambda)$ defined analogously as for the DMC, consider now the new AVC with input alphabet $\mathcal{U} = \{u_i : 1 \leq i \leq N\}$, output alphabet $D = \{D_i : 1 \leq i \leq N\}$, and class of channels $\left\{ (W(D_j|u_i, s^n))_{\substack{i=1,\dots,N \\ j=1,\dots,N}}, s^n \in \mathcal{S}^n \right\} = \mathcal{W}^*$. Here we can enlarge $\mathcal{W}^*$ to the row convex closure $\overline{\overline{\mathcal{W}}}^*$ with convex sets of transmission probabilities $T^*(u)$, $u \in \mathcal{U}$. *Property (5.40) implies that these sets are disjoint.* Therefore we can apply the theorem with the capacity formula

$$\max_{P \in \mathcal{P}(\mathcal{U})} \min_{W \in \overline{\overline{\mathcal{W}}}^*} T(P|W),$$

which again (as in (5.37)) is larger than

$$(1 - 3\lambda) \log N(n(\lambda), \lambda)$$

and a contradiction follows as before for the DMC.

Remark A proof had to be given because the original proof has a flaw.

Research Problem Can this result be derived without using information quantities?

5.4.6 Feedback Does Not Increase the Capacity of the DMC

For reasons of simplicity we shall prove the weak converse for the DMC with feedback only for sequential block codes. It can also be shown that for general coding procedures feedback does not increase the capacity of the DMC.

Theorem 35 *For the discrete memoryless channel with feedback we have*

$$\inf_{\lambda > 0} \overline{\lim}_{n \to \infty} \frac{1}{n} \log N_f(n, \lambda) \le C.$$

Let $\{(u_i(\cdot), D_i), 1 \le i \le N\}$ be a sequential encoding function with block length n and let U be a random variable uniformly distributed on $\{u_1, \ldots, u_N\}$. $V \triangleq Y^n = (Y_1, \ldots, Y_n)$ has the conditional distributions

$$P(Y^n = y^n | i) = \prod_{t=1}^{n} w\big(y_t | u_{it}(y_{t-1})\big)$$

for all $i \in \{1, \ldots, N\}$ and $y^n \in \mathcal{Y}^n$. Hence for all $y^n \in \mathcal{Y}^n$

$$P(Y^n = y^n) = \sum_{i=1}^{N} \frac{1}{N} \cdot \prod_{t=1}^{n} w\big(y_t | u_{it}(y_{t-1})\big)$$

If we can show that $I(U \wedge Y^n) \le n \cdot C$, we are done, since in this case we can again apply Fano's Lemma, which does not use the structure of the channel, and proceed as in the proof of the weak converse for the DMC.

Because of the feedback, however, $Y^n = U \cdot W$ does not hold here and so we cannot make use of the fundamental inequality. Therefore things are more complicated here and we have to introduce some further notation.

Definition 28 For discrete random variables T_1, T_2 and T_3 the conditional information between T_1 and T_2 for given T_3 is defined as

$$I(T_1 \wedge T_2 | T_3) \triangleq H(T_1 | T_3) - H(T_1 | T_2, T_3).$$

For the proof of the theorem we need the following lemma

Lemma 54 (Kolmogorov identity)

$$I(T_1 \wedge T_2 T_3) = I(T_1 \wedge T_2) + I(T_1 \wedge T_3|T_2)$$

$$= I(T_1 \wedge T_3) + I(T_1 \wedge T_2|T_3).$$

Proof Because of symmetry we only have to prove the first equality. By definition
$I(T_1 \wedge T_2 T_3) = H(T_1) - H(T_1|T_2 T_3)$,
$I(T_1 \wedge T_2) = H(T_1) - H(T_1|T_2)$,
and $I(T_1 \wedge T_3|T_2) = H(T_1|T_2) - H(T_1|T_2 T_3)$. □

Proof of the theorem By the above discussion, it suffices to show that $I(U \wedge Y^n) \leq n \cdot C$. Therefore we write

$$I(U \wedge Y^n) = I(U \wedge Y^{n-1} Y_n) = I(U \wedge Y^{n-1}) + I(U \wedge Y_n|Y^{n-1})$$

by the Kolmogorov identity. By induction we are done, if we can prove that

$$I(U \wedge Y_n|Y^{n-1}) \leq C.$$

Since $I(U \wedge Y_n|Y^{n-1}) = \sum_{y^{n-1}} P(Y^{n-1} = y^{n-1}) I(U \wedge Y_n|Y^{n-1} = y^{n-1})$, it is sufficient to show that

$$I(U \wedge Y_n|Y^{n-1} = y^{n-1}) \leq C \text{ for all } y^{n-1} \in \mathcal{Y}^{n-1}.$$

Therefore let $P(i, y_n, y^{n-1}) \triangleq P(U = i, Y_n = y_n, Y^{n-1} = y^{n-1})$. Now

$$I(U \wedge Y_n|Y^{n-1} = y^{n-1}) = \sum_{i=1}^{N} \sum_{y \in \mathcal{Y}} P(i, y|y^{n-1}) \cdot \log \frac{P(y|i, y^{n-1})}{P(y|y^{n-1})}$$

$$= \sum_{x \in \mathcal{X}} \sum_{i: u_i(y^{n-1}) = x} \sum_{y \in \mathcal{Y}} P(i, y|y^{n-1}) \cdot \log \frac{P(y|i, y^{n-1})}{P(y|y^{n-1})}.$$

Since the channel is memoryless, we have $P(y|i, y^{n-1}) = w(y|x)$ if $u_i(y^{n-1}) = x$.

Further $\sum_{i: u_i(y^{n-1}) = x} P(i, y|y^{n-1}) = P(x, y|y^{n-1})$. Therefore

$$I(U \wedge Y_n|Y^{n-1} = y^{n-1}) = \sum_{x \in \mathcal{X}} \sum_{y \in \mathcal{Y}} P(x, y|y^{n-1}) \log \frac{w(y|x)}{P(y|y^{n-1})}.$$

Now we can write

$$P(x, y | y^{n-1}) \quad = P(y|x)P(x|y^{n-1}) = w(y|x)P(x|y^{n-1}),$$

$$P(y|y^{n-1}) = \sum_{x \in \mathcal{X}} P(x, y|y^{n-1}) = \sum_{x \in \mathcal{X}} w(y|x)P(x|y^{n-1}),$$

and with $\overline{P}(x) \triangleq P(x|y^{n-1})$

$$I(U \wedge Y_n | Y^{n-1} = y^{n-1}) = \sum_{x \in \mathcal{X}} \sum_{y \in \mathcal{Y}} \overline{P}(x)w(y|x) \log \frac{w(y|x)}{\sum_x \overline{P}(x)w(y|x)} = I(X \wedge Y),$$

if X has distribution $\overline{P}$ and Y is defined by

$$P(Y = y, X = x) = w(y|x) \cdot \overline{P}(X = x) \text{ for all } x \in \mathcal{X}, y \in \mathcal{Y}.$$

Since obviously $I(X \wedge Y) \leq C = \max_X I(X \wedge Y)$, we are done. $\square$

Remark For sequential block codes Kemperman and Kersten also proved the strong converse, which does not hold in general. For channels with memory, feedback increases the capacity.

References

1. R. Ahlswede, A constructive proof of the coding theorem for discrete memoryless channels in case of complete feedback, Sixth Prague Conference on Inform. Theory, Stat. Dec. Fct's and Rand. Proc., Sept. 1971, Publ. House Czechosl. Academy of Sci. 1–22, (1973)
2. R. Ahlswede, G. Dueck, Every bad code has a good subcode: a local converse to the coding theorem. Z. Wahrscheinlichkeitstheorie und verw. Geb. **34**, 179–182 (1976)
3. R. Ahlswede, P. Gács, J. Körner, Correction to: "Bounds on conditional probabilities with applications in multi-user communication" Z. Wahrscheinlichkeitstheorie und verw. Gebiete **39**(4), 353–354 (1977)
4. S. Arimoto, On the converse to the coding theorem for discrete memoryless channels. IEEE Trans. Inform. Theory **19**, 357–359 (1973)
5. U. Augustin, Gedächtnisfreie Kanäle für diskrete Zeit. Z. Wahrscheinlichkeitstheorie und verw. Geb. **6**, 10–61 (1966)
6. J. Beck, A uniform estimate for the sum of independent R.V'.s with an application to Stein's lemma, problems control information theory/problemy upravlen. Teor. Inform. **7**(2), 127–136 (1978)
7. D. Blackwell, L. Breiman, A.J. Thomasian, The capacity of a class of channels. Ann. Math. Stat. **30**(4), 1229–1241 (1959)
8. R.M. Fano, *Class Notes for Transmission of Information, Course 6.574* (MIT Cambridge, Mas, 1952)
9. A. Feinstein, A new basic theorem of information theory. IRE Trans. Sect. Inform. Theory **PGIT-4**, 2–22 (1954)
10. J.H.B. Kemperman, *Studies in Coding Theory I*. Mimeo graphed lecture notes (1961)
11. J.H.B. Kemperman, *On the Optimum Rate of Transmitting Information, Probability and Information Theory*, (Springer, Berlin, 1969) pp. 126–169

12. J.H.B. Kemperman, Strong converses for a general memoryless channel with feedback, Trans. Sixth Prague Conf. on Inform. Theory, Stat. Dec. Fct's, and Rand. Proc., 375–409, 1973.
13. C.E. Shannon, A mathematical theory of communication. Bell Syst. Tech. J. **27**(379–423), 632–656 (1948)
14. J. Wolfowitz, The coding of messages subject to chance errors. Illinois J. Math. **1**, 591–606 (1957)

Chapter 6
Towards Capacity Functions

6.1 Lecture on Concepts of Performance Parameters for Channels

Among the mostly investigated parameters for noisy channels are code size, error probability in decoding, block length; rate, capacity, reliability function; delay, complexity of coding. There are several statements about connections between these quantities. They carry names like "coding theorem", "converse theorem" (weak, strong, ...), "direct theorem", "capacity theorem", "lower bound", "upper bound", etc. There are analogous notions for source coding.

This note has become necessary after the author noticed that Information Theory suffers from a lack of precision in terminology. Its purpose is to open a discussion about this situation with the goal to gain more clarity.

6.1.1 Channels

It is beyond our intention to consider questions of modelling, like what is a channel in reality, which parts of a communication situation constitute a channel etc. Shannon's mathematical description in terms of transmission probabilities is the basis for our discussion.

Also, in most parts of this note we speak about one-way channels, but there will be also comments on multi-way channels and compound channels.

Abstractly, let $\mathcal{I}$ be any set, whose elements are called input symbols and let $\emptyset$ be any set, whose elements are called output symbols.

An *(abstract) channel* $W : \mathcal{I} \to (\emptyset, \mathcal{E})$ is a set of probability distributions

$$W = \big\{ W(\cdot|i) : i \in \mathcal{I} \big\} \tag{6.1}$$

on $(\emptyset, \mathcal{E})$.

A. Ahlswede et al. (eds.), *Storing and Transmitting Data,*
Foundations in Signal Processing, Communications and Networking 10,
DOI: 10.1007/978-3-319-05479-7_6, © Springer International Publishing Switzerland 2014

So for every input symbol i and every (measurable) $E \in \mathcal{E}$ of output symbols $W(E|i)$ specifies the probability that a symbol in E will be received, if symbol i has been sent.

The set $\mathcal{I}$ does not have to carry additional structure.

Of particular interest are *channels with time-structure*, that means, symbols are words over an alphabet, say $\mathcal{X}$ for the inputs and $\mathcal{Y}$ for the outputs. Here $\mathcal{X}^n = \prod_{t=1}^{n} \mathcal{X}_t$ with $\mathcal{X}_t = \mathcal{X}$ for $t \in \mathbb{N}$ (the natural numbers) are the input words of (block)-length n and $\mathcal{Y}^n = \prod_{t=1}^{n} \mathcal{Y}_t$ with $\mathcal{Y}_t = \mathcal{Y}$ for $t \in \mathbb{N}$ are the output words of length n.

Moreover, again for the purpose of this discussion we can assume that a transmitted word of length n leads to a received word of length n. So we can define a (constant block length) channel by the set of stochastic matrices

$$\mathcal{K} = \{W^n : \mathcal{X}^n \to \mathcal{Y}^n : n \in \mathbb{N}\}. \tag{6.2}$$

In most channels with time-structure there are (compatibility) relations between these matrices.

We don't have to enter these delicate issues. Instead, we present now three channel concepts, which serve as key examples in this note.

DMC: The most familiar channel is the discrete memoryless channel, defined by the transmission probabilities

$$W^n(y^n|x^n) = \prod_{t=1}^{n} W(y_t|x_t) \tag{6.3}$$

for $W : \mathcal{X} \to \mathcal{Y}$, $x^n = (x_1, \ldots, x_n) \in \mathcal{X}^n$, $y^n = (y_1, \ldots, y_n) \in \mathcal{Y}^n$, and $n \in \mathbb{N}$.

NDMC: The *nonstationary* discrete memoryless channel is given by a sequence $(W_t)_{t=1}^{\infty}$ of stochastic matrices $W_t : \mathcal{X} \to \mathcal{Y}$ and the rule for the transmission of words

$$W^n(y^n|x^n) = \prod_{t=1}^{n} W_t(y_t, x_t). \tag{6.4}$$

Other names are "inhomogeneous channel", "non-constant" channel.

Especially, if $\quad W_t = \begin{cases} W & \text{for } t \text{ even} \\ V & \text{for } t \text{ odd} \end{cases}$

one gets a "periodic" channel of period 2 or a "parallel" channel. (c.f. [2, 37]).

ADMC: Suppose now that we have two channels $\mathcal{K}_1$ and $\mathcal{K}_2$ as defined in (6.2). Then following [3] we can associate with them an *averaged* channel

$$\mathcal{A} = \left\{ \left(\frac{1}{2}W_1^n + \frac{1}{2}W_2^n : \mathcal{X}^n \to \mathcal{Y}^n \right) : n \in \mathbb{N} \right\} \tag{6.5}$$

and when both constituents, $\mathcal{K}_1$ and $\mathcal{K}_2$ are DMC's (resp. NDMC's) we term it ADMC (resp. ANDMC).

It is a very simple channel with "strong memory", suitable for theoretical investigations. They are considered in [3] in much greater generality (any number of constituents, infinite alphabets) and have been renamed by Han and Verdu "mixed channels" in several chapters (see [29]).

We shall see below that channel parameters, which have been introduced for the DMC, where their meaning is without ambiguities, have been used for general time-structured channels for which sometimes their formal or operational meaning is not clear.

Nonstationary and *Memory*, incorporated in our examples of channels, are tests for concepts measuring channel performance.

6.1.2 Three Unquestioned Concepts: The Two Most Basic, Code Size and Error Probability, then Further Block Length

Starting with the abstract channel $W : \mathcal{I} \to (\emptyset, \mathcal{E})$ we define a *code*

$$\mathcal{C} = \big\{(u_i, D_i) : i \in I\big\} \text{ with } u_i \in \mathcal{I}, D_i \in \mathcal{E}$$

for $i \in I$ and pairwise disjoint D_i's.

$$M = |\mathcal{C}| \text{ is the code size} \tag{6.6}$$

$$e(\mathcal{C}) = \max_{i \in I} W(D_i^c | u_i) \tag{6.7}$$

is the (maximal) probability of error and

$$\bar{e}(\mathcal{C}) = \frac{1}{M} \sum_{i=1}^{M} W(D_i^c | u_i) \tag{6.8}$$

is the average probability of error.

One can study now the functions

$$M(\lambda) = \max_{\mathcal{C}} \big\{ |\mathcal{C}| : e(\mathcal{C}) \leq \lambda \big\} \text{ (resp. } \overline{M}(\lambda)) \tag{6.9}$$

and

$$\lambda(M) = \min_{\mathcal{C}} \big\{ e(\mathcal{C}) : |\mathcal{C}| = M \big\} \text{ (resp. } \overline{\lambda}(M)), \tag{6.10}$$

that is, finiteness, growth, convexity properties etc.

It is convenient to say that C is an (M, λ)-code, if

$$|C| \geq M \text{ and } e(C) \leq \lambda. \tag{6.11}$$

Now we add time-structure, that means here, we go to the channel defined in (6.2). The parameter n is called the *block length* or word length.

It is to be indicated in the previous definitions. So, if $u_i \in \mathcal{X}^n$ and $D_i \subset \mathcal{Y}^n$ then we speak about a code $C(n)$ and definitions (6.9), (6.10), and (6.11) are to be modified accordingly:

$$M(n, \lambda) = \max_{C(n)} \{ |C(n)| : e(C(n)) \leq \lambda \} \tag{6.12}$$

$$\lambda(n, M) = \min_{C(n)} \{ e(C(n)) : |C(n)| = M \} \tag{6.13}$$

$$C(n) \text{ is an } (M, n, \lambda)\text{-code, if } |C(n)| \geq M, e(C(n)) \leq \lambda. \tag{6.14}$$

Remark One could study blocklength as function of M and λ in smooth cases, but this would be tedious for the general model $\mathcal{K}$, because monotonicity properties are lacking for $M(n, \lambda)$ and $\lambda(M, n)$.

We recall next Shannon's fundamental statement about the two most basic parameters.

6.1.3 Stochastic Inequalities: The Role of the Information Function

We consider a channel $W : \mathcal{X} \to \mathcal{Y}$ with finite alphabets. To an input distribution P, that is a PD on $\mathcal{X}$, we assign the output distribution $Q = PW$, that is a PD on $\mathcal{Y}$, and the joint distribution $\tilde{P}$ on $\mathcal{X} \times \mathcal{Y}$, where $\tilde{P}(x, y) = P(x)W(y|x)$.

Following Shannon [47] we associate with (P, W) or $\tilde{P}$ the *information function (per letter)* $I : \mathcal{X} \times \mathcal{Y} \to \mathbb{R}$, where

$$I(x, y) = \begin{cases} \log \frac{\tilde{P}(x,y)}{P(x)Q(y)} \\ 0 \qquad\qquad \text{, if } \tilde{P}(x, y) = 0. \end{cases} \tag{6.15}$$

If X is an (input) RV with values in $\mathcal{X}$ and distribution $P_X = P$ and if Y is an (output) RV with values in $\mathcal{Y}$ and distribution $P_Y = Q$ such that the joint distribution P_{XY} equals $\tilde{P}$, *then $I(X, Y)$ is a RV.* Its distribution function will be denoted by F, so

$$F(\alpha) = \Pr\{I(X, Y) \leq \alpha\} = \tilde{P}(\{(x, y) : I(x, y) \leq \alpha\}). \tag{6.16}$$

We call an $(M, \overline{\lambda})$-code $\{(u_i, D_i) : 1 \leq i \leq M\}$ *canonical,* if $P(u_i) = \frac{1}{M}$ for $i = 1, \ldots, M$ and the decoding sets are defined by maximum likelihood decoding, which results in a (minimal) average error probability $\overline{\lambda}$.

Theorem 36 (Shannon 1957, [47]) *For a canonical $(M, \overline{\lambda})$-code and the corresponding information function there are the relations*

$$\frac{1}{2} F \left(\log \frac{M}{2} \right) \leq \overline{\lambda} \leq F \left(\log \frac{M}{2} \right). \tag{6.17}$$

Remarks

1. Shannon carries in his formulas a blocklength n, but this is nowhere used in the arguments. The bounds hold for abstract channels (without time structure). The same comment applies to his presentation of his random coding inequality: there exists a code of length M and average probability of error

$$\overline{\lambda} \leq F(\log M + \theta) + e^{-\theta}, \theta > 0.$$

2. Let us emphasize that all of Shannon's bounds involve the information function (per letter), which is highlighted also in Fano [22], where it is called mutual information. In contrast, Fano's inequality is *not a stochastic inequality.* It works with the *average* (or expected) mutual information $I(X \wedge Y)$ (also written as $I(X; Y)$), which is a constant. Something has been given away.

6.1.4 Derived Parameters of Performance: Rates for Code Sizes, Rates for Error Probabilities, Capacity, Reliability

The concept of rate involves a renormalisation in order to put quantities into a more convenient scale, some times per unit. Exponentially growing functions are renormalized by using the logarithmic function. In Information Theory the prime example is $M(n, \lambda)$ (see Chap. 2). Generally speaking, with any function $f : \mathbb{N} \to \mathbb{R}_+$ (or, equivalently, any sequence $(f(1), f(2), f(3), \ldots)$ of non-negative numbers) we can associate a rate function rate (f), where

$$\text{rate}\,(f(n)) = \frac{1}{n} \log f(n). \tag{6.18}$$

We also speak of the *rate at n,* when we mean

$$\text{rate}_n (f) \triangleq \text{rate}\,(f(n)) = \frac{1}{n} \log f(n). \tag{6.19}$$

This catches statements like "an increase of rate" or "rate changes".

In Information Theory f is related to the channel $\mathcal{K}$ or more specifically $f(n)$ depends on W^n. For example choose $f(n) = M(n, \lambda)$ for $n \in \mathbb{N}$, λ constant. Then rate(f) is a *rate function* for certain code sizes.

Now comes a *second step*: for many *stationary* systems like stationary channels (c.f. DMC) f behaves very regular and instead of dealing with a whole rate function one just wants to associate a *number* with it.

We state for the three channels introduced in Sect. 6.1 the results—not necessarily the strongest known—relevant for our discussion.

DMC: There is a constant $C = C(W)$ (actually known to equal $\max_P I(W|P)$) such that

(i) for every $\lambda \in (0, 1)$ and $\delta > 0$ there exists an $n_0 = n_0(\lambda, \delta)$ such that *for all* $n \geq n_0$ there exist

$$(n, e^{(C-\delta)n}, \lambda)\text{-codes,}$$

(ii) for every $\lambda \in (0, 1)$ and $\delta > 0$ there exists an $n_0 = n_0(\lambda, \delta)$ such that *for all* $n \geq n_0$ there does *not* exist an

$$(n, e^{(C+\delta)n}, \lambda)\text{-code.}$$

ADMC: There is a constant C (actually known to equal $\max_P \min_{i=1,2} I(W_i|P)$ [3]) such that

(i) holds

(ii) for every $\delta > 0$ there *exists* a $\lambda \in (0, 1)$ and an $n_0 = n_0(\lambda, \delta)$ such that *for all* $n \geq n_0$ there does *not* exist an

$$(n, e^{(C+\delta)n}, \lambda)\text{-code.}$$

NDMC: There is a sequence of numbers $\left(C(n) \right)_{n=1}^{\infty}$ (which actually can be chosen as $C(n) = \frac{1}{n} \sum_{t=1}^{n} \max_P I(W_t|P)$ [2]) such that

(i′) for every $\lambda \in (0, 1)$ and $\delta > 0$ there exists an $n_0 = n_0(\lambda, \delta)$ such that for all $n \geq n_0$ there exist
$$(n, e^{(C(n)-\delta)n}, \lambda)\text{-codes.}$$

(ii′) for every $\lambda \in (0, 1)$ and $\delta > 0$ there exists an $n_0 = n_0(\lambda, \delta)$ such that for all $n \geq n_0$ there does *not* exist an

$$(n, e^{(C(n)+\delta)n}, \lambda)\text{-code.}$$

(This is still true for infinite output alphabets, for infinite input alphabets in general not. There the analogue of (i), say (iii′) is often still true, but also not always.)

Notice that with every sequence $\left(C(n)\right)_{n=1}^{\infty}$ satisfying (i') and (ii') or (i') and (iii') also every sequence $\left(C(n)+o(1)\right)_{n=1}^{\infty}$ does. In this sense the sequence is not unique, whereas earlier the constant C is.

The *pair* of statements [(i), (ii)] has been called by Wolfowitz *Coding theorem with strong converse* and the number C has been called the *strong capacity* in [2]. For the ADMC there is no C satisfying (i) *and* (ii), so this channel *does not have* a strong capacity.

The pair of statements [(i), (ii)] have been called by Wolfowitz coding theorem with *weak converse* and the number C has been called in [2] the *weak capacity*. So the ADMC does have a weak capacity.

(For completeness we refer to two standard textbooks. On page 9 of Gallager [24] one reads "The converse to the coding theorem is stated and proved in varying degrees of generality in Chaps. 4, 7, and 8. In imprecise terms, it states that if the entropy of a discrete source, in bits per second, is greater than C, then independent of the encoding and decoding used in transmitting the source output at the destination cannot be less than some positive number which depends on the source and on C. Also, as shown in Chap. 9, if R is the minimum number of binary digits per second required to reproduce a source within a given level of average distortion, and if $R > C$, then, independent of the encoding and decoding, the source output cannot be transmitted over the channel and reproduced within that given average level of distortion."

In spite of its pleasant preciseness in most cases, there seems to be no definition of the weak converse in the book by Csiszár and Körner [17].)

Now the NDMC has in general no strong and no weak capacity (see our example in Sect. 6.1.7).

However, if we replace the concept of capacity by that of a capacity function $\left(C(n)\right)_{n=1}^{\infty}$ then the pair [(a'), (ii')] (resp. [(i'), (iii')]) may be called coding theorem with strong (resp. weak) converse and accordingly one can speak about *strong (resp. weak) capacity functions*, defined modulo $o(1)$.

These concepts have been used or at least accepted—except for the author— also by Wolfowitz, Kemperman, Augustin and also Dobrushin [18, 19], Pinsker [44]. The concept of information stability (Gelfand/Yaglom; Pinsker) defined for *sequences of numbers* and *not*—like some authors do nowadays—for a *constant only*, is in full agreement at least with the [(i), (iii)] or [(i'), (iii')] concepts. Equivalent formulations are

(i') $\inf_{\lambda>0} \lim_{n\to\infty} \left(\frac{1}{n}\log M(n,\lambda) - C(n)\right) \geq 0$

(ii') for all $\lambda \in (0, 1)$ $\overline{\lim_{n\to\infty}} \left(\frac{1}{n}\log M(n,\lambda) - C(n)\right) \leq 0$

(iii') $\inf_{\lambda>0} \overline{\lim_{n\to\infty}} \left(\frac{1}{n}\log M(n,\lambda) - C(n)\right) \leq 0.$

(For a constant C this gives (a), (b), (c).)

Remarks

1. A standard way of expressing (iii) is: for rates above capacity the error probability is bounded away from 0 for *all large n*. ([23], called *"partial converse"* on page 44.)
2. There are cases (c.f. [3]), where the uniformity in λ valid in (ii) or (ii') holds only for $\lambda \in (0, \lambda_1)$ with an absolute constant λ_1—a "medium" strong converse. It also occurs in "second order" estimates of [32] with $\lambda_1 = \frac{1}{2}$.
3. There are cases where (iii) [or (iii')] don't hold for constant λ's but for $\lambda = \lambda(n)$ going to 0 sufficiently fast, in one case [9] like $\frac{1}{n}$ and in another like $\frac{1}{n^4}$ [7]. In both cases $\lambda(n)$ decreases reciprocal to a polynomial and it makes sense to speak of polynomial-weak converses. The soft-converse of [12] is for $\lambda(n) = e^{o(n)}$. Any decline condition on λ_n could be considered.
4. For our key example in Sect. 6.1.7 [(i'), (iii')] holds, but not [(i), (iii)]. It can be shown that for the constant $C = 0$ and any $\delta > 0$ there is a $\lambda(\delta) > 0$ such that $(n, e^{(C+\delta)n})$-codes have error probability exceeding $\lambda(\delta)$ for *infinitely many n*. *By the first remark of this lecture, this is weaker than (c) and equivalent to*

$$\inf_{\lambda > 0} \lim_{n \to \infty} \frac{1}{n} \log M(n, \lambda) = C.$$

Now comes a seemingly small twist. Why bother about "weak capacity", "strong capacity" etc. and their existence—every channel should have a capacity.

Definition 29 $\underline{C}$ is called the (pessimistic) capacity of a channel $\mathcal{K}$, if it is the supremum over all numbers C for which (i) holds. Since $C = 0$ satisfies (a), the number $\underline{C} = \underline{C}(\mathcal{K})$ exists. Notice that there are no requirements concerning (ii) or (iii) here.

To every general $\mathcal{K}$ a constant performance parameter has been assigned!
What does it do for us?
First of all the name "pessimistic" refers to the fact that another number $\overline{C} = \overline{C}(\mathcal{K})$ can be introduced, which is at least a large as $\underline{C}$.

Definition 30 $\overline{C}$ is called the (optimistic) capacity of a channel $\mathcal{K}$, if it is the supremum over all numbers C for which in (a) the condition "for all $n \geq n_0(\lambda, \delta)$" is replaced by "for infinitely many n" or equivalently

$$\overline{C} = \inf_{\lambda > 0} \overline{\lim_{n \to \infty}} \frac{1}{n} \log M(n, \lambda).$$

Here it is measured whether for every λ $R < \overline{C}$ this "rate" is occasionally, but infinitely often achievable.

(Let us briefly mention that "the reliability function" $E(R)$ is commonly defined through the values

$$\underline{E}(R) = -\lim_{n \to \infty} \frac{1}{n} \log \lambda(e^{Rn}, n)$$

$$\overline{E}(R) = -\varlimsup_{n \to \infty} \frac{1}{n} \log \lambda(e^{Rn}, n)$$

if they coincide. Again further differentiation could be gained by considering the sequence

$$E_n(R_n) = -\frac{1}{n} \log \lambda(e^{R_n n}, n), \ n \in \mathbb{N},$$

for sequences of rates $(R_n)_{n=1}^{\infty}$. But that shall not be pursuit here.)

In the light of old work [2] we were shocked when we learnt that these two definitions were given in [17] and that the pessimistic capacity was used throughout that book. Since the restriction there is to the DMC-situation it makes actually no difference. However, several of our Theorems had just been defined away. Recently we were even more surprised when we learned that these definitions were not new at all and have indeed been standard and deeply rooted in the community of information theorists (the pessimistic capacity $\underline{C}$ is used in [16, 22, 50], and the optimistic capacity $\overline{C}$ is used in [17] on page 223 and in [38]).

Fano [22] uses $\underline{C}$, but he at the same time emphasizes throughout the book that he deals with "constant channels".

After quick comments about the optimistic capacity concept in the next section we report on another surprise concerning $\underline{C}$.

6.1.5 A Misleading Orientation at the DMC: the Optimistic Rate Concept Seems Absurd

Apparently for the DMC the optimistic as well as the pessimistic capacities, $\overline{C}$ and $\underline{C}$, equal $C(W)$. For multi-way channels and compound channels $\{W(\cdot|\cdot, s) : s \in \mathcal{S}\}$ the optimistic view suggests a dream world.

1. Recently Cover explained that under this view for the broadcast channel ($W : \mathcal{X} \to \mathcal{Y}, V : \mathcal{X} \to \mathcal{Z}$) the rate pair $(R_\mathcal{Y}, R_\mathcal{Z}) = (C(W), C(V))$ is in the capacity region, which in fact equals $\{(R_\mathcal{Y}, R_\mathcal{Z}) : 0 \leq R_\mathcal{Y} \leq C(W), 0 \leq R_\mathcal{Z} \leq C(W)\}$. Just assign periodically time intervals of lengths $m_1, n_1, m_2, n_2, m_3, n_3, \ldots$ to the DMC's W and V for transmission. Just choose every interval very long in comparison to the sum of the lengths of its predecessors. Thus again and again every channel comes in its rate close, and finally arbitrary close, to its capacity. The same argument applies to the MAC, TWC etc.— so in any situation where the communicators have a choice of the channels for different time intervals.
2. The reader may quickly convince himself that $\overline{C} = \min_{s \in \mathcal{S}} C(W(\cdot|\cdot, s)) \geq \max_P \min_s I(W(\cdot|\cdot, s)|P)$ for the compound channel. For the sake of the argument choose $\mathcal{S} = \{1, 2\}$. The sender not knowing the individual channel transmits for

channel $W(\cdot|\cdot, 1)$ on the m-intervals and for channel $W(\cdot|\cdot, 2)$ on the n-intervals. The receiver *can test* the channel and knows in which intervals to decode!

3. As a curious Gedankenexperiment: Is there anything one can do in this context for the AVC?

For the semicontinuous compound channel, $|\mathcal{S}| = \infty$, the ordinary weak capacity [[(i),(iii)] hold] is unknown. We guess that optimism does not help here, because it does seem to help if there are infinitely many proper cases.

The big issue in all problems here is of course delay. It ought to be incorporated (Space-time coding).

6.1.6 A "Paradox" for Products of Channels

Let us be given s channels $(W_j^n)_{n=1}^\infty$, $1 \le j \le s$. Here $W_j^n : \mathcal{X}_j^n \to \mathcal{Y}_j^n$, $1 \le j \le s$. The product of these channels $(W^{*n})_{n=1}^\infty$ is defined by

$$W^{*n} = \prod_{j=1}^s W_j^n : \prod_{j=1}^s \mathcal{X}_j^n \to \prod_{j=1}^s \mathcal{Y}_j^n.$$

A chapter by Wyner [50] is very instructive for our discussion. We quote therefore literally the beginning of the chapter (page 423) and also its Theorem with a sketch of the proof (page 425), because it is perhaps instructive for the reader to see how delicate things are even for leading experts in the field.

"In this chapter we shall consider the product or parallel combination of channels, and show that (1) the *capacity of the product channel is the sum of the capacities of the component channels*, and (2) the "strong converse" holds for the product channel if it holds for each of the component channels. The result is valid for any class of channels (with or without memory, continuous or discrete) provided that the capacities exist. "Capacity" is defined here *as the supremum of those rates for which arbitrarily high reliability is achievable with block coding for sufficiently long delay*.

Let us remark here that there are two ways in which "channel capacity" is commonly defined. The first definition takes the channel capacity to be the supremum of the "information" processed by the channel, where "information" is the difference of the input "uncertainty" and the "equivocation" at the output. *The second definition, which is the one we use here, takes the channel capacity to be the maximum "error free rate"*. For certain classes of channels (e.g., memoryless channels, and finite state indecomposable channels) it has been established that these two definitions are equivalent. In fact, this equivalence is the essence of the Fundamental Theorem of Information Theory. For such channels, (1) above follows directly. The second definition, however, is applicable to a broader class of channels than the first. One very important such class are time-continuous channels."

Theorem 37 *(i) Let C^* be the capacity of the product of s channels with capacities $C_1, C_2, \ldots, C_s$ respectively. Then*

$$C^* = \sum_{j=1}^{s} C_j. \tag{6.20}$$

(ii) If the strong converse holds for each of these s channels, then it holds for the product channel.

The proof of (i) is divided into two parts. In the first (the "direct half") we will show that any $R < \sum_{j=1}^{s} C_j$ is a permissible rate. This will establish that $C^* \geq \sum_{j=1}^{s} C_j$. In the second ("weak converse") we will show that no $R > \sum_{j=1}^{s} C_j$ is a permissible rate, establishing that $C^* \leq \sum_{j=1}^{s} C_j$. The proof of (ii) parallels that of the weak converse.

It will suffice to prove the theorem for the product of two channels ($s = 2$), the result for arbitrary s following immediately by induction."

Let's first remark that $C^* \geq \sum_{j=1}^{s} C_j$ for the pessimistic capacities (apparently used here) follows immediately from the fact that by taking products of codes the errors at most behave additive. By proving the reverse inequality the weak converse, statement (iii) in Sect. 6.1.4 is *tacitly assumed* for the component channels and from there on everything is okay. The point is that this assumption does not appear as a hypothesis in the Theorem! Indeed our key example of Sect. 6.1.7 shows that (6.20) is in general not true. The two factor channels used in the example do not have a weak converse (or weak capacity for that matter).

The reader is reminded that having proved a weak converse for the number $\underline{C}$, the pessimistic capacity, is equivalent to having shown that the weak capacity exists.

6.1.7 The Pessimistic Capacity Definition: An Information-Theoretic Perpetuum Mobile

Consider the two matrices $V^1 = \begin{pmatrix} 1 & 0 \\ 0 & 1 \end{pmatrix}$ and $V^0 = \begin{pmatrix} \frac{1}{2} & \frac{1}{2} \\ \frac{1}{2} & \frac{1}{2} \end{pmatrix}$. We know that $C(V^1) = 1$ and $C(V^0) = 0$.

Consider a NDMC $\mathcal{K}$ with $W_t \in \{V^0, V^1\}$ for $t \in \mathbb{N}$ and a NDMC $\mathcal{K}^*$ with t-th matrix W_t^* also from $\{V^0, V^1\}$ but *different* from W_t. Further consider the product channel $(\mathcal{K}, \mathcal{K}^*)$ specified by $W_1 W_1^* W_2 W_2^*$—again a NDMC.

With the choice $(m_1, n_1, m_2, n_2, \ldots)$, where for instance $n_i \geq 2^{m_i}$, $m_{i+1} \geq 2^{n_i}$ we define channel $\mathcal{K}$ completely by requiring that $W_t = V^1$ in the m_i-length intervals and $W_t = V^0$ in the n_i-length intervals. By their growth properties we have for the pessimistic capacities $\underline{C}(\mathcal{K}) = \underline{C}(\mathcal{K}^*) = 0$. However, apparently $\underline{C}(\mathcal{K}, \mathcal{K}^*) = 1$.

6.1.8 A Way Out of the Dilemma: Capacity Functions

If $M(n, \lambda)$ fluctuates very strongly in n and therefore also $\mathrm{rate}_n(M)$, then it does not make much sense to describe its growth by one number $\underline{C}$. At least one has to be aware of the very limited value of theorems involving that number.

For the key example in Sect. 6.1.7 $\underline{C}(\mathcal{K}) = \underline{C}(\mathcal{K}^*) = 0$ and on the other hand $\overline{C}(\mathcal{K}) = \overline{C}(\mathcal{K}^*) = 1$. In contrast we can choose the sequence $(c_n)_{n=1}^{\infty} = \left(\frac{1}{n} \sum_{t=1}^{n} C(W_t)\right)_{n=1}^{\infty}$ for channel $\mathcal{K}$ and $(c_n^*)_{n=1}^{\infty} = \left(\frac{1}{n} \sum_{t=1}^{n} C(W_t^*)\right)_{n=1}^{\infty}$ for channel $\mathcal{K}^*$, who are always *between* 0 and 1.

They are (even strong) capacity functions and for the product channel $\mathcal{K} \times \mathcal{K}^*$ we have the capacity function $(c_n + c_n^*)_{n=1}^{\infty}$, which equals identically 1, what it should be. Moreover thus also in general the "perpetuum mobile of information" disappears. We have been able to prove the

Theorem 38 *For two channels $\mathcal{K}_1$ and $\mathcal{K}_2$*

(i) *with weak capacity functions their product has the sum of those functions as weak capacity function*
(ii) *with strong capacity functions their product has the sum of those functions as strong capacity function.*

We hope that we have made clear that capacity functions in conjunction with converse proofs carry in general more information—perhaps not over, but *about channels*—than optimistic or pessimistic capacities. This applies even for channels without a weak capacity function because they can be made this way at least as large $\underline{C}$ and still satisfy (i).

Our conclusion is, that

1. When speaking about capacity formulas in non standard situations one must clearly state which definition is being used.
2. There is no "true" definition nor can definitions be justified by authority.
3. Presently weak capacity functions have most arguments in their favour, also in comparison to strong capacity functions, because of their wide validity and the primary interest in direct theorems. To call channels without a strong capacity "channels without capacity" ([49]) is no more reasonable than to name an optimistic or a pessimistic capacity "the capacity".
4. We must try to help enlightening the structure of channels. For that purpose for instance $\underline{C}$ can be a useful bound on the weak capacity function, because it may be computable whereas the function isn't.
5. Similar comments are in order for other quantities in Information Theory, rates for data compression, reliability functions, complexity measures.

6.1.9 Some Concepts of Performance from Channels with Phases

In this section we explore other capacity concepts involving the phase of the channel, which for stationary systems is not relevant, but becomes an issue otherwise. Again the NDMC $(W_t)_{t=1}^{\infty}$ serves as a genuine example. In a phase change by m we are dealing with $(W_{t+m})_{t=1}^{\infty}$. "Capacity" results for the class of channels $\{(W_{t+m})_{t=1}^{\infty} : 0 \leq m < \infty\}$ in the spirit of a compound channel, that is, for codes which are good simultaneously for all m are generally unknown. The AVC can be produced as a special case and even more so the zero-error capacity problem.

An exception is for instance the case where $(W_t)_{t=1}^{\infty}$ is almost periodic in the sense of Harald Bohr. Because these functions have a mean also $\big(C(W_t)\big)_{t=1}^{\infty}$ has a mean and it has been shown that there is a strong capacity [2].

Now we greatly simplify the situation and look only at $(W_t)_{t=1}^{\infty}$ where

$$W_t \in \left\{ \begin{pmatrix} 1 & 0 \\ 0 & 1 \end{pmatrix}, \begin{pmatrix} \frac{1}{2} & \frac{1}{2} \\ \frac{1}{2} & \frac{1}{2} \end{pmatrix} \right\}$$

and thus $C(W_t) \in \{0, 1\}$. Moreover, we leave error probabilities aside and look only at $0-1$-sequences $(C_1, C_2, C_3, \dots)$ and the associated $C(n) = \frac{1}{n}\sum_{t=1}^{n} C_t \in [0, 1]$.

So we just play with $0-1$-sequences $(a_n)_{n=1}^{\infty}$ and associated Cesaro-means $A_n = \frac{1}{n}\sum_{t=1}^{n} a_t$ and $A_{m+1,m+n} = \frac{1}{n}\sum_{t=m+1}^{m+n} a_t$.

First of all there are the familiar

$$\underline{A} = \varliminf_{n\to\infty} A_n \text{ (the pessimistic mean)} \tag{6.21}$$

$$\overline{A} = \varlimsup_{n\to\infty} A_n \text{ (the optimistic mean).} \tag{6.22}$$

We introduce now a new concept

$$\underline{\underline{A}} = \lim_{n\to\infty} \inf_{m\geq 0} A_{m+1,m+n} \text{ (the pessimistic phase independent mean).} \tag{6.23}$$

The "inf" reflects that the system could be in any phase (known to but not controlled by the communicators). Next we assume that the communicators can choose the phase m for an intended n and define

$$\overline{\overline{A}} = \varlimsup_{n\to\infty} \sup_{m\geq 0} A_{m+1,m+n} \text{ (super optimistic mean).} \tag{6.24}$$

We shall show first

Lemma 55

$$\varlimsup_{n\to\infty} \inf_{m\geq 0} A_{m+1,m+n} = \underline{\underline{A}} \qquad (6.25)$$

$$\lim_{n\to\infty} \sup_{m\geq 0} A_{m+1,m+n} = \overline{\overline{A}} \qquad (6.26)$$

Proof We prove only (6.25), the proof for (6.26) being "symmetrically" the same. We have to show that

$$\underline{\underline{A}} = \varliminf_{n\to\infty} \inf_{m\geq 0} A_{m+1,m+n} \geq \varlimsup_{n\to\infty} \inf_{m\geq 0} A_{m+1,m+n}. \qquad (6.27)$$

$\square$

For every n let $m(n)$ give minimal $A_{m+1,m+n}$. The number exists because these means take at most $n+1$ different values. Let n^* be such that $A_{m(n^*)+1,m(n^*)+n^*}$ is within ε of $\underline{\underline{A}}$ and choose a much bigger N^* for which $A_{m(N^*)+1,m(N^*)+N^*}$ is within ε of the expression at the right side of (6.27) and $N^* \geq \frac{1}{\varepsilon}n^*$ holds.

Choose r such that $rn^* + 1 \leq N^* \leq (r+1)n^*$ and write

$$N^* A_{m(N^*)+1,m(N^*)+N^*} = \sum_{s=0}^{r-1} \sum_{t=m(N^*)+sn^*+1}^{m(N^*)+(s+1)n^*} a_t + \sum_{t=m(N^*)+rn^*+1}^{m(N^*)+N^*} a_t$$

$$\geq r \cdot n^* A_{m(n^*)+n^*} \geq r \cdot n^* (\underline{\underline{A}} - \varepsilon)$$

$$\geq (N^* - n^*)(\underline{\underline{A}} - \varepsilon) \geq N^*(1 - \varepsilon)(\underline{\underline{A}} - \varepsilon).$$

Finally, by changing the order of operations we get four more definitions, however, they give nothing new. In fact,

$$\inf_{m} \varliminf_{n\to\infty} A_{m+1,m+n} = \sup_{m} \varliminf_{n\to\infty} A_{m+1,m+n} = \underline{A} \qquad (6.28)$$

$$\inf_{m} \varlimsup_{n\to\infty} A_{m+1,m+n} = \sup_{m} \varlimsup_{n\to\infty} A_{m+1,m+n} = \overline{A}, \qquad (6.29)$$

because for an m_0 close to an optimal phase the first m_0 positions don't affect the asymptotic behaviour.

The list of quantities considered is not intended to be complete in any sense, but serves our illustration.

We look now at $\underline{\underline{A}} \leq \underline{A} \leq \overline{A} \leq \overline{\overline{A}}$ in four examples to see what constellations of values can occur.

We describe a 0–1-sequence $(a_n)_{n=1}^{\infty}$ by the lengths of its alternating strings of 1's and 0's: $(k_1, \ell_1, k_2, \ell_2, k_3, \dots)$

Examples

1. $k_t = k$, $\ell_t = \ell$ for $t = 1, 2, \ldots$; a periodic case:

$$\underset{=}{A} = \underline{A} = \overline{A} = \overset{=}{A} = \frac{k}{k + \ell}.$$

2. $k_t = \ell_t = t$ for $t = 1, 2, \ldots$. Use $\sum_{t=1}^{n} k_t = \sum_{t=1}^{n} \ell_t = \frac{n(n+1)}{2}$ and verify

$$0 = \underset{=}{A} < \frac{1}{2} = \underline{A} = \overline{A} < 1 = \overset{=}{A}.$$

3. $k_t = \sum_{s=1}^{t-1} k_s$, $\ell_t = \sum_{s=1}^{t-1} \ell_s$ for $t = 1, 2, \ldots$

$$0 = \underset{=}{A} < \frac{1}{2} = \underline{A} < \frac{2}{3} = \overline{A} < 1 = \overset{=}{A}.$$

Here all four values are different.

4. $k_t = \sum_{s=1}^{t-1} k_s$, $\ell_t = t$ for $t = 2, 3, \ldots, k_1 = 1$

$$0 = \underset{=}{A} < 1 = \underline{A} = \overline{A} = \overset{=}{A}.$$

All four quantities say something about $(A_n)_{n=1}^{\infty}$, they all say less than the *full record*, the sequence itself (corresponding to our capacity function).

6.1.10 Some Comments on a Formula for the Pessimistic Capacity

A noticeable observation of Verdú and Han [48] is that $\underline{C}$ can be expressed for every channel $\mathcal{K}$ in terms of a stochastic limit (per letter) mutual information.

The renewed interest in such questions originated with the Theory of Identification, where converse proofs for the DMC required that output distributions of a channel, generated by an arbitrary input distribution (randomized encoding for a message), be "approximately" generated by input distributions of controllable sizes of the carriers. Already in [12] it was shown that essentially sizes of $\sim e^{Cn}$ would do and then in [31, 32], the bound was improved (strong converse) by a natural random selection approach. They termed the name "resolvability" of a channel for this size problem.

The approximation problem (like the rate distortion problem) is a "covering problem" as opposed to a "packing problem" of channel coding, but often these problems are very close to each other, actually ratewise identical for standard channels like the DMC. To establish the strong second order identification capacity for more general

channels required in the approach of [31] that resolvability must equal capacity and
for that the strong converse for $\mathcal{K}$ was needed.

This led them to study the ADMC [3], which according to Han [28] plaid a key
role in the further development. Jacobs has first shown that there are channels with a
weak converse, but without a strong converse. In his example the abstract reasoning
did not give a channel capacity formula. This is reported in [37] and mentioned in
[3], from where the following facts should be kept in mind.

1. The ADMC has no strong converse but a weak converse (see Sect. 6.1.4 for precise
   terminology).
2. The term weak capacity was introduced.
3. The weak capacity (and also the λ-capacity were determined for the ADMC by
   linking it to the familiar max min -formula for the compound channel in terms of
   (per letter)-mutual information.
4. It was shown that $\lim_{n\to\infty} \frac{1}{n} \max_{X^n} I(X^n \wedge Y^n)$ does not describe the weak capacity in general. Compare this with Wyner's first capacity definition in Sect. 6.1.6.
5. It was shown that Fano's inequality, involving only the *average* mutual information $I(X^n \wedge Y^n)$, fails to give the weak converse for the ADMC.

The observation of [48] is again natural, one should use the information function
of the ADMC directly rather than the max min -formula. They defined for general $\mathcal{K}$
the *sequence* of pairs

$$(\mathbf{X}, \mathbf{Y}) = (X^n, Y^n)_{n=1}^{\infty} \tag{6.30}$$

and

$$\underline{I}(\mathbf{X} \wedge \mathbf{Y}) = \sup \left\{ \alpha : \varliminf_{n\to\infty} \Pr \left\{ (x^n, y^n) : \frac{1}{n} I(x^n, y^n) \leq \alpha \right\} = 0 \right\}. \tag{6.31}$$

Their general formula asserts

$$\underline{C} = \sup_{\mathbf{X}} \underline{I}(\mathbf{X} \wedge \mathbf{Y}). \tag{6.32}$$

The reader should be aware that

α. The stochastic inequalities used for the derivation (10.3) are both (in particular
   also Theorem 4 of [48]) not new.
β. Finally, there is a very important point. In order to show that a certain quantity
   K (for instance $\sup_{\mathbf{X}} \underline{I}(\mathbf{X} \wedge \mathbf{Y})$) equals $\underline{C}$ one has to show $K \geq \underline{C}$ and then (by
   definition of $\underline{C}$) that $K + \delta$, any $\delta > 0$, is not a rate achievable for arbitrary small
   error probabilities or equivalently, that $\inf_{\lambda} \varliminf_{n\to\infty} \log M(n, \lambda) < K + \delta$. For
   this one does *not need* the *weak* converse (ii) $\inf_{\lambda} \varlimsup_{n\to\infty} \log M(n, \lambda) \leq K$, but
   only

$$\inf_{\lambda} \varliminf_{n\to\infty} \log M(n, \lambda) \leq K \tag{6.33}$$

(see also Sect. 6.1.4) The statement may be termed the "weak-weak converse" or the "weak-converse" or "occasional-converse" or whatever. Keep in mind that the fact that the weak converse does not hold for the factors led to the "information theoretic perpetuum mobile". The remark on page 1153 "Wolfowitz ... referred to the conventional capacity of Definition 1 (which is always defined) as *weak capacity*" is not only wrong, because Wolfowitz never used the term "weak capacity", it is—as we have explained—very misleading. After we have commented on the drawbacks of the pessimistic capacity, especially also for channel NDMC, we want to say that on the other hand the formula $\sup_X \underline{I}(\mathbf{X} \wedge \mathbf{Y})$ and also its dual $\sup_X \overline{I}(\mathbf{X} \wedge \mathbf{Y})$ are helpful in characterizing or bounding quantities of interest not only in their original context, Theory of Identification. Han has written a book [29] in which he introduces these quantities and their analogues into all major areas of Information Theory.

6.1.11 Pessimistic Capacity Functions

We think that the following concept suggests itself as one result of the discussion.

Definition 31   A sequence $(C_n)_{n=1}^{\infty}$ of non-negative numbers is *a* capacity sequence of $\mathcal{K}$, if

$$\inf_{\lambda>0} \varliminf_{n\to\infty} \left(\frac{1}{n} \log M(n, \lambda) - C_n \right) = 0.$$

The sequence $(\underline{C}, \underline{C}, \underline{C}, \dots)$ is a capacity sequence, so by definition there are always capacity sequences.

Replacing α by α_n in (6.31) one can characterize capacity sequences in term of sequences defined in terms of (per letter) information functions. Every channel $\mathcal{K}$ has a class of capacity sequences $\mathcal{C}(\mathcal{K})$.

It can be studied. In addition to the constant function one may look for instance at the class of functions of period m, say $\mathcal{C}(\mathcal{K}, m) \subset \mathcal{C}(\mathcal{K})$. More generally complexity measures μ for the sequences may be used and accordingly one gets say $\mathcal{C}(\mathcal{K}, \mu \le \rho)$, a space of capacity functions of μ-complexity less than ρ.

This seems to be a big machinery, but channels $\mathcal{K}$ with no connections between W^n and $W^{n'}$ required in general constitute a *wild* class of channels. The capacity sequence space $\mathcal{C}(\mathcal{K})$ characterizes a channel in time like a capacity region for multiway channels characterizes the possibilities for the communicators.

Its now not hard to show that for the product channel $\mathcal{K}_1 \times \mathcal{K}_2$ for any $f \in \mathcal{C}(\mathcal{K}_1 \times \mathcal{K}_2)$ there exist $f_i \in \mathcal{C}(\mathcal{K}_i)$; $i = 1, 2$; such that $f_1 + f_2 \ge f$. The component channels together can do what the product channel can do. This way, both, the nonstationarity and perpetuum mobile problem are taken care of.

We wonder how all this looks in the light of "quantum parallelism".

We finally quote statements by Shannon. In [46] he writes "Theorem 4, of course, is analogous to known results for the ordinary capacity C, *where the product channel has the sum of the ordinary capacities* and the sum channel has an equivalent number of letters equal to the sum of the equivalent numbers of letters for the individual channels. We conjecture, but have not been able to prove, that the equalities in Theorem 4 hold in general—not just under the conditions given". Both conjectures have been disproved (Haemers and Alon).

6.2 Lecture on Comments to "On Concepts of Performance Parameters for Channels"

This contribution has appeared in the preprint series of the SFB 343 at Bielefeld University in the year 2000 [6].

Its declared purpose was to open a discussion about basic concepts in Information Theory in order to gain more clarity. This had become necessary after the work [12] for identification gave also new interpretation to classical theory of transmission leading in particular to "A general formula for channel capacity" by S. Verdú and T.S. Han, which however, does not cover the results of [2].

The final answer is not found until now.

We give now a brief reaction to some comments and criticism we received.

1. Strong objections were made against statement α) after (6.32), which implies the claim that the inequality

$$\overline{\lambda} \geq F(\log M - \Theta) - e^{-\Theta}, \ \Theta > 0 \tag{6.34}$$

is essentially not new. Particularly it has been *asserted that this inequality is not comparable to Shannon's*

$$\overline{\lambda} \geq \frac{1}{2}F\left(\log \frac{M}{2}\right) \tag{6.35}$$

in (6.17) of Shannon's Theorem, because it is stronger.

Therefore we have to justify our statement by a proof. Indeed, just notice that Shannon worked with the constant $\frac{1}{2}$ for simplicity in the same way as one usually extracts from a code with average error probability $\overline{\lambda}$ a subcode of size $\frac{M}{2}$ with maximal probability of error not exceeding $2\overline{\lambda}$. However, again by the pigeon-hole principle for any $\beta \in (0, \frac{1}{2})$ there are $M\beta$ codewords with individual error probabilities $\leq \frac{\overline{\lambda}}{1-\beta}$. (This argument was used in [1, 2, 13]).

Now just replace in Shannon's proof $\frac{1}{2}$ by β to get the inequality

$$\overline{\lambda} \geq (1 - \beta)F(\log M\beta). \tag{6.36}$$

Equating now $F(\log M - \Theta)$ with $F(\log M \beta)$ we get $\beta = e^{-\Theta}$ and it suffices to show that

$$(1 - e^{-\Theta})F(\log M - \Theta) \geq F(\log M - \Theta) - e^{-\Theta}.$$

Indeed $e^{-\Theta} \geq e^{-\Theta}F(\log M - \Theta)$, because F is a distribution function. So (6.36) is even slightly stronger than (6.34).   □

For beginners we carry out the details.

Introduce $W^*(u|y) = \frac{\tilde{P}(u,y)}{Q(y)}$ and notice that $I(u, y) \leq \log M\beta$ is equivalent with $\frac{W^*(u|y)}{P(u)} \leq M\beta$ or, since $P(u) = M^{-1}$,

$$W^*(u|y) \leq \beta. \tag{6.37}$$

Now concentrate attention on those pairs (u, y) for which (6.37) holds.

Consider the bipartite graph $G = (\mathcal{U}, \mathcal{Y}, \mathcal{E})$ with vertix sets $\mathcal{U} = \{u_1, \ldots, u_M\}$, $\mathcal{Y}$, and edge set $\mathcal{E} = \{(u, y) : W^*(u|y) \leq \beta\}$

Clearly,

$$\tilde{P}(\mathcal{E}) = F(\log M\beta) \tag{6.38}$$

We partition now $\mathcal{Y}$ into

$$\mathcal{Y}_+ = \{y \in \mathcal{Y} : \text{ exists } u \text{ with } W^*(u|y) > \beta\}, \ \mathcal{Y}_- = \mathcal{Y}\backslash\mathcal{Y}_+ \tag{6.39}$$

and correspondingly we partition $\mathcal{E}$ into

$$\mathcal{E}_+ = \{(u, y) \in \mathcal{E} : y \in \mathcal{Y}_+\}, \ \mathcal{E}_- = \mathcal{E}\backslash\mathcal{E}_+. \tag{6.40}$$

Clearly $\tilde{P}(\mathcal{E}) = \tilde{P}(\mathcal{E}_+) + P(\mathcal{E}_-)$. For $y \in \mathcal{Y}_+$ ML-decoding chooses a u with $W^*(u|y) > \beta$, but (u, y) is not in $\mathcal{E}$ and not in $\mathcal{E}_+$. Therefore all $(u', y) \in \mathcal{E}_+$ contribute to the error probability. The total contribution is $\tilde{P}(\mathcal{E}_+)$.

The contribution of the edges in $\mathcal{E}_-$ to the error probability is, if f is the ML-decoder,

$$\sum_{y \in \mathcal{Y}_-} Q(y)(1 - W^*(f(y)|y)) \geq \tilde{P}(\mathcal{E}_-)(1 - \beta)$$

and hence

$$\bar{\lambda} \geq \tilde{P}(\mathcal{E}_+) + \tilde{P}(\mathcal{E}_-)(1 - \beta)$$

(even slightly stronger than 6.36), written explicitly

$$\bar{\lambda} \geq \tilde{P}(\{(u, y) : \log \frac{W(y|u)}{Q(y)} \leq \log M - \Theta\}) - e^{\Theta} \tilde{P}(\{(u, y) :$$

$$\textit{for all } u' \in \mathcal{U} \ \log \frac{W(y|u')}{Q(y)} \leq \log M - \Theta\}).$$

For those who do not accept it as Shannon's result it would be only consequential to name it then the Shannon/Ahlswede inequality.

2. We also have some good news. In Sect. 6.1.5 we argued that the optimistic capacity concept seems absurd and we provided convincing examples.

Investigations [10] in Cryptography made us aware that this concept, a dual to the pessimistic capacity, finds a natural place here, because *one wants to protect also against enemies having fortunate time for themselves in using their wire-tapping channel!*

3. Finally, being concerned about performance criteria, we should not forget that in Information Theory, similarly as in Statistics, asymptotic theory gives a first coarse understanding, but never should be the last word.

In particular with the availability of a lot of computing power not only small, but even medium size samples call for algorithmic procedures with suitable parameters. This was the message from Ziv in his contribution [52].

4. S. Dodunekov in [20] explained that for linear codes with parameters block-length n, dimension k and minimal distance d, fixing two parameters and optimizing the third of the quantities (a) $N_q(k, d)$, (b) $K_q(n, d)$, (c) $D_q(n, k)$ *the first one gives the most accurate description.*

5. It has been pointed out that there are different concepts of capacity, but that usually in a chapter it is clearly explained which is used and a definition is given. It can be felt from those reactions that Doob's famous (and infamous with respect to Shannon and his work) criticism [21] about the way in which Information Theory proceeds from Coding Theorem n to Coding Theorem $n + 1$ keeps people alert.

That mostly researchers know which concepts they are using and that they sometimes even give definitions, that is still not enough.

For instance we have been reminded that our statement "there seems to be no definition of the weak converse in the book [17]" is wrong and that a look at the index leads to the problem session, where a definition is given. This is correct.

However, it is not stated there on page 112 that this definition of a converse, referred to by us as "weak weak converse, ... or ...", is not the definition, which was established in Information Theory at least since the book [49] by Wolfowitz came out in 1961 and was used at least by most of the Mathematicians and Statisticians working in the field.

Unfortunately this happened even though since 1974 we had many discussions and joint work with Körner and Csiszár, a great part of which entered the book and influenced the shape of the later parts of the book. It was also given to us for reading prior to publication and we have to apologize for being such a poor reader. Otherwise we would have noticed what we only noticed in 1998.

It is clear now why [2] and [3] are not cited in the book, because they don't fit into the frame.

This frame became the orientation for people starting to learn Information Theory via typical sequences. Shannon's stochastic inequalities ([47]) perhaps were not taught anymore.

6. We know that the weak capacity has the additivity property for parallel channels. We draw attention to the fact that *Shannon (and also later Lovász) conjectured* this property to hold also for his zero-error capacity (which was disproved by Haemers [27]). Apparently, Shannon liked to have this property!

We all do, often naively, *time-sharing*, which is justified, if there is an additivity property!

We like to add that without thinking in terms of time-sharing we never would have discovered and proved our characterization of the (weak) capacity region for the MAC in [4] (with a different proof in [5]).

So our message is "Shannon also seems to think that additivity is an important property" and not "Shannon made a wrong conjecture".

The additivity property for quantum channels is of great interest to the community of Quantum Informationtheorists. This led M. Horodecki to his question quoted in [36]. The answer is positive for degraded channels, but not in general!

7. Once the situation is understood it is time to improve it. We suggest below a unified description of capacity concepts with conventions for their notations.

In every science it is occasionally necessary to agree on some standards—a permanent fight against the second law of thermodynamics. We all know how important the settings of such standards have been in Physics, Chemistry, Biology etc. Every advice to the standards proposed here is welcome.

We start here with the case corresponding to B) under 4. above and restrict ourself to one-way channels.

We consider a general channel $\mathcal{K}$ with time structure, which is defined in (6.2) of [6]. Recall that for a positive integer n and a non-negative real number λ. $N(n, \lambda)$ denotes for $\mathcal{K}$ the maximal cardinality of a code of block-length n with an error probability not exceeding λ.

Often λ depends on n and we write $\lambda(n)$. The sequence $\{\lambda(n)\}$ is typically of one of the following forms

(i) $\lambda(n) = 0$ for all $n \in \mathbb{N}$
(ii) $\lambda(n) = \lambda, 0 < \lambda \leq 1$, for all $n \in \mathbb{N}$
(iii) $\lambda(n) = n^{-\alpha}, 0 < \alpha$, for all $n \in \mathbb{N}$
(iv) $\lambda(n) = e^{-En}, 0 < E$, for all $n \in \mathbb{N}$

We speak about zero, constant, polynomial, and exponential error probabilities.

With the sequence $\{N(n, \lambda(n))\}$ we associate a very basic and convenient performance parameter for the channel $\mathcal{K}$ the

rate-error function $R : \Lambda \rightarrow \mathbb{R}_+,$

where Λ is the space of all non-negative real-valued sequences and for every $\{\lambda(n)\} \in \Lambda$.

$R(\{\lambda(n)\})$ is the largest real number with $\frac{1}{n} \log N(n, \lambda(n)) \geq R(\{\lambda(n)\}) - \delta$ for every $\delta > 0$ and all large n.

How does it relate to capacities? In the four cases described we get for $R(\{\lambda(n)\})$ the values

  (i') $C(0)$, that is, Shannon's zero error capacity
  (ii') $C(\lambda)$, that is, the λ-capacity introduced in [13]
  (iii') $C(\alpha)$, that is, a novel α-capacity
  (iv') $C(E)$, that is, the E-capacity introduced by Evgueni Haroutunian in [33].

A word about notation is necessary. The functions $C(0)$, $C(\lambda)$, $C(\alpha)$, and $C(E)$ are distinguished only by their arguments, these will always appear explicitly. All our results have to be interpreted with this understanding.

This convention was made already in [13] where not only the maximal error probability λ but also the average error probability $\overline{\lambda}$, the maximal error probability λ_R for randomized encoding, and the average error probability $\overline{\lambda_R}$ for randomized encoding were considered. For example, one of our theorems in [13] says that

$$C(\lambda_R) = C(\overline{\lambda}) = C(\overline{\lambda_R}).$$

under certain conditions where $\lambda_R = \overline{\lambda} = \overline{\lambda_R}$. Taken literally this is a trivial statement. In the light of our notation it means that these functions coincide for certain values of the argument. This notation result is no confusion or ambiguity, and has the advantage of suggestiveness new and typographical simplicity.

An important point about the introduced rate-error function and the capacities is their existence for every channel $\mathcal{K}$.

The same is the case for the (ordinary) capacity

$$C = \inf_{0 < \lambda \leq 1} C(\lambda).$$

Our rate-error function may be called pessimistic and it has an optimistic twin $\overline{R}(\{\lambda(n)\})$, the largest real number with

$$\overline{\lim}_{n \to \infty} \frac{1}{n} \log N(n, \lambda(n)) \geq \overline{R}(\{\lambda(n)\}).$$

Correspondingly we get the optimistic capacities $\overline{C}(0)$, $\overline{C}(\lambda)$, $\overline{C}(\alpha)$, $\overline{C}(E)$, and $\overline{C}$. Of course for a DMC $\overline{C}(0) = C(0)$ etc.

They are relevant, for example, if a wire-tapper chooses the times of an attack.

Again all these quantities exist. Moreover, the error criteria $\overline{\lambda}$, λ_R, $\overline{\lambda_R}$ lead to analoga of the capacities in (iii)–(iv'), namely, $C(\overline{\alpha})$, $C(\alpha_R)$, $C(\overline{\alpha_R})$, $C(\overline{E})$, $C(E_R)$, and $C(\overline{E_R})$ and similarly for $\overline{C}$. In [13] compound channels, a certain class of channels, were considered. The concepts are even more relevant for the more sophisticated AVC, for example.

8. Well-known is Shannon's *rate-distortion function* in source coding. It is amazing that our preceding analogue for channel coding was not introduced. However, it must be said that Shannon introduced his function "informationally" (and so does Toby Berger in [15]) and not "operational" as we did. In channel coding he gave two definitions for the channel capacity, an informational and an operational one. This is very well discussed in the explanation of Aaron Wyner, which we cite in [6]. Unfortunately, some well-known text books like [16] or [51] give the informational one. But $\{\max_{P^n} I(W^n|P^n)\}$ describes the operational capacity C only in special cases like the DMC and is too large for instance for averaged channels (memory!). Here lies one of the main roots for conceptional confusions about channel capacity!

9. As we have explained earlier a nice property to have is additivity of a capacity for parallel channels. This is the case for the ordinary capacity C, if $C = \overline{C}$ and this is exactly in the case where the *weak converse* holds. We also say in this case that the *weak capacity* exists (see [3]). *So this quantity does not exist automatically.* This is sometimes overlooked and even more true for the strong capacity introduced in [3], when the strong converse holds.

This must be kept in mind when we compare results. Generality is of course easier to obtain for capacities with weaker properties than for those with stronger properties.

In proving upper bounds like the weak converse it is helpful to prove first bounds, which can be obtained easier, like polynomial or exponential (see also soft converse in [12]) weak converses discussed in [7].

10. We just indicate that in the spirit of [6] one should introduce also

rate sequence − error functions

• saying in particular much more about non-stationary channels than rate-error functions,

classes of rate sequence − error functions

• catching tighter descriptions suggested in Sect. 11 of [6].

Of course associated with these performance criteria are capacity concepts.

11. The discussion should be continued and should some time include other performance criteria like the analoga to (i) and (iii) above. Also analoga of (i), (ii), (iii) for combinatorial channel models and also criteria for sources are to be classified.

It should be specific about distinctions stemming from multi-way channels—feedback situations—non-block codes—delay—synchronization.

Beyond capacities for Shannon's transmission there are those for identification (second order)—common randomness—general information transfer (first and second order).

In combinatorial channel models to be distinguished are the various error concepts:

failure of error detection or wrong detection–defects—insertions–deletions
—localized errors–unidirectional errors—etc.

The case of feedback just brings us to search (see Chap. 6) with the recently
studied lies with cost constraints etc.

Also J. Körner's models of a combinatorial universe with information aspects,
starting with ambiguous alphabets and going along trifferent paths, definitely must
be included (Sperner Capacity).

Recent work by G. Katona and K. Tichler in search deserves immediate attention.
There a test is a partition of the search space $\mathcal{X}$ into 3 sets $(\mathcal{Y}, \mathcal{N}, \mathcal{A})$. If the object
x searched for satisfies $x \in \mathcal{Y}$ the answer is Yes, if it satisfies $x \in \mathcal{N}$ the answer is
No, it is for $x \in \mathcal{A}$ arbitrary Yes or No.

Also to be classified are the performance criteria in the very important work
on codes introduced by Kautz/Singleton [39] and studied by Lindström, Dyachkov,
Erdös and many others (see survey [26]).

The results found an application in the probabilistic model of K-identification
in [7].

Finally, analogous performance criteria are to be defined in Statistics in particular
in the interplay between Multi-user Source Coding Theory and Hypothesis Testing
or Estimation starting with work [8, 11], with Csiszar and Burnashev and continued
by many others (see survey [30] by Han and Amari).

6.3 Lecture on the "Capacity" of the Product Channel

6.3.1 The Strong Converse of the Coding Theorem

Central in the proof of the strong converse for the DMC is the following lemma. As
Fano's Lemma it does not make use of the time structure, i.e., the block length n is
not involved and can hence be chosen w.l.o.g. as $n = 1$.

Let $\mathcal{X}$ and $\mathcal{Y}$ denote finite sets, $W = \left(w(y|x)\right)_{x \in \mathcal{X}, y \in \mathcal{Y}}$ a stochastic matrix and let
Q be an arbitrary probability distribution on $\mathcal{Y}$. Further for all $x \in \mathcal{X}$ we have some
$\Theta_x > 0$ and define

$$ B_x(\Theta_x, Q) \triangleq \left\{ y \in \mathcal{Y} : \frac{w(y|x)}{Q(y)} \geq 2^{\Theta_x} \right\}. $$

Lemma 56 *If for an (N, λ)-code $\left\{(u_i, D_i) : i = 1, \ldots, N\right\}$ ($u_i \in \mathcal{X}, D_i \subset \mathcal{Y}$ for
all i, $D_i \cap D_j = \varnothing$ for all $i \neq j$) there exists a probability distribution Q on $\mathcal{Y}$ with
$\max_{u_i} \sum_{y \in B_{u_i}(\Theta_{u_i}, Q)} w(y|u_i) < \gamma$ and $\lambda + \gamma < 1$, then*

$$ N < (1 - \lambda - \gamma)^{-1} 2^{\frac{1}{N} \sum_{i=1}^{N} \Theta_{u_i}}. $$

Proof We shortly write Θ_i instead of Θ_{u_i} and set for $i = 1, \ldots, N$

$$A_i \triangleq \left\{ y \in D_i : \frac{w(y|u_i)}{Q(y)} < 2^{\Theta_i} \right\} = D_i \cap \left(B_{u_i}(\Theta_i, Q) \right)^c. \qquad \square$$

For $y \in A_i$ then by definition $2^{\Theta_i} Q(y) > w(y|u_i)$ and hence for all i

$$2^{\Theta_i} Q(D_i) \geq 2^{\Theta_i} Q(A_i) > w(A_i|u_i) = w(D_i|u_i) - w(D_i \setminus A_i|u_i)$$

$$\geq w(D_i|u_i) - B_{u_i}(\Theta_i, Q) \geq 1 - \lambda - \gamma$$

by the assumptions. Division by $Q(D_i)$ and taking logarithm on both sides yields for all $i = 1, \ldots, n$ $\Theta_i \geq \log\left(\frac{1-\lambda-\gamma}{Q(D_i)}\right)$ and hence

$$\frac{1}{N} \sum_{i=1}^{N} \Theta_i \geq \frac{1}{N} \sum_{i=1}^{N} \log\left(\frac{1-\lambda-\gamma}{Q(D_i)}\right) = -\sum_{i=1}^{N} \log Q(D_i) + \log(1-\lambda-\gamma)$$

$$> -\sum_{i=1}^{N} \frac{1}{N} \log \frac{1}{N} + \log(1-\lambda-\gamma) = \log N + \log(1-\lambda-\gamma),$$

since the relative entropy $\sum_{i=1}^{N} \frac{1}{N} \log\left(\frac{\frac{1}{N}}{Q(D_i)}\right) > 0$.

Exponentiation on both sides yields the statement of the lemma. $\qquad \square$

Theorem 39 (Strong Converse for the DMC) *For all $\lambda \in (0, 1)$*

$$\varlimsup_{n \to \infty} \frac{1}{n} \log N(n, \lambda) \leq C.$$

The strong converse for the DMC was first proved by Wolfowitz. The following proof is due to Kemperman.

Proof With the notation of the preceding lemma we now choose $\mathcal{X} = \mathcal{X}^n$, $\mathcal{Y} = \mathcal{Y}^n$ and $W = W^n$ and let $\{(u_i, D_i) : i = 1, \ldots, N\}$ be an (n, N, λ)-code. We partition $\mathcal{U} \triangleq \{u_1, \ldots, u_N\}$ into subcodes $\mathcal{U}(P)$, where each code word in $\mathcal{U}(P)$ is of type $P \in \mathcal{P}$. $\qquad \square$

Let the type P^* be such that $\mathcal{U}(P^*)$ is a subcode of maximum length $M \triangleq |\mathcal{U}(P^*)|$. Then $N \leq (n+1)^{|\mathcal{X}|} \cdot M$.

Further let $Q \triangleq P^* \cdot W$. In each code word $u = (u_{(1)}, \ldots, u_{(n)}) \in \mathcal{U}(P^*)$ by definition every $x \in \mathcal{X}$ has frequency exactly $< u|x >= P^*(x) \cdot n$. We denote by $\mathbb{E}_P$ and Var_P the expected value and the variance of a random variable on a set $\mathcal{X}$ with probability distribution P.

$$\mathbb{E}_{w(\cdot|u)} \log \frac{w(\cdot|u)}{Q^n(\cdot)} = \sum_{t=1}^{n} \mathbb{E}_{w(\cdot|u_{(t)})} \log \frac{w(\cdot|u_{(t)})}{Q(\cdot)}$$

$$= \sum_{t=1}^{n} \sum_{y \in \mathcal{Y}} w(y|u_{(t)}) \log \frac{w(y|u_{(t)})}{Q(y)}$$

$$= \sum_{x \in \mathcal{X}} \sum_{y \in \mathcal{Y}} n P^*(x) w(y|x) \cdot \log \frac{w(y|x)}{Q(y)}$$

$$= n \cdot I(P^*, W).$$

Next we shall show that $\mathrm{Var}_{w(\cdot|u)} \cdot \log \frac{w(\cdot|u)}{Q^n(\cdot)}$ is bounded.

$$\mathrm{Var}_{w(\cdot|u)} \log \frac{w(\cdot|u)}{Q^n(\cdot)} = \sum_{t=1}^{n} \mathrm{Var}_{w(\cdot|u_{(t)})} \log \frac{w(\cdot|u_{(t)})}{Q(\cdot)}$$

$$\leq \sum_{t=1}^{n} \mathbb{E}_{w(\cdot|u_{(t)})} \left(\log \frac{w(\cdot|u_{(t)})}{Q(\cdot)} + \log P^*(u_{(t)}) \right)^2$$

$$\text{(since } \mathrm{Var}X \leq \mathbb{E}(X+a)^2 \text{ for all } a \in \mathbb{R})$$

$$= \sum_{t=1}^{n} \sum_{y \in \mathcal{Y}} w(y|u_{(t)}) \cdot \left(\log \frac{P^*(u_{(t)}) w(y|u_{(t)})}{Q(y)} \right)^2$$

$$= \sum_{y \in \mathcal{Y}} n P^*(u_{(t)}) w(y|u_{(t)}) \cdot \left(\log \frac{P^*(u_{(t)}) w(y|u_{(t)})}{Q(y)} \right)^2$$

$$= n \sum_{y \in \mathcal{Y}} Q(y) \frac{P^*(u_{(t)}) \cdot w(y|u_{(t)})}{Q(y)} \left(\log \frac{P^*(u_{(t)}) w(y|u_{(t)})}{Q(y)} \right)^2.$$

Now observe that the function $x \cdot \log^2 x$ is uniformly continuous and therefore bounded by some constant. Hence the function $\sum_{i=1}^{b} x_i \log^2 x_i$ is also bounded for every fixed b. It can be shown that $\sum_{i=1}^{b} x_i \log^2 x_i \leq \max\{\log^2(3), \log^2(b)\}$, when $\sum_{i=1}^{b} x_i = 1$. Hence

$$\mathrm{Var}_{w(\cdot|u)} \log \frac{w(\cdot|u)}{Q^n(u)} \leq n \cdot c, \text{ where } c \triangleq \max\{\log^2 3, \log^2 |y|\}.$$

Now choose for all $i = 1, \ldots, N$ $\Theta_i = \Theta \triangleq nI(P^*, W) + \sqrt{\frac{2n}{1-\lambda}} c$ and $Q \triangleq Q^n$. Then by Chebyshev's Inequality

$$w\big(B_u(\Theta, Q)|u\big) = w\left(\frac{w(\cdot|u)}{Q^n(\cdot)} \geq 2^{\Theta}|u\right)$$

$$= w\left(\log\frac{w(\cdot|u)}{Q^n(\cdot)} - \mathbb{E}_{w(\cdot|u)}\log\frac{w(\cdot|u)}{Q^n(\cdot)} \geq \sqrt{\frac{2nc}{1-\lambda}}\bigg|u\right)$$

$$\leq \frac{nc}{\left(\sqrt{\frac{2nc}{1-\lambda}}\right)^2} = \frac{1-\lambda}{2}.$$

If we use the preceding lemma with $\gamma \triangleq \frac{1-\lambda}{2}$ we obtain

$$N \leq \left(1 - \lambda - \frac{1-\lambda}{2}\right)^{-1} 2^{nI(P^*,W) + \sqrt{\frac{2nc}{1-\lambda}}} = \frac{2}{1-\lambda}\exp\left\{nI(P^*,W) + \sqrt{\frac{2nc}{1-\lambda}}\right\}.$$

From this we can conclude that

$$N \leq \exp\left\{Cn + \sqrt{\frac{2nc}{1-\lambda}} + a\log(n+1) + \log\frac{2}{1-\lambda}\right\}.$$

This proves the strong converse.   □

6.3.2 On the "Capacity" of the Product of Channels

1. Wyner's Approach

As in Lecture on Chap. 4 we define a parallel channel $W_1 \times \cdots \times W_k$. Each component is abstract, with or without memory, continuous or discrete, but with time structure.

Definition 32 A real number R is said to be a permissible rate of transmission for W if for every $\delta > 0$ and for n sufficiently large, there exists a code with parameter n with $M = \exp Rn$ codewords and error probability $\overline{\lambda} \leq \delta$. Since $R = 0$ is a permissible rate, the set of permissible rates is not empty. We define the channel capacity C as the supremum of permissible rates (see [CK])!

Definition 33 The strong converse holds if for any fixed $R > C \lim_{n\to\infty} \lambda(n, e^{Rn}) = 1$. Equivalently, for any $R > C$ and $\delta < 1$ and n sufficiently large, any (n, e^{Rn}) code has $\lambda \geq \delta$.

Remark

1. Augustin [14] has proved that these channels have a coding theorem with a weak converse with capacity C^*. However, they do not always have a strong converse. So there may be a lack of uniformity for $0 \leq \varepsilon < 1$, however we know now that there is always this uniformity for $0 \leq \varepsilon \leq \varepsilon_0$.

Actually this occured under different circumstances for averaged channels and for compound channels for average errors.

2. Interesting is also a quotation from [50], where Wyner writes

> Let us remark here that there are two ways in which "channel capacity" is commonly defined. *The first definition* takes the channel capacity to be the supremum of the "information" processed by the channel, where "information" is the difference of the input "uncertainty" and the "equivolation" at the output. The second definition, which is the one we use here, takes the channel capacity to be the maximum "error free rate". For certain classes of channels (e.g., memoryless channels, and finite state indecomposable channels) it has been established that these two definitions are equivalent. In fact, this equivalence is the essence of the Fundamental Theorem of Information Theory...

Wyner states that

Theorem 40 *(i) The capacity of the product channel (or parallel channel, or special nonstationary channel) is the sum of the capacities of the component channels.*

(ii) The "strong converse" holds for the product channel, if it holds for each of the component channels.

Proof Inductively it suffices to consider two factors

(i) To show that any $R < C_1 + C_2$ is permissible choose $R_1 + R_2 = R$, $R_i < C_i (i = 1, 2)$ and concatenate existing $(n, e^{R_i n})$ codes

$$\{(u_j^i, D_j^i) : 1 \leq j \leq M_i\}, \quad M_i = e^{R_i n}, i = 1, 2,$$

with error probabilities λ_i and get

$$\{(u_j^1, u_k^2, D_j^1 \times D_k^2) : 1 \leq j \leq M_1, 1 \leq k \leq M_2\}$$

with error probability $\lambda \leq \lambda_1 + \lambda_2$.

Thus $C^2 \geq C_1 + C_2$. For the "converses" use the fact that from a listcode one can select at random a subcode (as in 5.3.3) such that the following relations hold.   □

Lemma 57 *Let W^n be a channel with input space $\mathcal{X}^n = \prod_1^n \mathcal{X}$ and output space $\mathcal{Y}^n = \prod_1^n \mathcal{Y}$, then for $M = e^{Rn}$, $L = e^{R_L n}$ and $R_1 < R - R_L$, $M_1 = e^{R_1 n}$*

$$\lambda(1, M_1, n) \leq \lambda(L, M, n) + \varepsilon(n), \tag{6.41}$$

where $\lim_{n \to \infty} \varepsilon(n) = 0$. (In particular, if the strong converse holds, then also $\lambda(e^{R_2 n}, e^{Rn}, n) \to 1$ as $n \to \infty$, if $R - R_L > C$.)

Additionally we use Lemma 60 in Lecture 7.1 and get in particular for the product channel

$$\lambda(1, M, n) \geq \lambda^{(1)}(L, M, n)\lambda^{(2)}(1, L, n). \tag{6.42}$$

Let now for the product channel be a code with rate $R > C_1 + C_2$.
Set $\eta = \frac{R - (C_1 + C_2)}{2} > o$, $R_L = C_2 + \eta$ and $L = e^{R_L n}$. From (6.42)

$$\lambda \geq \lambda(1, e^{Rn}, n) \geq \lambda^{(1)}(e^{R_L n}, e^{Rn}, n)\lambda^{(2)}(1, e^{R_L n}, n). \tag{6.43}$$

Since $R_L > C_2$, the capacity of the second channel, necessarily $\lambda^{(2)}(1, e^{R_L n}, n)$ is "bounded away from 0". Also by Lemma 57

$$\lambda^{(1)}(e^{R_L n}, e^{Rn}, n) \geq \lambda^{(1)}(1, e^{R_L n}, n) + \varepsilon(n) \tag{6.44}$$

with $\varepsilon(n) \to 0$ as $n \to \infty$ and $R_2 < R - R_L$.
   If we set $R_2 = C_1 + \frac{\eta}{2} = C_1 + \frac{R - C_1 - C_2}{4} = R - R_L - \eta/2 < R - R_L$ (6.44) is applicable.
   Since $R_2 > C_1$, the capacity of channel 1, $\lambda^{(1)}(1, e^{R_2 n}, n)$ is "also bounded away from zero". Hence by (6.44) $\lambda^{(1)}(e^{R_L n}, e^{Rn}, n)$ is also "bounded away from zero", so that by (6.43) λ is bounded away from zero and the weak converse is established.

(ii)   In this reasoning replacing "bounded away from zero" by "bounded below by a constant $\gamma < 1$". This holds for every $\gamma < 1$, the strong converse holds again.

Remarks

1. We actually proved (ii), but where is the flaw in the proof of (i)? (The product of two functions, each taking values 0 and 1 can be identical 0!)
2. "Usual" definitions of weak converse $\inf_{\lambda > 0} \overline{\lim}_{n \to \infty} |\frac{1}{n} \log M(n, \lambda) = C$ "cannot stay away from 0" again and again!

Example $C^2 > C_1 + C_2$ $W = \begin{pmatrix} \frac{1}{2} & \frac{1}{2} \\ \frac{1}{2} & \frac{1}{2} \end{pmatrix}$ $W' = \begin{pmatrix} 1 & 0 \\ 0 & 1 \end{pmatrix}$

$$
\begin{array}{ll}
\underline{W \ldots W \; W' \ldots W'} & \text{Channel 1} \\[4pt]
\underline{W' \ldots W' \; W \ldots W} & \text{Channel 2} \\[2pt]
\quad n_1 \qquad\quad n_2 & \\[4pt]
\qquad C^2 = 1 & n_{t+1} = n_t^2 \\[4pt]
C_1 = 0, \; C_2 = 0 &
\end{array}
$$

Shows error in Wyner's chapter (i). No strong converses rate $\frac{1}{2}$ has again and again error probability smaller than ε. Also no "weak converses" (Wyner's sense) $\frac{1}{2} > R > o$ $\exists \lambda(R) > o$ but $\lambda_n(R)$ fluctuates between 0^+ 1^-.
The capacity theorem is wrong!
More important: The "folklore" definition is paradoxical: Two channels have capacity 0, but combined they have capacity 1.

3. In my definition (for nonstationary channels) capacity functions of parallel channels *are additive* (modulo $o(n)$)!

2. Augustin's Approach

In [14] the formulation is this:

For an abstract channel $\mathcal{K}$ let $M(\mathcal{K}, \varepsilon)$ be the maximal size of an $(1 - \varepsilon)$-maximal probability of error code. Actually $M(\mathcal{K}, \varepsilon) = \sup\{M : M \text{ is length of an } (1-\varepsilon)\text{-code}$ for $\mathcal{K}\}$. Clearly, $1 \leq M(\mathcal{K}, \varepsilon) = \infty$. Let now $\mathcal{K}_1, \mathcal{K}_2$ be abstract channels and $\mathcal{K}^2 = \mathcal{K}_1 \times \mathcal{K}_2$.

Then obviously by concatenation

$$M(\mathcal{K}^2, \varepsilon_1 \cdot \varepsilon_2) \geq M(\mathcal{K}_1, \varepsilon_1) M(\mathcal{K}_2, \varepsilon_2). \tag{6.45}$$

Next we apply Augustin's version of using Fano*-sources for the packing lemma as for nonstationary channels. Formally, let $\{(u_i, D_i) : 1 \leq i \leq M\}$ be an $(1-\varepsilon)$-code for $\mathcal{K}_1 \times \mathcal{K}_2$ and let for $u_i = a_i b_i$

$$q \triangleq q_1 \times q_2 \triangleq \left(\frac{1}{M} \sum_{i=1}^{M} W_1(\cdot|a_i) \right) \times \left(\frac{1}{M} \sum_{i=1}^{M} W_2(\cdot|b_i) \right). \tag{6.46}$$

Then for arbitrary $\theta_1, \theta_2 > 0$

$$\theta_1 \theta_2 \, q(D_i) \geq W\left(\left\{ \frac{dW(\cdot|u_i)}{dq} \leq \theta_1 \theta_2 \right\} \cap D_i | u_i \right)$$

$$\geq q(D_i) - W\left(\left\{ \frac{dW(\cdot|u_i)}{dq} > \theta_1 \theta_2 \right\} | u_i \right)$$

$$> \varepsilon - W\left(\left\{ \frac{dW(\cdot|u_i)}{dq} > \theta_1 \theta_2 \right\} | u_i \right)$$

$$> \varepsilon - W_1\left(\left\{ \frac{dW_1(\cdot|a_i)}{dq_1} > \theta_1 \right\} | a_i \right) - W_2\left(\left\{ \frac{dW_2(\cdot|b_i)}{dq_2} > \theta_2 \right\} b_i \right).$$

Hence

$$\theta_1 \theta_2 \frac{1}{M} \geq \theta_1 \theta_2 \frac{1}{M} \sum_{i=1}^{M} q(D_i) > \varepsilon - \frac{1}{M} \sum_{i=1}^{M} W_1\left(\left\{ \frac{dw_1(\cdot|a_i)}{dq_1} > \theta_1 \right\} \right)$$

$$- \frac{1}{M} \sum_{i=1}^{M} W_2\left(\left\{ \frac{dw_2(\cdot|b_i)}{dq_2} > \theta_2 \right\} \right). \tag{6.47}$$

The maximal code estimate (see Chap. 5) gives

$$M(\mathcal{K}_j, \varepsilon_j^*) > \theta_j \left[\frac{1}{M_j} \sum_{i=1}^{M} W_j \left(\left\{ \frac{dW(\cdot|a_i)}{dq_j} > \theta_j \right\} \right) - \varepsilon_j^* \right] \tag{6.48}$$

$0 < \varepsilon_j^* < 1$.

Combining (6.47) and (6.48) gives

$$\theta_1 \theta_2 \frac{1}{M} > \varepsilon - (\varepsilon_1^* + \varepsilon_2^*) - \frac{M(\mathcal{K}_1, \varepsilon_1^*)}{\theta_1} - \frac{M(\mathcal{K}_2, \varepsilon_2^*)}{\theta_2}. \tag{6.49}$$

Assume now $\varepsilon_1^* + \varepsilon_2^* = (1 - c)\varepsilon$ for $0 < c < 1$ and $M(\mathcal{K}_1, \varepsilon_1^*) \cdot M(\mathcal{K}_2, \varepsilon_2^*) < \infty$, because otherwise (3) below is trivially true, and set

$$\theta_j = \frac{3M(\mathcal{K}_j, \varepsilon_j^*)}{c\varepsilon} \qquad (j = 1, 2). \tag{6.50}$$

Then (6.49) yields

$$\frac{q}{c^2 \varepsilon^2} M(\mathcal{K}_1, \varepsilon_1^*) M(\mathcal{K}_2, \varepsilon_2^*) > \frac{c\varepsilon}{3} M(\mathcal{K}^2, \varepsilon) \tag{6.51}$$

and on the other hand for $\varepsilon \leq \varepsilon_1 \varepsilon_2$ by (6.45)

$$M(\mathcal{K}_1, \varepsilon_1) M(\mathcal{K}_2, \varepsilon_2) \leq M(\mathcal{K}^2, \varepsilon). \tag{6.52}$$

The two inequalities together give the

Theorem 41 *For the product channel* $\mathcal{K}_1 \times \mathcal{K}_2$

$$M(\mathcal{K}_1, \varepsilon_1) M(\mathcal{K}_2, \varepsilon_2) \leq M(\mathcal{K}^2, \varepsilon) \leq \left(\frac{3}{c\varepsilon} \right)^3 M(\mathcal{K}_1, \varepsilon_1^*) M(\mathcal{K}_2, \varepsilon_2^*) \tag{6.53}$$

where $\varepsilon \leq \varepsilon_1 \varepsilon_2$ *and* $\varepsilon_1^* + \varepsilon_2^* < (1 - c)\varepsilon$ $(0 < \varepsilon_1, \varepsilon_2, \varepsilon, \varepsilon_1^*, \varepsilon_2^* < 1)$.

Introducing time structure we see that for instance the assumptions for $j = 1, 2$ imply

$$\lim_{n \to \infty} \frac{1}{n} \log M(\mathcal{K}_j, \lambda, n) = C_j \ (0 < \lambda < 1)$$

$$\lim_{n \to \infty} \frac{1}{n} \log M(\mathcal{K}^2, \lambda, n) = C_1 + C_2 \ (0 < \lambda < 1).$$

Strong Converses and Coding Theorem (in this sense) are preserved under the product.

Remarks

1. In (6.45) we have a trivial lower bound on $M(\mathcal{K}^2, \varepsilon)$. The upper bound comes, as usual, from the Lagerungs Lemma, however, in terms of unknown information quantities, which we relate to $M(\mathcal{K}_j, \varepsilon)$ $(j = 1, 2)$ by the conjunction with the maximal code lemma involving the same unknown quantities, to be eliminated.
2. We discuss this theorem in connection with our general discussion in Lectures 6.1 and 6.2 about rates, error exponents and capacity concepts.

6.4 Lecture on Every Channel with Time Structure has a Capacity Sequence

6.4.1 The Concept

On August 25th 2008 we lectured at the workshop in Budapest which honored I. Csiszár in the year of his 70th birthday, on the result that for every AVC under maximal error probability pessimistic capacity and optimistic capacity are equal. This strongly motivated us to think again about performance criteria and we came back to what we called already a long time ago [2] a (weak) capacity function (now sequence!). But this time we were bold enough to conjecture the theorem below. Its proof was done in hours. In the light of this shining observation we omit now the word "weak" which came from the connection with the weak converse and make the following

Definition 34 For a channel with time structure $\mathcal{K} = (W^n)_{n=1}^{\infty}$ $C : \mathbb{N} \to \mathbb{R}_+$ is a capacity sequence, if for maximal code size $M(n, \lambda)$, where n is the block length or time and λ is the permitted error probability, and the corresponding rate $R(n, \lambda) = \frac{1}{n} \log M(n, \lambda)$

$$\inf_{\lambda > 0} \varliminf_{n \to \infty} (R(n, \lambda) - C(n)) \geq 0 \tag{6.54}$$

$$\inf_{\lambda > 0} \varlimsup_{n \to \infty} (R(n, \lambda) - C(n)) \leq 0. \tag{6.55}$$

Recall that the pessimistic capacity is $\underline{C} = \inf_{\lambda > 0} \varliminf_{n \to \infty} R(n, \lambda)$ and the optimistic capacity is $\overline{C} = \inf_{\lambda > 0} \varlimsup_{n \to \infty} R(n, \lambda)$.

6.4.2 The Existence Result and Its Proof

Theorem 42 *Every channel with time structure has a capacity sequence, if $\overline{C} > \infty$. Moreover, if $(C, C, C, \dots)$ is a capacity sequence, then $C = \underline{C} = \overline{C}$.*

Proof We use only that $R(n, \lambda)$ is not decreasing in λ. $\qquad\qquad\qquad\square$

Let $(\delta_l)_{l=1}^\infty$ be a null-sequence of positive numbers and let $(\lambda_l)_{l=1}^\infty$ be such that $\lambda_l \in (0, 1)$ and

$$\underline{C} + \delta_l \geq \lim_{n\to\infty} \tfrac{1}{n} R(n, \lambda) \geq \underline{C} \qquad\qquad (6.56)$$

$$\overline{C} + \delta_l \geq \overline{\lim_{n\to\infty}} \tfrac{1}{n} R(n, \lambda) \geq \overline{C}. \qquad\qquad (6.57)$$

Moreover, let $(n_l)_{l=1}^\infty$ be a monotone increasing sequence of natural numbers such that for all $n \geq n_l$

$$\overline{C} + \delta_l \geq R(n, \lambda_l) \geq \underline{C} - \delta_l. \qquad\qquad (6.58)$$

For $d_l = \left\lceil \frac{\overline{C} - \underline{C}}{\delta_l} \right\rceil$ define

$$A_l(i) = \{n : n_l \leq n, \overline{C} - (i-1)\delta_l \geq R(n, \lambda_l) \geq \overline{C} - i\delta_l\}, \text{ if } 2 \leq i \leq d_{l-1},$$
$$A_l(1) = \{n : n_l \leq n, \overline{C} + \delta_l \geq R(n, \lambda_l) \geq \overline{C} - \delta_l\},$$
$$A_l(d_l) = \{n : n_l \leq n, \overline{C} - (d_{l-1})\delta_l \geq R(n, \lambda_l) \geq \underline{C} - \delta_l\}.$$

Define by using lower end points

$$C(n) = \begin{cases} \overline{C} - i\delta_l & \text{for } n \in A_l(i), n < n_{l+1} \text{ and } 1 \leq i \leq d_l - 1 \\ \underline{C} - \delta_l & \text{for } n \in A_l(d_l), n < n_{l+1}. \end{cases}$$

Now for any $\lambda \in (0, 1)$ and $\lambda_l < \lambda$

$$R(n, \lambda) - C(n) \geq R(n, \lambda_{i+j}) - C(n) \geq 0$$

for $n_{l+j} \leq n < n_{l+j+1}$ and $j = 0, 1, 2, \ldots$ and thus

$$\lim_{n\to\infty} R(n, \lambda) - C(n) \geq 0$$

and (6.54) follows.

Finally, for any $\lambda < \lambda_l$ by monotonicity of $R(n, \lambda)$

$$\overline{\lim_{n\to\infty}} R(n, \lambda) - C(n) \leq \overline{\lim_{n\to\infty}} R(n, \lambda_l) - C(n) \leq 2\delta_l$$

and

$$\inf_{\lambda\in(0,1)} \overline{\lim_{n\to\infty}} R(n, \lambda) - C(n) \leq \lim_{l\to\infty} 2\delta_l = 0$$

and (6.55) holds. $\qquad\qquad\qquad\qquad\qquad\qquad\qquad\qquad\qquad\qquad\qquad\square$

Example The channel with

$$\frac{1}{n}\log M(n, \lambda) = \begin{cases} \lambda n & \text{for even } n \\ 0 & \text{for odd } n \end{cases} \text{ has } \underline{C} = 0, \overline{C} = \infty$$

and no capacity sequence.

6.4.3 Parallel Channels or the Product of Channels

Theorem 43 *For two channels with time structure* $\mathcal{K}_1 = (W^n)_{n=1}^{\infty}$ *and* $\mathcal{K}_2 = (V^n)_{n=1}^{\infty}$, *which have capacity sequences* C_1 *and* C_2, *the product channel* $\mathcal{K}_1 \times \mathcal{K}_2$ *has capacity sequence*

$$C_{12} = C_1 + C_2.$$

This theorem was not easy to discover, because Wyner proved additivity of Wolfowitz's strong capacity and so did Augustin (subsequently) with an elegant proof based on the following lemma. Wyner's result of additivity of (pessimistic) capacities is false. We could save this part under the assumption $\underline{C} = \overline{C}$ for both channels, that is, if the weak converse holds in the usual sense with a constant. However, often it does not hold like for the channel $\mathcal{K} = AA \ldots AB \ldots$, where $A = \begin{pmatrix} 1 & 0 \\ 0 & 1 \end{pmatrix}$, $B = \begin{pmatrix} \frac{1}{2} & \frac{1}{2} \\ \frac{1}{2} & \frac{1}{2} \end{pmatrix}$, and $\overline{C}(\mathcal{K}) = 1$, $\underline{C}(\mathcal{K}) = 0$, if alternating strings of A's and B's are sufficiently long (see [6]). We call it AB-channel.

Actually, switching the A's and the B's gives a nonstationary DMC (see [6]) $\mathcal{K}'$ again with $\overline{C}(\mathcal{K}') = 1$, $\underline{C}(\mathcal{K}') = 0$. However, $C(\mathcal{K} \times \mathcal{K}') = 1 > 0 + 0$!

On the other hand additivity is a desirable property as can be seen from the fact that Shannon and later also Lóvasz conjectured it for the zero-error capacity (that means, I think, wanted it to hold). But it was disproved by Haemers.

Recently Hastings [34] disproved additivity for the capacity in the HSW-Theorem for quantum channels ([35, 45])

Lemma 58 (Augustin [14]) *For the product channel* $\mathcal{K}_1 \times \mathcal{K}_2$ *of abstract (time free, discrete) channels* $\mathcal{K}_1$ *and* $\mathcal{K}_2$

$$N(\mathcal{K}_1, \epsilon_1)N(\mathcal{K}_2, \epsilon_2) \leq N(\mathcal{K}_1 \times \mathcal{K}_2, \epsilon) \leq \left(\frac{3}{c\epsilon}\right)^3 N(\mathcal{K}_1, \epsilon_1^*)N(\mathcal{K}_2, \epsilon_2^*) \quad (6.59)$$

where $\epsilon \leq \epsilon_1 \epsilon_2$ *and* $\epsilon_1^* + \epsilon_2^* < (1 - c)\epsilon$ $(0 < \epsilon_1, \epsilon_2, \epsilon, \epsilon_1^*, \epsilon_2^* < 1)$ *and* $N(\cdot, \lambda) = M(\cdot, 1 - \lambda)$.

Remarks

1. In [6] we also wanted additivity to hold and found a complicated way by assigning *classes of functions* to a signle channel as performance criterion. The present discovery is oriented at the "weak converse" and shows that a capacity sequence always exists under the mild assumption $\overline{C} < \infty$. So now the class of functions can be reduced to one modulo $o(W)$.
2. In [2] the strong converse was defined for the capacity sequence $(C_n)_{n=1}^{\infty}$ by

$$\frac{1}{n}\log M(n, \lambda) \le C_n + \frac{o(n)}{m}. \tag{6.60}$$

3. Also in [2] the weak converse was defined by

$$\frac{1}{n}\log M(n, \lambda) \le f(\lambda)C_n$$

for all large n and $\inf_\lambda f(\lambda) = 1$.

Proof of Theorem 43 Apply Lemma 58 for W^n, V^n, $\epsilon_1 = \epsilon_2 = 1 - \lambda$, $c = \frac{1}{2}$, $\epsilon_1^* = \epsilon_2^* = \frac{1-c}{2}(1 - \lambda)^2$ and conclude that

$$R(W^n, \lambda) + R(V^n, \lambda) \le R(W^n \times V^n, 2\lambda - \lambda^2)$$
$$\le \frac{1}{n}\log\frac{3}{c(1 - \lambda)^2} + R(W^n, \lambda') + R(V^n, \lambda')$$

where $\lambda' = 1 - \frac{1-c}{2}(1 - \lambda)^2$.

$$\lim_{n\to\infty} (R(W^n \times V^n, \lambda) - C_1(n) - C_2(n))$$
$$\ge \lim_{n\to\infty} (R(W^n, \lambda) - C_1(n)) + \lim_{n\to\infty} (R(V^n, \lambda) - C_2(n)) \ge 0$$

also

$$\inf_\lambda (R(W^n \times V^n, \lambda) - C_1(n) - C_2(n)) \ge 0$$

$$\inf_\lambda \overline{\lim_{n\to\infty}} (R(W^n \times V^n, \lambda) - C_1(n) - C_2(n))$$

$$\le \inf_\lambda \left(\overline{\lim_{n\to\infty}} (R(W^n, \lambda) - C_1(n)) + \overline{\lim_{n\to\infty}} (R(V^n, \lambda) - C_2(n)) \right) \le \lim_{l\to\infty} 2\delta_l = 0.$$

$\square$

6.4.4 Zero-error Capacity

Replacing inf and $\underline{\lim}$ (resp. $\overline{\lim}$) we get a definition of $C_0 : \mathbb{N} \to \mathbb{R}_+$ by the conditions

$$\overline{\lim_{n\to\infty}} R(n, 0) - C_0(n) \ge 0$$

and

$$\overline{\lim_{n \to \infty}} R(n, 0) - C_0(n) \leq 0.$$

Since

$$M(n_1 + n_2, 0) \geq M(n_1, 0)M(n_2, 0) \tag{6.61}$$

for the DMC we get superadditivity for parallel channels. We do not have (6.61) for general channels, however, we have

$$M_{12}(n, 0) \geq M_1(n, 0)M_2(n, 0)$$

and therefore superadditivity for capacity sequences. Why has every channel with this structure a zero-error capacity sequence? Because we can choose $C_0(n) = R(n, 0)$.

6.5 Lecture on the Capacity Functions $C(\lambda), C(\overline{\lambda})$

6.5.1 Explanations Via Compound Channels

With some abuse of notations we use as in [13] for capacity functions for maximal error probability λ and average error probability $\overline{\lambda}$ the same symbol C (for the different functions).

Recall that $N(N, \lambda)$ (resp. $N(n, \overline{\lambda})$) is the maximal size of n-block codes for a channel in question with maximal error probabilities $\leq \lambda$ (resp. average error probability $\leq \overline{\lambda}$).

Since $0 \leq \log N(n, \lambda) \leq n \log a$,

$$C^+(\lambda) = \overline{\lim_{n \to \infty}} \frac{1}{n} \log N(n, \overline{\lambda}) \tag{6.62}$$

and

$$C^-(\lambda) = \lim_{n \to \infty} \frac{1}{n} \log N(n, \overline{\lambda}) \tag{6.63}$$

are well-defined for all $\lambda \in (0, 1)$. If for a λ $C^*(\lambda) = C^-(\lambda)$, then $C(\lambda)$ exists. Analogous definitions are given for $\overline{\lambda}$.

It was shown in [13] that *for CC with finitely many states $C(\overline{\lambda})$ is a non-decreasing step functions with finitely many jumps at specified arguments* $\overline{\lambda}_1, \ldots, \overline{\lambda}_{k(|\mathcal{S}|)}$.

Remark 1 It is essential that we required $\leq \overline{\lambda}$ rather than $< \overline{\lambda}$, which would make the capacity functions continuous from the left (or lower semicontinuous) and thus it would be defined and known in the jump points.

The same is the case for instance for averaged channels with finitely many constituents.

In the definition zero-error codes play a role. The principal observation is easiest to explain for the case $|S| = 2$, where we have two DMC, $W(\cdot | \cdot | 1)$ and $W(\cdot | \cdot | 2)$, with corresponding capacities C_1 and C_2 with $C_1 \geq C_2$ w.l.o.g.

A further quantity is C_{12}, the compound capacity. Here

$$C(\overline{\lambda}) = \begin{cases} C_{12} \text{ for } \overline{\lambda} < \frac{1}{2} \\ C_{12} \vee (C_{01} \wedge C_{02}) \text{ for } \overline{\lambda} = \frac{1}{2} \\ C_2 = C_1 \wedge C_2 \text{ for } \overline{\lambda} > \frac{1}{2}, \end{cases}$$

where C_{0i} is the zero-error capacity of the ith channel.

Obviously the new definition with $< \overline{\lambda}$ would have the second line replaced by "C_{12} for $\overline{\lambda} = \frac{1}{2}$".

$$(C_{12} \vee C_{13} \vee C_{23}) \wedge C_1 \wedge C_2 \wedge C_3 = (C_{12} \wedge C_3) \vee (C_{13} \wedge C_2) \vee (C_{23} \wedge C_1)$$

Suppose $C_{12} \wedge C_3 > C_{13} \wedge C_2, C_{23} \wedge C_1$ then $C(\overline{\lambda}) = C_{12} \wedge C_3$.

Permuting the indices we would get $C(\overline{\lambda}) \neq C_{12} \wedge C_3$.

Given $W = \{W(\cdot | \cdot | s) : 1 \leq s \leq k\}$ define $C_{lr} \cdots = \max_P \inf_{s=l,r} I(P, W(\cdot | \cdot | s))$.

From these basic formulas we can build $C(\overline{\lambda})$.

Theorem 44 *Except perhaps for finitely many points* $\lambda_1, \ldots, \lambda_{K^*(k)}$ *for every* $\overline{\lambda} \in (0, 1)$

$$C(\overline{\lambda}) = C_S = \max_P \inf_{s \in S} I(P, W(\cdot | \cdot | s))$$

for some $S \subset S = \{1, 2, \ldots, k\}$.

Example $k = 3$

$$C(\overline{\lambda}) = \begin{matrix} C_{123} \text{ for } 0 < \overline{\lambda} < \frac{1}{3} \\ C_{12} \wedge C_{13} \wedge C_{23} \text{ for } \frac{1}{3} < \overline{\lambda} < \frac{1}{2} \\ (C_{12} \vee C_{13} \vee C_{23}) \wedge C_1 \wedge C_2 \wedge C_3 \text{ for } \frac{1}{2} < \overline{\lambda} < \frac{2}{3} \\ C_1 \wedge C_2 \wedge C_3 \text{ for } \frac{2}{3} < \overline{\lambda} < 1 \end{matrix}$$

Remark 2 This example shows that $C(\overline{\lambda})$ cannot be expressed by $C(\lambda)$.

6.5.2 Explanations Via Abstract Channels with Time Structure

For a channel with time structure $\mathcal{K} = (W^n)_{n=1}^{\infty}$ the function $R : \mathbb{N} \times (0, 1) \to \mathbb{R}_+$ defined by

$$R(n, \lambda) = \frac{1}{n} \log M(n, \lambda) \qquad (6.64)$$

is the rate function-sequence, $C^+(\overline{\lambda})$ is the optimistic, $C^-(\overline{\lambda})$ is the pessimistic, and if $C^-(\overline{\lambda}) = C^+(\overline{\lambda})$ then $C(\overline{\lambda}) = C^-(\overline{\lambda}) = C^+(\overline{\lambda})$ is the error capacity function. For fixed $\overline{\lambda}$ $C(\overline{\lambda})$ is the $\overline{\lambda}$-capacity.

Remarks

3. For the AB-channel we have constant optimistic and pessimistic capacity functions. $\overline{C}(\lambda) = 1$, $\underline{C}(\lambda) = 0$ for all $\lambda \in (0, 1)$.
4. Notabene: In the terminology of *capacity sequences* there is no problem in defining "error capacity sequences". The rate function-sequence R takes care of it by (6.64).

6.5.3 Further Clarifications for Error Capacity Functions

We have introduced and analyzed time-capacity sequences in Lecture 6.4 and error-capacity functions $C(\overline{\lambda})$ above. They were first introduced by Ahlswede/Wolfowitz [13]. It must be emphasized that here $C(\overline{\lambda})$ exists, if $\overline{C}(\overline{\lambda}) = \underline{C}(\overline{\lambda})$. Much earlier in 1963 Parthasarathy introduced for maximal error probability ε, $\varepsilon \in (0, 1)$, what he called ε-capacity

$$C(\varepsilon) = \lim_{n \to \infty} n^{-1} \log N(n, \varepsilon)$$

(see also Kieffer [43]: the "optimum asymptotic rate"), we write this quantity as above in the form $\underline{C}(\varepsilon)$. It is monotone non-decreasing, and therefore it has at most countably many discontinuities, which are all jump discontinuities.

The right hand and the left hand limits of the pessimistic ε-capacity functions are

$$\underline{C}(\varepsilon^+) = \lim_{\varepsilon' \downarrow \varepsilon} \underline{C}(\varepsilon'), \quad 0 \le \varepsilon < 1.$$

$$\underline{C}(\varepsilon^-) = \lim_{\varepsilon' \uparrow \varepsilon} \underline{C}(\varepsilon'), \quad 0 < \varepsilon < 1.$$

Clearly $\underline{C}(\varepsilon^-) \le \underline{C}(\varepsilon) \le \underline{C}(\varepsilon^+)$, $0 < \varepsilon < 1$. Note that $\underline{C}(0^+)$ is the *pessimistic capacity*.

Kieffer determines $\underline{C}(\varepsilon)$ for any $\varepsilon \in (0, 1)$ for which the gap $\underline{C}(\varepsilon^+) - \underline{C}(\varepsilon^-)$ is sufficiently small for a BSAC. The problem of computing $\underline{C}(\varepsilon)$ for general DC is still open and is generally much harder than computing $\underline{C}(\varepsilon^+)$.

There has been work on "information quantities" by Kieffer [40], Gray/Ornstein [25], Kieffer [41], and Verdú/Han [48]. This is discussed in [43], where the belief is expressed that the techniques used extend also to some other channels.

We are convinced that it is easier to follow the definition under Remark 1 and cannot see any *practical reason* for a very, very small loosening of the error constraint (from $< \lambda$ to $\leq \lambda$).

Unfortunately, we have to point out some errors in terminology of some of our experts.

Remarks

1. Verdú/Han defined in [48], page 1149, the pessimistic $\underline{C}(\varepsilon)$ as "ε-capacity" and the (pessimistic) capacity $C = \lim_{\varepsilon \downarrow 0} \underline{C}(\varepsilon)$. Kieffer erroneously wrote on page 1268 of [42] that Verdú/Han called $\limsup_{n \to \infty} n^{-1} \log N(n, \varepsilon)$ the ε-capacity! This is our $\overline{C}(\varepsilon)$.

2. On page 1153 of [48] the authors write that Wolfowitz referred to the conventional capacity (the pessimistic in the previous remark) as weak capacity. Actually, this term was introduced for a constant for which coding theorem and weak converse hold, however, more importantly it was claimed that the weak capacity equals the (pessimistic) capacity above. Of course the latter exists by definition, but the former need not exist. If it does, then the quantities are equal and in fact they also equal the (optimistic) $\overline{C}$.

6.5.4 Analogous Concepts for Sources

In addition to input alphabet $\mathcal{X}$ and output alphabet $\mathcal{Y}$ there is now a source alphabet $\mathcal{Z}$. To channels with time structure correspond sources with time structure $Q = (P^n)_{n=1}^{\infty}$ where P^n is a PD on Z^n.

Definition 35 For a source with time structure $Q = (P^n)_{n=1}^{\infty}$ the function $T : \mathbb{N} \to \mathbb{R}_+$ is a compression sequence if for minimal code size $M(n, \varepsilon)$, where n is the block length or time and ε is the permitted error probability, and the corresponding rate $R(n, \varepsilon) = \frac{1}{n} \log M(n, \varepsilon)$

$$\sup_{\varepsilon > 0} \varlimsup_{n \to \infty} (R(n, \varepsilon) - T(n)) \leq 0 \tag{6.65}$$

and

$$\sup_{\varepsilon > 0} \varliminf_{n \to \infty} (R(n, \varepsilon) - T(n)) \geq 0. \tag{6.66}$$

We call

$$\overline{T} = \sup_{\varepsilon > 0} \varlimsup_{n \to \infty} R(n, \varepsilon) \tag{6.67}$$

the pessimistic compression and so

$$\underline{T} = \sup_{\varepsilon>0} \lim_{n\to\infty} R(n, \varepsilon) \tag{6.68}$$

the optimistic compression.

6.5.5 The Analogous Result and Its Proof

Theorem 45 *Every source with time structure has a compression function if $\underline{T} < \infty$. Moreover, if $(T, T, T, \dots)$ is a compression sequence, then $T = \overline{T} = \underline{T}$.*

Proof We use only that $R(n, \varepsilon)$ is not increasing in ε.   □

Let $(\delta_l)_{l=1}^\infty$ be a null-sequence of positive numbers and let $(\varepsilon_l)_{l=1}^\infty$ be such that $\varepsilon_l \in (0, 1)$ and

$$\overline{T} + \delta_l \geq \overline{\lim_{n\to\infty}} R(n, \varepsilon_n) \geq \overline{T} \tag{6.69}$$

$$\underline{T} + \delta_l \geq \lim_{n\to\infty} R(n, \varepsilon_n) \geq \underline{T} \tag{6.70}$$

Moreover, let $(n_l)_{l=1}^\infty$ be a monotone increasing sequence of natural numbers such that for all $n \geq n_l$

$$\overline{T} + \delta_l \geq R(n, \varepsilon_l) \geq \underline{T} - \delta_l. \tag{6.71}$$

Replacing $\underline{C}, \overline{C}, \lambda_l$ by $\underline{T}, \overline{T}, \varepsilon_l$ in the definition we get again sets $A_l(i)$ for $1 \leq i \leq d_l$ and a sequence $(T(n))_{n=1}^\infty$ instead of $(C(n))_{n=1}^\infty$.

We get now for any $\varepsilon \in (0, 1)$ and $\varepsilon_l < \varepsilon$

$$R(n, \varepsilon) - T(n) \leq \mathcal{K}(n, \varepsilon_{l+j}) - T(n) \leq s \tag{6.72}$$

for $n_{l+j} \leq n < n_{l+j+1}$ and $j = 0, 1, 2, \dots$ and thus

$$\overline{\lim_{n\to\infty}} R(n, \varepsilon) - T(n) \leq 0 \tag{6.73}$$

and thus (6.65).

Finally, for any $\varepsilon < \varepsilon_l$ by $R(n, \varepsilon)$ monotonically decreasing in ε

$$\lim_{n\to\infty} R(n, \varepsilon) - T(n) \geq \lim_{n\to\infty} R(n, \varepsilon_l) - T(n) \geq 2\delta_l \tag{6.74}$$

and

$$\sup_{\varepsilon \in (0,1)} \lim_{n \to \infty} R(n, \varepsilon) - T(n) \geq \lim_{l \to \infty} 2\delta_l = 0 \qquad (6.75)$$

and (6.63) holds.

Example No compression function.

6.5.6 Parallel Sources or the Product of Sources

Theorem 46 *For two sources with time structure $Q_1 = (P_1^n)_{n=1}^{\infty}$ and $Q_2 = (P_2^n)_{n=1}^{\infty}$, which have compression sequences T_1 and T_2, the product source $Q_1 \times Q_2$ has compression sequence $T_{12} = T_1 + T_2$.*

There may be a proof like the one for channels. However, we suggest another approach. Obviously for $\varepsilon = \varepsilon_1, \ldots, \varepsilon_n$

$$N(Q_1, \varepsilon_1)N(Q_2, \varepsilon_2) \geq N(Q_1 \times Q_2), \varepsilon_1, \ldots.$$

Establishing an appropriate lower bound leads to a nice combinatorial (or analytic) problem. For PD's $P = (p_1, p_2, \ldots, p_a)$ and $Q = (q_1, \ldots, q_b)$ and any integer $s \in [a, b]$ consider

$$F(a, b, s) = \max \left\{ \sum_{(i,j) \in S} p_i q_j \text{ for } S \in [a] \times [b] \text{ with } |S| = s \right\}.$$

Conjecture

$$F(a, b, s) \geq \frac{1}{2} \max \left\{ \sum_{i \in A} p_i \sum_{j \in B} q_j : A \subset [a], B \subset [b], |A||B| \leq S \right\}.$$

It may relate to Chebyshev's inequality. Put for $c \leq \min(a, b)$

$$r_k = \max\{p_i q_k, \alpha_k q_i : 1 \leq i \leq k\}, \quad 1 \leq k \leq c,$$

then

$$\sum_{i=1}^{c} p_i \sum_{j=1}^{c} q_j \leq c \sum_{k=1}^{c} r_k.$$

In particular, if $p_1 \leq p_2 \leq \cdots \leq p_c$ and

$$\sum_{i=1}^{c} c p_i \sum_{j=1}^{c} q_j \leq c \sum_{i=1}^{c} p_i q_i.$$

Further Reading

1. R. Ahlswede, On two-way communication channels and a problem by Zarankiewicz. in Sixth Prague Conference on Information Theory, Statistical Decision Function's and Random Process (Publishing House of the Czech Academy of Sciences, 1973), pp. 23–37
2. R. Ahlswede, Channel capacities for list codes. J. Appl. Probab. **10**, 824–836 (1973)
3. R. Ahlswede, Elimination of correlation in random codes for arbitrarily varying channels. Z. Wahrscheinlichkeitstheorie und verw. Geb. **44**, 159–175 (1978)
4. R. Ahlswede, Coloring hypergraphs: a new approach to multi-user source coding I. J. Comb. Inf. Syst. Sci. **4**(1), 76–115 (1979)
5. R. Ahlswede, Coloring hypergraphs: a new approach to multi-user source coding II. J. Comb. Inf. Syst. Sci. **5**(3), 220–268 (1980)
6. R. Ahlswede, A method of coding and its application to arbitrarily varying channels. J. Comb. Inf. Syst. Sci. **5**(1), 10–35 (1980)
7. R. Ahlswede, An elementary proof of the strong converse theorem for the multiple-access channel. J. Comb. Inf. Syst. Sci. **7**(3), 216–230 (1982)
8. R. Ahlswede, V. Balakirsky, Identification under random processes, problemy peredachii informatsii (special issue devoted to M.S. Pinsker). Probl. Inf. Transm. **32**(1), 123–138
9. R. Ahlswede, V. Balakirsky, Identification under random processes, problemy peredachii informatsii (special issue devoted to M.S. Pinsker). Probl. Inf. Transm. **32**(1), 144–160 (1996)
10. R. Ahlswede, I. Csiszár, Common randomness in information theory and cryptography, part I: secret sharing. IEEE Trans. Inform. Theor. **39**(4), 1121–1132 (1993)
11. R. Ahlswede, I. Csiszár, Common randomness in information theory and cryptography, part II: CR capacity. Preprint 95–101, SFB 343 Diskrete Strukturen in der Mathematik, Universität Bielefeld. IEEE Trans. Inform. Theor. **44**(1), 55–62 (1998)
12. R. Ahlswede, G. Dueck, Every bad code has a good subcode: a local converse to the coding theorem. Z. Wahrscheinlichkeitstheorie und verw. Geb. **34**, 179–182 (1976)
13. R. Ahlswede, G. Dueck, Good codes can be produced by a few permutations. IEEE Trans. Inform. Theor. IT-28(3), 430–443 (1982)
14. R. Ahlswede, G. Dueck, Identification in the presence of feedback–a discovery of new capacity formulas. IEEE Trans. Inform. Theor. **35**(1), 30–39 (1989)
15. R. Ahlswede, P. Gács, J. Körner, Bounds on conditional probabilities with applications in multiuser communication. Z. Wahrscheinlichkeitstheorie und verw. Geb. **34**, 157–177 (1976)
16. R. Ahlswede, T.S. Han, On source coding with side information via a multiple-access channel and related problems in multi-user information theory. IEEE Trans. Inform. Theor. IT-29(3), 396–412 (1983)
17. R. Ahlswede, B. Verboven, On identification via multi-way channels with feedback. IEEE Trans. Inform. Theor. **37**(5), 1519–1526 (1991)
18. R. Ahlswede, Z. Zhang, New directions in the theory of identification via channels. Preprint 94–010, SFB 343 Diskrete Strukturen in der Mathematik, Universität Bielefeld. IEEE Trans. Inform. Theor. **41**(4), 1040–1050 (1995)

19. S. Arimoto, On the converse to the coding theorem for the discrete memoryless channels. IEEE Trans. Inform. Theor. IT-19, 357–359 (1973)
20. R. Ash, *Information Theory, Interscience Tracts in Pure and Applied Mathematics*, vol. 19 (Wiley, New York, 1965)
21. G. Aubrun, S. Szarek, E. Werner, Hastings' additivity counterexample via Dvoretzky's theorem. Commun. Math. Phys. **305**, 85–97 (2011)
22. R.E. Blahut, Hypothesis testing and information theory, IEEE Trans. Inform. Theor. IT-20, 405–417 (1974)
23. R.E. Blahut, Composition bounds for channel block codes. IEEE Trans. Inform. Theor. IT-23, 656–674 (1977)
24. I. Csiszár, J. Körner, Graph decomposition: a new key to coding theorems. IEEE Trans. Inform. Theor. IT-27, 5–12 (1981)
25. I. Csiszár, J. Körner, K. Marton, *A new look at the error exponent of a discrete memoryless channel (preprint)* (IEEE Intern. Symp. Inform. Theory, Ithaca, NY, 1977)
26. R.L. Dobrushin, S.Z. Stambler, Coding theorems for classes of arbitrarily varying discrete memoryless channels. Prob. Peredachi Inform. **11**, 3–22 (1975)
27. G. Dueck, *Omnisophie: über richtige, wahre und natürliche Menschen* (Springer, Berlin Heidelberg, 2003)
28. G. Dueck, J. Körner, Reliability function of a discrete memoryless channel at rates above capacity. IEEE Trans. Inform. Theor. IT-25, 82–85 (1979)
29. H. Dudley, The vocoder. Bell. Lab. Rec. **18**, 122–126 (1939)
30. A. Feinstein, A new basic theorem of information theory. IRE Trans. Inform. Theor. **4**, 2–22 (1954)
31. R.G. Gallager, A simple derivation of the coding theorem and some applications. IEEE Trans. Inform. Theor. IT-11, 3–18 (1965)
32. R.G. Gallager, Source coding with side information and universal coding (preprint). in IEEE International Symposium Information Theory (Ronneby, Sweden, 1976)
33. V.D. Goppa, Nonprobabilistic mutual information without memory. Prob. Contr. Inform. Theor. **4**, 97–102 (1975)
34. A. Haroutunian, Estimates of the error exponent for the semi-continuous memoryless channel. Prob. Peredachi Inform. **4**, 37–48 (1968)
35. H. Kesten, Some remarks on the capacity of compound channels in the semicontinuous case. Inform. Contr. **4**, 169–184 (1961)
36. V.N. Koselev, On a problem of separate coding of two dependent sources. Prob. Peredachi Inform. **13**, 26–32 (1977)
37. J.K. Omura, A lower bounding method for channel and source coding probabilities. Inform. Contr. **27**, 148–177 (1975)
38. C.E. Shannon, A mathematical theory of communication. Bell Syst. Tech. J. **27**(379–423), 632–656 (1948)
39. C.E. Shannon, R.G. Gallager, E.R. Berlekamp, Lower bounds to error probability for coding on discrete memoryless channels I-II. Inform. Contr. **10**(65–103), 522–552 (1967)
40. D. Slepian, J.K. Wolf, Noiseless coding of correlated information sources, IEEE Trans. Inform. Theor. IT-19, 471–480 (1973)
41. J. Wolfowitz, The coding of messages subject to chance errors. Ill. J. Math. **1**, 591–606 (1957)

References

1. R. Ahlswede, Certain results in coding theory for compound channels. in Proceedings of the Colloquium Information Theory (Debrecen, Hungary, 1967), pp. 35–60
2. R. Ahlswede, Beiträge zur Shannonschen Informationstheorie im Fall nichtstationärer Kanäle, Z. Wahrscheinlichkeitstheorie und verw. Geb. 10, 1–42 (1968) (Diploma Thesis Nichtstationäre Kanäle, Göttingen 1963).
3. R. Ahlswede, The weak capacity of averaged channels. Z. Wahrscheinlichkeitstheorie und verw. Geb. 11, 61–73 (1968)
4. R. Ahlswede, Multi-way communication channels, in Proceedings of 2nd International Symposium on Information Theory, Thakadsor, Armenian SSR, September 1971 (Akademiai Kiado, Budapest, 1973), pp. 23–52
5. R. Ahlswede, The capacity region of a channel with two senders and two receivers. Ann. Probab. 2(5), 805–814 (1974)
6. R. Ahlswede, in textitOn Concepts of Performance Parameters for Channels, General Theory of Information Transfer and Combinatorics. Lecture Notes in Computer Science, vol. 4123 (Springer, 2006), pp. 639–663
7. R. Ahlswede, General theory of information transfer, Preprint 97–118, SFB 343 "Diskrete Strukturen in der Mathematik", Universität Bielefeld, 1997. General theory of information transfer: updated, General Theory of Information Transfer and Combinatorics, a Special Issue of Discrete Applied Mathematics 156(9), 1348–1388 (2008)
8. R. Ahlswede, M.V. Burnashev, On minimax estimation in the presence of side information about remote data. Ann. Statist. 18(1), 141–171 (1990)
9. R. Ahlswede, N. Cai, Z. Zhang, Erasure, list, and detection zero-error capacities for low noise and a relation to identification. Preprint 93–068, SFB 343 Diskrete Strukturen in der Mathematik, Universität Bielefeld. IEEE Trans. Inform. Theor. 42(1), 55–62 (1996)
10. R. Ahlswede, N. Cai, Z. Zhang, in Secrecy Systems for Identification Via Channels with Additive-Like Instantaneous Block Enciphers, General Theory of Information Transfer and Combinatorics. Lecture Notes in Computer Science, vol. 4123 (Springer, 2006), pp. 285–292
11. R. Ahlswede, I. Csiszár, Hypothesis testing under communication constraints. IEEE Trans. Inform. Theor. IT-32(4), 533–543 (1986)
12. R. Ahlswede, G. Dueck, Identification via channels. IEEE Trans. Inform. Theor. 35(1), 15–29 (1989)
13. R. Ahlswede, J. Wolfowitz, The structure of capacity functions for compound channels. in Proceedings of the International Symposium on Probability and Information Theory, April 1968 (McMaster University, Canada, 1969), pp. 12–54
14. U. Augustin, Gedächtnisfreie Kanäle für diskrete Zeit. Z. Wahrscheinlichkeitstheorie u. verw. Geb. 6, 10–61 (1966)
15. T. Berger, *Rate Distortion Theory: A Mathematical Basis for Data Compression* (Prentice-Hall, Englewood Cliffs, 1971)
16. T.M. Cover, J. Thomas, *Elements of Information Theory* (Wiley, New York, 1991)
17. I. Csiszár, J. Körner, *Information Theory: Coding Theorems for Discrete Memoryless Systems* (Academic, New York, 1981)
18. R.L. Dobrushin, General formulation of Shannon's main theorem of information theory. Usp. Math. Nauk. 14, 3–104 (1959)
19. R.L. Dobrushin, General formulation of Shannon's main theorem of information theory. Am. Math. Soc. Trans. 33, 323–438 (1962)
20. S.M. Dodunekov, Optimization problems in coding theory. in Workshop on Combinatorial Search, Budapest, 23–26 April 2005
21. J.L. Doob, Review of "A mathematical theory of communication". Math. Rev. 10, 133 (1949)
22. R.M. Fano, *Transmission of Information, A Statistical Theory of Communication* (Wiley, New York, 1961)
23. A. Feinstein, *Foundations of Information Theory* (McGraw-Hill, New York, 1958)

24. R.G. Gallager, *Information Theory and Reliable Communication* (Wiley, New York, 1968)
25. R.M. Gray, D.S. Ornstein, Block coding for discrete stationary $\bar{d}$-continuous noisy channels. IEEE Trans. Inf. Theor. IT-25(3), 292–306 (1979)
26. S. Györi, Coding for a multiple access OR channel: a survey. General Theory of Information Transfer and Combinatorics, Special Issue of, Discrete Applied Mathematics, 156(9), 2008
27. W. Haemers, On some problems of Lovasz concerning the Shannon capacity of a graph. IEEE Trans. Inform. Theor. **25**(2), 231–232 (1979)
28. T.S. Han, Oral, communication in 1998
29. T.S. Han, Information-Spectrum Methods in Information Theory, April 1998 (in Japanese).
30. T.S. Han, S.I. Amari, Statistical inference under multiterminal data compression, information theory: 1948–1998. IEEE Trans. Inform. Theor. **44**(6), 2300–2324 (1998)
31. T.S. Han, S. Verdú, Approximation theory of output statistics. IEEE Trans. Inf. Theor. IT-39(3), 752–772 (1993)
32. T.S. Han, S. Verdú, New results in the theory of identification via channels. IEEE Trans. Inform. Theor. **39**(3), 752–772 (1993)
33. E.A. Haroutunian, Upper estimate of transmission rate for memoryless channel with countable number of output signals under given error probability exponent, (in Russian). in 3rd All-Union Conference on Theory of Information Transmission and Coding (Publication house of Uzbek Academy of Sciences, Uzhgorod, Tashkent, 1967), pp. 83–86
34. M.B. Hastings, Superadditivity of communication capacity using entangled inputs. Letters (2009)
35. A.S. Holevo, The capacity of quantum channel with general signal states. IEEE Trans. Inf. Theor. **44**, 269–273 (1998)
36. M. Horodecki, Is the Classical Broadcast Channel Additive? Oral Communication (Cambridge England, 2004)
37. K. Jacobs, Almost periodic channels. in Colloquium on Combinatorial Methods in Probability Theory, Matematisk Institute, Aarhus University, 1–10 Aug 1962, pp. 118–126
38. F. Jelinek, *Probabilistic Information Theory* (McGraw-Hill, New York, 1968)
39. W. Kautz, R. Singleton, Nonrandom binary superimposed codes. IEEE Trans. Inform. Theor. **10**, 363–377 (1964)
40. J.C. Kieffer, A general formula for the capacity of stationary nonanticipatory channels. Inf. Contr. **26**, 381–391 (1974)
41. J.C. Kieffer, Block coding for weakly continuous channels. IEEE Trans. Inf. Theor. IT-27(6), 721–727 (1981)
42. J.C. Kieffer, Epsilon-capacity of a class of nonergodic channels. in Proceedings of IEEE International Symposium Information Theory (Seattle, WA, 2006), pp. 1268–1271.
43. J.C. Kieffer, ε-capacity of binary symmetric averaged channels. IEEE Trans. Inf. Theor. **53**(1), 288–303 (2007)
44. M.S. Pinsker, *Information and Stability of Random Variables and Processes* (Izd-vo Akademii Nauk, Moscow, 1960)
45. B. Schumacher, M.D. Westmoreland, Sending classical information via noisy quantum channels. Phys. Rev. A **56**(1), 131–138 (1997)
46. C.E. Shannon, The zero error capacity of a noisy channel. IRE Trans. Inform. Theor. **2**, 8–19 (1956)
47. C.E. Shannon, Certain results in coding theory for noisy channels. Inform. Contr. **1**, 6–25 (1957)
48. S. Verdú, T.S. Han, A general formula for channel capacity. IEEE Trans. Inform. Theor. **40**(4), 1147–1157 (1994)
49. J. Wolfowitz, Coding theorems of information theory. in Ergebnisse der Mathematik und ihrer Grenzgebiete, vol. 31, 3rd edn., *(Springer* (Englewood Cliffs, Berlin-Göttingen-Heidelberg, Prentice-Hall, 1978), p. 1961
50. A.D. Wyner, The capacity of the product channel. Inform. Contr. **9**, 423–430 (1966)

51. R.W. Yeung, *A First Course in Information Theory, Information Technology: Transmission, Processing and Storage* (Kluwer Academic/Plenum Publishers, New York, 2002)
52. J. Ziv, Back from Infinity: a constrained resources approach to information theory. IEEE Inform. Theor. Soc. Newslett. **48**(1), 30–33 (1998)

Chapter 7
Error Bounds

7.1 Lecture on Lower Bounds on Error Probabilities

7.1.1 The Sphere Packing Bound: Its Origin for the BSC

A binary symmetric channel, abbreviated by BSC, is a DMC with a transmission matrix

$$W = \begin{pmatrix} 1 - \varepsilon & \varepsilon \\ \varepsilon & 1 - \varepsilon \end{pmatrix}, \quad 0 \leq \varepsilon \leq \frac{1}{2}$$

For any (n, M) code $\{(u_i, D_i) : 1 \leq i \leq 1\}$ for the BSC the average probability of correct decoding is

$$\overline{\lambda}_c = \frac{1}{M} \sum_{i=1}^{M} \sum_{y^n \in D_i} W(y^n | u_i). \tag{7.1}$$

Let $A_{m,i}$ be the *number* of words y^n that are decoded into u_i and have Hamming distance m from u_i. Then we can rewrite (7.1) as

$$\overline{\lambda}_c = \frac{1}{M} \sum_{i=1}^{M} \sum_{m=0}^{n} A_{m,i} \, \varepsilon^m (1 - \varepsilon)^{n-m}. \tag{7.2}$$

Since $1 = \sum_{m=0}^{n} \binom{n}{m} \varepsilon^m (1 - \varepsilon)^{n-m}$ we can write he average probability of error $\overline{\lambda} = 1 - \overline{\lambda}_c$ as

$$\overline{\lambda} = \frac{1}{M} \sum_{i=1}^{M} \sum_{m=0}^{n} \left[\binom{n}{m} - A_{m,i} \right] \varepsilon^m (1 - \varepsilon)^{n-m}. \tag{7.3}$$

A. Ahlswede et al. (eds.), *Storing and Transmitting Data*,
Foundations in Signal Processing, Communications and Networking 10,
DOI: 10.1007/978-3-319-05479-7_7, © Springer International Publishing Switzerland 2014

There are $\binom{n}{m}$ words at distance m from u_i (for every i), $\binom{n}{m} - A_{m,i}$ of them cause decoding errors if u_i is transmitted. We free ourselves now from the decoding structure and just choose the cardinalities $A_{m,i}$ in an optimal way (even if not realizable). The constraints we use are

$$A_{m,i} \leq \binom{n}{m} \quad \text{for } m, i. \tag{7.4}$$

$$\sum_{i=1}^{M} \sum_{m=0}^{n} A_{m,i} \leq 2^n \tag{7.5}$$

(since there 2^n output words; each decoded into at most one codeword).

The minimum of (7.3), subject to these constraints, is achieved by selecting for all i

$$A_{m,i} = \begin{cases} \binom{n}{m} & \text{for } 0 \leq m \leq k-1 \\ 0 & \text{for } k+1 \leq m \leq n \end{cases} \tag{7.6}$$

and $A_{k,i}$ such that

$$M \sum_{m=0}^{k-1} \binom{n}{m} + \sum_{i=1}^{M} A_{k,i} = 2^k; \; 0 \leq \sum_{i=1}^{M} A_{k,i} \leq M \binom{n}{k}. \tag{7.7}$$

Substitution into (7.3) yields

$$\overline{\lambda}(n, M) > \left[\binom{n}{k} - \frac{1}{M} \sum_{i=1}^{M} A_{k,i} \right] \varepsilon^k (1-\varepsilon)^{n-k} + \sum_{m=k+1}^{n} \binom{n}{m} \varepsilon^m (1-\varepsilon)^{n-m}. \tag{7.8}$$

This bound is known as the sphere packing bound, because we can view the words of distance k or less from a codeword as a sphere of radius k around that codeword (with incomplete boundary). The bound (7.8) is the error probability that would result if such a partition of $\mathcal{Y}^n$ where possible. ("Sphere" packed codes, compare also perfect codes.)

We put now (7.8) into an analytical form. There are several ways of doing this with varying precisions.

Theorem 47 (Sphere packing bound for the BSC) *For the BSC* $W = \begin{pmatrix} 1-\varepsilon & \varepsilon \\ \varepsilon & 1-\varepsilon \end{pmatrix}$, $0 \leq \varepsilon < \frac{1}{2}$, let $\delta \in [\varepsilon, \frac{1}{2}]$, arbitrary otherwise.
 If the number of codewords satisfies

$$M \geq \sqrt{8(n+1)} \exp\{n[1 - h(\delta)]\} \tag{7.9}$$

then

$$\overline{\lambda}(n, M) \geq \frac{\varepsilon}{(1 - \varepsilon)\sqrt{8(n + 1)}} \exp\{n[h(\delta) + D((\delta, 1 - \delta)\|(\varepsilon, 1 - \varepsilon))]\}. \quad (7.10)$$

Proof Clearly any increase in any of the sums $\sum_{i=1}^{M} A_{m,i}$ over the values in (7.6), (7.7) will further lower the bound on $\overline{\lambda}$. For $m' = \lceil \delta n \rceil$ choose

$$\sum_{i=1}^{M} A_{m',i} = s^n$$

and for all $m \neq m'$ choose

$$\sum_{i=1}^{M} A_{m,i} = \binom{n}{m} M.$$

Thus we clearly upperbound those in (7.6), (7.7) and obtain

$$\overline{\lambda}(n, M) \geq \left[\binom{n}{m'} - \frac{n^2}{M}\right] \varepsilon^{m'} (1 - \varepsilon)^{n - m'}. \quad (7.11)$$

By Stirling's formula (see 4.36)

$$\binom{n}{m'} \geq \frac{1}{\sqrt{2n}} \exp\left\{nh\left(\frac{m'}{n}\right)\right\}. \quad (7.12)$$

If $m' \leq \frac{n}{2}$ we have $h\left(\frac{m'}{m}\right) \geq h(\delta)$. Also $\frac{1}{n} \geq \frac{1}{m}$, so that

$$\binom{n}{m'} \geq \frac{1}{\sqrt{2(n + 1)}} \exp\{nh(\delta)\}. \quad (7.13)$$

Since $\delta < \frac{1}{2}$ and $m' = \lceil \delta n \rceil$ the only possible value of m' greater than $\frac{n}{2}$ is $\frac{n+1}{2}$. In the case we use the special bound

$$\binom{n}{\frac{n+1}{2}} \geq \frac{1}{\sqrt{2(n + 1)}} 2^n.$$

Since $2^n \geq \exp\{nh(\delta)\}$, the bound in (7.13) is valid for all possible m'. Also since m' exceeds δn by at most 1, we have

$$\left(\frac{\varepsilon}{1 - \varepsilon}\right)^{m'} \geq \frac{\varepsilon}{1 - \varepsilon} \left(\frac{\varepsilon}{1 - \varepsilon}\right)^{\delta n}.$$

Substituting this and (7.13) into (7.11) and using (7.9) we have (7.10).

Remark Notice the ratewise coincidence with the result in Lecture 3.1.

7.1.2 Beck's Uniform Estimate for Stein's Lemma
and a Medium Converse

Lemma 59 (Beck) *Let* $\mu_1, \ldots, \mu_n$ *be PD's on a measurable space* $(\mathcal{Y}, \mathcal{B})$ *and let* $\mu^n = \mu_1 \times \cdots \times \mu_n$ *denote the product measure on* $(\mathcal{Y}^n, \mathcal{B}^n)$. *If*

$$\mu^n(B) \geq 1 - \delta(\varepsilon) \text{ for } B \in \mathcal{B}, \tag{7.14}$$

where $\delta(\varepsilon) = \left(\min \left\{ \frac{1}{4}, \varepsilon \right\} \right)^4 40^{-5}, 0 < \varepsilon < \varepsilon_0$ *(an absolute constant), then for*

$$J = \max_t D(\mu_t \| \lambda) \tag{7.15}$$

$$\lambda^n(B) \geq \delta(\varepsilon) n^{-\varepsilon} \exp\{-(n + \varepsilon)J\}. \tag{7.16}$$

A theorem of Kemperman asserts that Shannon's maximal mutual information C^* for a channel W can be written in the form

$$C^* = \inf_{\nu \in \mathcal{P}(\mathcal{Y}, \mathcal{B})} \max_{x \in \mathcal{X}} D(W(\cdot | x) \| \nu) \tag{7.17}$$

and moreover, if $C^* < \infty$, then there exists a unique PD $\tilde{\nu}$ such that

$$D(W(\cdot | x) \| \nu) \leq C^* \text{ for all } x \in \mathcal{X}$$

(c.f. Lecture for $\mathcal{Y}$ finite).

This has an immediate consequence for codes.

Theorem 48 (BC) *If there exists an* $(n, M, \delta(\varepsilon))$ *code for a stationary memoryless channel, then*

$$M \leq \frac{n^\varepsilon}{\delta(\varepsilon)} \exp\{(c + \varepsilon)C^*\}.$$

Proof Let $\{(u_i, D_i) : 1 \leq i \leq M\}$ be such a code. By Lemma 59, since $W^n(D_i | u_i) > 1 - \delta(\varepsilon)$,

$$\tilde{\nu}^n(D_i) \geq \delta(\varepsilon) n^{-\varepsilon} \exp\{-(n + \varepsilon)C^*\}$$

and thus

$$1 \geq \sum_{i=1}^M \tilde{\nu}^n(D_i) \geq M\delta(\varepsilon) n^{-\varepsilon} \exp\{-(n + \varepsilon)C^*\}.$$

The weak converse holds, if for any $R > C$ there exists a constant $\lambda(K)$ such that for large n every (n, e^{Rn}) code satisfies $\lambda \geq \lambda(R)$.

It is a way to establish that a number is the capacity, the optimal admissible rate. To show that a number C is the capacity means to show that for every $\delta > 0$ $C - \delta$ is admissible rate and $C + \delta$ is not admissible, that is,

$$\overline{\lim_{n \to \infty}} \lambda(n, e^{(C+\delta)n}) > 0.$$

7.1.3 From the Strong Converse to the Sphere Packing Bound

For a DMC $W : \mathcal{X} \to \mathcal{Y}$ Haroutunian [12] formulated the sphere packing exponent function

$$E_{sp}(R, P, W) = \min_{V : I(P|V) \leq R} D(V \| W | P), \tag{7.18}$$

where $P \in \mathcal{P}(\mathcal{X})$ and $V : \mathcal{X} \to \mathcal{Y}$ is a DMC and the sphere packing exponent

$$E_{sp}(R, W) = \max_P E_{sp}(R, P, W). \tag{7.19}$$

Theorem 49 (Sphere packing bounds) *(i) For every $R > 0$, $\delta > 0$ every constant composition code of type P with parameters (M, n, λ), where*

$$\frac{1}{n} \log M \geq R + \delta,$$

satisfies

$$\lambda \geq \frac{1}{2} \exp\{-n E_{sp}(R, P, W)(1 + \delta)\}$$

for $n \geq n_0(|\mathcal{X}|, |\mathcal{Y}|, \delta)$.
(ii) Every (M, n, λ)-code for W, where $\frac{1}{n} \log M \geq R + \delta$, satisfies for $n \geq n_0(|\mathcal{X}|, |\mathcal{Y}|, \delta)$

$$\lambda \geq \frac{1}{2} \exp\{-n E_{sp}(R, W)(1 + \delta)\}.$$

Proof (i) By the strong converse in Wolfowitz's (or by the packing lemma in Kemperman's) form the relation $I(V|P) \leq R$ implies that our code has under V a maximum probability of error at least $1 - \frac{\delta}{2}$. This means that for some i

$$V^n(\mathcal{D}_i^c | u_i) \geq 1 - \frac{\delta}{2}. \tag{7.20}$$

We conclude that $W^n(\mathcal{D}_i^c | u_i)$ cannot be too small. By the log-sum inequality for any $E \subset \mathcal{Z}$ and $Q_1, Q_2 \in \mathcal{P}(\mathcal{Z})$

$$Q_1(E) \log \frac{Q_1(E)}{Q_2(E)} + Q_1(E^c) \log \frac{Q_1(E^c)}{Q_2(E^c)} \leq D(Q_1 \| Q_2).$$

Therefore

$$Q_1(E) \log \frac{1}{Q_2(E)} \leq D(Q_1 \| Q_2) + h\big(Q_1(E)\big)$$

and

$$Q_2(E) \geq \exp\left\{-\frac{D(Q_1 \| Q_2) + h\big(Q_1(E)\big)}{Q_1(E)}\right\}. \tag{7.21}$$

Choose $\mathcal{Z} = \mathcal{Y}^n$, $E = \mathcal{D}_i^c$, $Q_1 = V^n(\cdot | u_i)$, and $Q_2 = W^n(\cdot | u_i)$, then we get from (7.20)

$$\lambda \geq W^n(\mathcal{D}_i^c | u_i) \geq \exp\left\{-\left(n\,D(V \| W | P) + h\left(1 - \frac{\delta}{2}\right)\right)\left(1 - \frac{\delta}{2}\right)^{-1}\right\}$$

$$\geq \frac{1}{2} \exp\{-n\,D(V \| W | P)(1 + \delta)\},$$

if δ is so small that $h\left(1 - \frac{\delta}{2}\right) < 1 - \frac{\delta}{2}$.

Choosing a V assuming the minimum in (7.18) this gives

$$\lambda \geq \frac{1}{2} \exp\{-n\,E_{sp}(R, P, W)(1 + \delta)\}.$$

(ii) Partition the code into subcodes of fixed composition and use type counting. Alternatively, use Augustin's form of the strong converse with the Fano*-input distribution P^{*n}.

Then

$$\frac{1}{M} \sum_{i=1}^{M} V^n(\mathcal{D}_i^c | u_i) \geq 1 - \frac{\delta}{2}, \text{ if } I(V^n | P^{*n}) \leq Rn, \text{ and by (7.21)}$$

$$\frac{1}{n} \sum_{i=1}^{M} W^n(\mathcal{D}_i^c | u_i) \geq \frac{1}{M} \sum_{i=1}^{M} \frac{1}{2} \exp\{-(D(V^n \| W^n | u_i)(1 + \delta)\}$$

$$\geq \frac{1}{2} \exp\left\{-\frac{1}{M} \sum_{i=1}^{M} D(V^n \| W^n | u_i)(1 + \delta)\right\}$$

(by convexity of "exp").

Finally,

$$\frac{1}{M} \sum_{i=1}^{M} D(V^n \| W^n | u_i) = \frac{1}{M} \sum_{i=1}^{M} \sum_{t=1}^{n} D(V \| W | u_{it})$$

$$= \sum_{t=1}^{n} \frac{1}{M} \sum_{i=1}^{M} D(V \| W | u_{it})$$

$$= \sum_{t=1}^{n} D(V \| W | P_t^*), \tag{7.22}$$

($\leq n \, D(V \| W | P^*)$, where P^* is AD-distribution).
Now

$$\max_{P^{*n}} \min_{V: \sum_{t=1}^{n} I(V | P_t^*) \leq Rn} \sum_{t=1}^{n} D(V \| W | P_t^*)$$

$$\leq \sum_{t=1}^{n} \max_{P_t^*} \min_{V: I(V|_t P^*) \leq R} D(V \| W | P_t^*) = n \, E_{sp}(R, W).$$

Remark Actually one can choose a non-stationary comparison channel $(V_t)_{t=1}^{\infty}$. Then

$$\max_{P^n = P_1 \times \cdots \times P_n} \min_{(V_t)_{t=1}^{\infty}: \sum_{t=1}^{n} I(V_t | P_t) \leq Rn} \sum_{t=1}^{n} D(V_t \| W | P_t) \leq n \, E_{sp}(R, W).$$

Does equality hold?

Generally, can consider $P^n \in \mathcal{P}(\mathcal{X}^n)$ and $V^n : \mathcal{X}^n \to \mathcal{Y}^n$.
(Compare: Random coding bound, sphere packing bound)
From here get immediately sphere packing bound for non-stationary memoryless channel NDMC in the form:

$$- \max_{P^n} \min_{(V_t)_{t=1}^{n}: \sum I(V_t | P_t) \leq Rn} D(V^n | W^n | P^n)$$

$$\geq - \max_{P^n = P_1 \times \cdots \times P_n} \min_{(R_1, \ldots, R_n) \sum_{t=1}^{n} R_t = R \cdot n} \sum_{t=1}^{n} \min_{I(V_t | P_t) \leq R_t} D(V_t \| W_t | P_t),$$

or

$$-\min_{\substack{\sum\limits_{t=1}^{n} R_t = R \cdot n}} \sum_{t=1}^{n} \max_{P_t} \min_{I(V_t|P_t) \leq R_t} D(V_t \| W_t | P_t)$$

$$= -\min_{\substack{\sum\limits_{t=1}^{n} R_t = R \cdot n}} \sum_{t=1}^{n} E_{sp}(W_t, R_t).$$

See 7.2.3.1 for comparison.

7.1.4 The Straight Line Lower Bound

Next to upper bounds, namely random coding and expurgated bounds discussed in the next lecture, and sphere packing as a lower bound there is a fourth important bound:

the straight line lower bound on optimal error probability, due to Shannon, Gallager, Berlekamp [17, 18].

It can be based on an inequality for list codes.

Lemma 60 *For arbitrary positive integers n_1, n_2, M, and L*

$$\overline{\lambda}(n_1 + n_2, M) \geq \overline{\lambda}(n_1, M, L)\lambda_{\max}(n_2, L + 1) \tag{7.23}$$

holds for every DMC W.

The guiding idea is this. For a codebook $\mathcal{U} \subset \mathcal{X}^{n_1+n_2}$ you must vote for a list of L codewords after having observed n_1 letters. This enforces a certain average error probability that the true word is not on the list. With the last n_2 letters the $L + 1$ remaining options are to be removed. The probabilities in (7.23) are related to these events.

Proof Let $u_i = u_{i,1} \cdot u_{i,2}$ with prefix $u_{i,1}$ and suffix $u_{i,2}$ and lengths n_1 and n_2. Similarly the received word y^n is separated into prefix y_1^n and suffix y_2^n. We rewrite now

$$\overline{\lambda} = \frac{1}{M} \sum_{i=1}^{M} \sum_{y^n \in \mathcal{D}_i^c} W(y^n | u_i) \tag{7.24}$$

by defining for each prefix $y_1^{n_1}$

$$\mathcal{D}_{i,2}(y_1^{n_1}) = \{y_2^{n_2} \text{ suffix with } y_1^{n_1} y_2^n \in \mathcal{D}_i\}. \tag{7.25}$$

Now

$$\bar{\lambda} = \frac{1}{M} \sum_{i=1}^{M} \sum_{y_1^{n_1}} W^{n_1}(y_1^{n_1}|u_{i,1}) \sum_{y_2^{n_2} \in \mathcal{D}_{i,2}(y_1^{n_1})^c} W(y_2^{n_2}|u_{i,2}). \qquad (7.26)$$

The system of pairs $\{(u_{i,2}, \mathcal{D}_{i,2}(y_1^{n_1})) : 1 \leq i \leq M\}$ is a code with error probabilities

$$\lambda_i(y_1^{n_1}) = \sum_{y_2^{n_2} \in \mathcal{D}_{i,2}(y_1^{n_1})^c} W(y_2^{n_2}|u_{i,2}), \ 1 \leq i \leq M. \qquad (7.27)$$

Let $i_1(y_1^{n_1})$ be the i for which $\lambda_i(y_1^{n_1})$ is smallest, $i_2(y_1^{n_1})$ for which it is next smallest, and so on. We maintain that for all i except perhaps $i_1(y_1^{n_1}), \dots, i_L(y_1^{n_1})$

$$\lambda_i(y_1^{n_1}) \geq \lambda_{\max}(n_2, L+1). \qquad (7.28)$$

If this were not so, then the set of $L+1$ codewords $\{u_{i,2} : i = i_1(y_1^{n_1}), \dots, i = i_{L+1}(y_1^{n_1})\}$ with their decoding regions $\{D_{i,2}(y_1^{n_1} : i = i_1(y_1^{n_1}), \dots, i_{L+1}(y_1^{n_1})\}$ would all have error probabilities less than $\lambda_{\max}(n_2, L+1)$, which is a contradiction. So we have the lower bound

$$\sum_{y_2^{n_2} \in \mathcal{D}_{i,2}^c(y_1^{n_1})} W(y_2^{n_2}|u_{i,2}) \geq \begin{cases} 0 & \text{for } i = i_1(y_1^{n_1}), \dots, i_L(y_1^{n_1}) \\ \lambda_{\max}(n_2, L+1) & \text{other } i. \end{cases} \qquad (7.29)$$

Interchanging summation between i and $y_1^{n_1}$ in (7.26) and substituting (7.29), we get

$$\bar{\lambda} \geq \frac{1}{M} \sum_{y_1^{n_1}} \sum_{\substack{i_\ell(y_1^{n_1}) \\ \ell > L}} W(y_1^{n_1}|u_{i_\ell(y_1^{n_1})}) \cdot \lambda_{\max}(n_2, L+1). \qquad (7.30)$$

Finally, we can consider the set of prefixes $\{u_{i,1} : 1 \leq i \leq M\}$ as codebook of block length n_1 with M words and we can use $u_{i_\ell}(y_1^{n_1}), 1 \leq \ell \leq L$, as a list decoding rule for this set of codewords. Thus

$$\frac{1}{M} \sum_{y_1^{n_1}} \sum_{\substack{i_\ell(y_1^{n_1}) \\ \ell > L}} W(y_1^{n_1}|u_{i_\ell(y_1^{n_1}),1}) > \bar{\lambda}(n_1, M, L). \qquad (7.31)$$

Inequality (7.23) follows from (7.30) and (7.31).

Theorem 50 *Let (R_1, E_1) be on the sphere packing curve and (R_2, E_2), $R_2 < R_1$ be below the sphere packing curve, known to be a lower bound on error exponent at rate R_2, then the pairs $(R, E') = \alpha(R_1, E_1) + (1 - \alpha)(R_2, E_2)$ give a lower bound for $0 \leq \alpha \leq 1$.*

Proof It is important that (R_1, E_1) is on the sphere packing curve, because it is then (see Lecture) also a lower bound for list codes with rate $\left(\frac{M}{L}\right) \geq R_1$. Apply now the Lemma for $n_1 = \alpha n$, $n_2 = (1 - \alpha)n$ and rate$M = \alpha R_1 (1 - \alpha)R_2 = R$, then

$$\frac{1}{n}\ell n\overline{\lambda}(n, e^{R_n}) \geq \frac{1}{n}\ell n\overline{\lambda}(\alpha n, e^{R_n}, e^{(1-\alpha)R_2}) + \frac{1}{n}\ell n\lambda_{\max}\left((1 - \alpha)n, e^{(1-\alpha)R_2}\right)$$

and therefore

$$-E \geq -\alpha E_1 - (1 - \alpha)E_2 = -E' \text{ and } E \leq E'.$$

7.1.4.1 The Marvelous Point $\left(0, \mathrm{E}_{\mathbf{ex}}(0)\right)$

The previous result is nice, but where do we find a point (R_2, E_2) with the desired properties?

Actually $\left(0, E_{ex}(0)\right)$ is such a point, as and we therefore have the

Corollary 4 *The straight line connecting* $\left(0, E_{ex}(0)\right)$ *with any point on the sphere packing curve, in particular the line tangent to this curve, provides a lower bound on the error exponent.*

This was proved by Shannon, Gallager, and Berlekamp for general DMC and will be proved here for the BSC.

Remark In all these estimates, like those in Lecture 3.2, there are technical terms of ratewise disappearing magnitudes. In CK they are avoided in the formulas by general phrases like "for $n \geq n_0(|\mathcal{X}|, |\mathcal{Y}|, \delta)$". In G they are carried out—a useful experience for beginners.

We need two more auxiliary results beyond the sphere packing bound, Lemma 1, and the list code inequality, Lemma 60, namely

Lemma 61 *Plotkin bound* [15] *The minimum distance of a codebook* $\mathcal{U} = \{u_1, \ldots, u_M\} \subset \{0, 1\}^n$ *satisfies*

$$d_{\min} \leq \frac{nM}{2(M - 1)}. \tag{7.32}$$

and a very low rate result

Lemma 62 *For the BSC* $W = \begin{pmatrix} 1 - \varepsilon & \varepsilon \\ \varepsilon & 1 - \varepsilon \end{pmatrix}$, $\varepsilon = 1/2$ *and for* $M > n + 2$

$$\lambda_{\max}(n, M) \geq \frac{\sqrt{n}\varepsilon}{(n + 4)(1 - \varepsilon)} \exp\{-n\, E_{ex}(0)\}. \tag{7.33}$$

Here

$$E_{ex}(0) = -1/4\ell n4\varepsilon \cdot (1 - \varepsilon)^{1)} \tag{7.34}$$

is the expurgated exponent at $R = 0$

Proof By Plotkin's bound for $M \geq n + 2$ $d_{\min} \leq n/2$. Suppose that two code-words u, u' are at distance $d = d(u, u') = d_{\min}$ from each other in an (n, M)-code $\{(u_i, \mathcal{D}_{u_i}) : 1 \leq i \leq M\}$. Then

$$\lambda(u) + \lambda(u') = \sum_{\lambda^n \in \mathcal{D}_u^c} W(y^n|u) + \sum_{y^n \in \mathcal{D}_{u'}^c} W(y^n|u'). \tag{7.35}$$

This can be lower bounded by replacing the two decoding sets by ML-decoding. Then

$$\frac{\lambda(u) + \lambda(u')}{2} \geq \sum_{i > \frac{d}{2}} \binom{d}{i} \varepsilon^i (1 - \varepsilon)^{d-i}. \tag{7.36}$$

For d even, this is lower bounded by

$$\begin{aligned}
\frac{\lambda(u) + \lambda(u')}{2} &\geq \binom{d}{\frac{d}{2} + 1} \varepsilon^{d/2+1} (1 - \varepsilon)^{d/2-1} \\
&= \frac{d\varepsilon}{(d + 2)(1 - \varepsilon)} \binom{d}{d/2} \varepsilon \\
&\geq \frac{d\varepsilon}{(d + 2)(1 - \varepsilon)\sqrt{2d}} \exp\left\{\frac{d}{2} \ell n[4\varepsilon(1 - \varepsilon)]\right\} \tag{7.37}
\end{aligned}$$

by Stirling's bound in 4.3.

This decreases in d for even d and since $d \leq \frac{n}{2}$ gives

$$\frac{\lambda(u) + \lambda(u')}{2} \geq \frac{\sqrt{n}\varepsilon}{(n + 4)(1 - \varepsilon)} \exp\left\{\frac{n}{4} \ell n[4\varepsilon(1 - \varepsilon)]\right\}. \tag{7.38}$$

For odd d we use Stirling's bound (4.36) and get

$$\frac{\lambda(u) + \lambda(u')}{2} \geq \sqrt{\frac{\varepsilon}{(1 - \varepsilon)(n + 2)}} \exp\left\{\frac{n}{4} \ell n[4\varepsilon(1 - \varepsilon)]\right\}, \tag{7.39}$$

which is larger than the bound in (7.38).

Interpretation. The rate of a code with $M = n + 2$ words goes to 0 as n goes to ∞. The exponent in the error bound is $E_{ex}(0)$, defined as $\lim_{R \to 0} E_{ex}(R)$. The bound in

(7.33) clearly applies to $\lambda_{\max}(n, e^{Rn})$ for every $R > 0$ and thus $E_{ex}(0)$ is the true error exponent at $R = 0$!

An important aspect of the lemma is that it brings to light the relevance of minimum distance for determining error probabilities for low rate codes. This relevance was noticed in the expurgated bound, where codewords too close to each other were eliminated. On the other hand for high rates close to capacity the minimum distances are relatively unimportant because it can be seen that in the random-coding ensemble most codes have a very small minimum distance.

By expurgation of the ensemble minimum distance is greatly increased, but at high rates the ensemble average error probability since this average is close to the sphere-packing bound to begin with.

It should also be emphasized that the assumption $M \geq n + 2$ is very essential for the bound (7.33) not only technically, because Plotkin's bound becomes effective, but also in reality:

For $M < n + 2$ and M fixed

$$\lim_{n \to \infty} \frac{-\ell n \lambda(n, M)}{n} = \frac{M}{M-1} E_{ex}(0).$$

(See [SGB].)

We are now prepared to go for the straight-lim exponent for the BSC $W = \begin{pmatrix} 1 - \varepsilon & \varepsilon \\ \varepsilon & 1 - \varepsilon \end{pmatrix}$, $\varepsilon < 1/2$. Let $\delta, \varepsilon < \delta < 1/2$, arbitrary otherwise and define

$$R_1 = \ell n 2 - h(\delta) \tag{7.40}$$

$$\begin{aligned} E_{sp}(R_1) &= -h(\delta) - \delta \ell n \varepsilon - (1 - \delta)\ell n(1 - \varepsilon) \\ &= D\big((\delta, 1 - \delta)\|(\varepsilon, 1 - \varepsilon)\big). \end{aligned} \tag{7.41}$$

For an (n, M) code with $M = e^{Rn}$ define the number α by

$$R = \alpha R_1 + \frac{3\ell n[2(n+1)]}{2n}. \tag{7.42}$$

We consider rates R in the range

$$-\frac{\varepsilon}{2n}\ell n[2(n+1)] \leq R \leq R_1 \tag{7.43}$$

so that $0 \leq \alpha < 1$.

Now define $n_1 = \lfloor \alpha n \rfloor$ and $n_2 = N - n$. By (7.42) the number of codewords $M = \lceil \exp nR \rceil$ satisfies

$$\frac{M}{n+1} \geq \sqrt{8(n+1)} \exp\{n\lambda R_1\} \tag{7.44}$$

$$\geq \sqrt{8(n_1 + 1)} \exp\{n_1 R_1\} \tag{7.45}$$

By the list code sphere packing bound (!) (Lemma 1)

$$\overline{\lambda}(n_1, M, n + 1) \geq \frac{\varepsilon}{(1-\varepsilon)\sqrt{8(n_1+1)}} \exp\{-n, E_{sp}(R_1)\}$$

$$\geq \frac{\varepsilon}{(1-\varepsilon)\sqrt{8(n+1)}} \exp\{-\lambda n E_{sp}(R_1)\} \tag{7.46}$$

and from Lemma 62

$$\lambda_{\max}(n_2, n + 2) \geq \frac{\sqrt{n_2}\varepsilon}{(n_2 + 4)(1 - \varepsilon)} \exp\{-n_2 E_{ex}(0)\}$$

$$\geq \frac{\varepsilon}{(n + 4)(1 - \varepsilon)} \exp\{-[(1 - \lambda)n + 1]E_{ex}(0)\}$$

$$\geq \frac{\sqrt{2}\varepsilon^{5/4}}{(n + 4)(1 - \varepsilon)^{3/4}} \exp\{-(1 - \lambda)n E_{ex}(0)\}. \tag{7.47}$$

Combining (7.46) and (7.47) with Lemma 60 yidles

$$\overline{\lambda}(n, M) \geq \frac{\varepsilon^{9/4}}{2(1 - \varepsilon)^{7/4}\sqrt{n + 1}(n + 4)}$$
$$\cdot \exp\{-n[\lambda E_{sp}(R_1) + (1 - \lambda)E_{ex}(0)]\}. \tag{7.48}$$

Finally using (7.42) and bringing the coefficient inside the exponent, we get

$$\overline{\lambda}(n, M) \geq \exp\left\{-n\left[E_{ex}(0) - R\left[\frac{E_{ex}(0) - E_{sp}(R_1)}{R_1}\right] + o_1(n)\right]\right\} \tag{7.49}$$

where

$$o_1(n) = \frac{1}{n}\left[\frac{E_{ex}(0) - E_{sp}(R_1)}{R_1}\frac{3}{2}\ell n[2(n + 1)] - \ell n\frac{\varepsilon^{9/4}}{2(1 - \varepsilon)^{7/4}\sqrt{n + 1}(n + 4)}\right]. \tag{7.50}$$

The exponent in (7.49) is linear in R describing a line from $(o, E_{ex}(0))$ to $(R_1, E_{sp}(R_1))$. The tightest bound arises by choosing R_1 (that is δ) to minimize this linear function.

Theorem 51 *For a BSC* $W = \begin{pmatrix} 1 - \varepsilon & \varepsilon \\ \varepsilon & 1 - \varepsilon \end{pmatrix}$, $0 < \varepsilon < 1/2$, *let* R_1 *be the rate at which a straight line through the point* $(R, E) = (0, E_{ex}(0))$ *is tangent to the curve* $(R, E_{sp}(R))$. *Then for any* $n \geq 1$ *and any* R *satisfying (7.43),* $\overline{\lambda}(n, e^{Rn})$ *satifies (7.49).*

7.1.5 Reliability Function of a Discrete Memoryless Channel at Rates Above Capacity

In [6] Dueck and Körner gave asymptotically coincident upper and lower bounds on the exponent of the largest possible probability of the correct decoding of block codes for all rates above capacity. The lower bound sharpens Omura's bound. The upper bound is proved by a simple combinatorial argument.

7.1.5.1 Introduction

For a given discrete memoryless channel (DMC) let $P_c(R, n)$ be the largest possible average probability of correct decoding achievable by block codes of length n and rate at least R. In [3] Arimoto proved an upper bound for $P_c(R, n)$ by applying Gallager-type estimates [9] together with an elegant idea of symmetrization, noticing that his result implies Wolfowitz's strong converse to the noisy channel coding theorem [20]. More recently Omura [14] established a lower bound on $P_c(R, n)$ using the "dummy channel" lower bounding method introduced by Haroutunian [12] and Blahut [4], [5]. The two bounds are not directly comparable; nevertheless it can be shown that they asymptotically coincide in *some range of rates* near capacity.

Here we give asymptotically coincident upper and lower bounds on $P_c(R, n)$ for all rates above capacity. The proof of the lower bound (i.e., the existence result) is an improved version of Omura's argument. The upper bound is established by a very elementary combinatorial argument. Though it can be shown to be equivalent to Arimoto's upper bound, the main point is that unlike Arimoto's development, the present proof gives the exponent of the upper bound in the same form as that of the lower bound. Moreover the new proof of the upper bound is simpler.

7.1.5.2 Definitions and Statement of the Result

The transmission probabilities of a DMC $W : \mathcal{X} \to \mathcal{Y}$ with finite alphabets $\mathcal{X}$ and $\mathcal{Y}$ are given for n-length sequences $x = (x_1, \dots, x_n) \in X^n, y = (y_1, \dots, y_n) \in Y^n$ by

$$w^n(y|x) \triangleq \prod_{i=1}^{n} w(y_i|x_i),$$

where w is a stochastic matrix with $|\mathcal{X}|$ rows and $|\mathcal{Y}|$ columns. An *n-length block code* for this DMC is a pair of mappings $f : M \to \mathcal{X}^n, \varphi : \mathcal{Y}^n \to M'$, where M and M' are arbitrary finite sets, M being the set of messages to be transmitted and M' the set of possible decisions of the decoder. We suppose that M' contains M. The *average probability of correct decoding* of the code (f, φ) is

$$\bar{c}(w, f, \varphi) \triangleq \frac{1}{|M|} \sum_{m \in M} w^n \left(\varphi^{-1}(m) | f(m) \right).$$

Write $P_c(R, n) \triangleq \max \bar{c}(w, f, \varphi)$, where the maximum is taken over all n-length block codes (f, φ) with message set M satisfying

$$\frac{1}{n} \log |M| \geq R.$$

Let P be the set of all probability distributions on X and let V be the set of all stochastic matrices $V : X \to Y$. For every $P \in \mathcal{P}$ and $\mathcal{V} \in V$ we define the output distribution PV of $\mathcal{V}$ corresponding to input distribution P by

$$PV(y) \triangleq \sum_{x \in X} P(x) V(y|x), \quad \text{for all } y \in Y.$$

Let $H(P)$ denote the *entropy* of $P \in \mathcal{P}$

$$H(P) \triangleq - \sum_{x \in X} P(x) \log P(x)$$

and let $H(V|P)$ be the *conditional entropy* of the output of channel V given an input of distribution P, i.e.

$$H(V|P) \triangleq - \sum_{x \in X} \sum_{y \in Y} P(x) V(y|x) \log V(y|x).$$

The corresponding *mutual information* is

$$I(V|P) \triangleq H(PV) - H(V|P)$$

$$= \sum_{x \in X} \sum_{y \in Y} P(x) V(y|x) \log \frac{V(y|x)}{PV(y)}.$$

Finally, let $D(V \| W | P)$ be the average informational divergence between the rows of the matrices V and w given an input of distribution P, i.e.,

$$D(V \| w | P) \triangleq \sum_{x \in X} \sum_{y \in Y} P(x) V(y|x) \log \frac{V(y|x)}{w(y|x)}.$$

We shall prove the following theorem.

Theorem 52 *For the DMC $w : X \to Y$ we have*

$$\lim_{n \to \infty} \left(-\frac{1}{n} \log P_c(R, n) \right) = \min_{P \in \mathcal{P}} \min_{V \in \mathcal{V}} \{ D(V \| w | P) + (R - I(P, V))^+ \}$$

for all rates $R \geq 0$, where $i^+ \triangleq \max\{0, t\}$.

7.1.5.3 Proof of the Theorem

We establish the theorem by means of several lemmas. First we state some well-known elementary combinatorial identities and inequalities which are needed in the sequel. The proof of the direct part of the theorem is in Lemma 64. The converse part is Lemma 67, the proof of which relies on the combinatorial Lemma 65.

For every $c \in X^n$ and $x \in X$ let $N(x|x)$ be the number of occurences of x among the coordinates of the sequence $x \in X^n$. The empirical distribution, in short E-distribution of a sequence $x \in X^n$ is the distribution $P_x \in \mathcal{P}$ on X whose probabilities are the numbers

$$P_x(x) \triangleq \frac{1}{n} N(x|x);$$

similarly the joint E-distribution $P_{x,y}$ is defined for a pair $x \in X^n$, $y \in Y^n$ by

$$P_{x,y}(x, y) \triangleq \frac{1}{n} N(x, y|x, y).$$

For every integer n let P_n denote the family of those distributions on X which are possible E-distributions of sequences $x \in X^n$. Similarly, for every $P \in \mathcal{P}_n$ let us denote by $V_n(P)$ the set of all stochastic matrices $V : X \to Y$ such that for some pair (x, y), $x \in X^n$, $y \in Y^n$, we have $P_{x,y}(x, y) = P(x)V(y|x)$. For every E-distribution $P \in \mathcal{P}$ write

$$T_P \triangleq \{x \in X^n | P_x = P\}.$$

For every $x \in T_P$ and $V \in \mathcal{V}_n(P)$ we define

$$T_V(x) \triangleq \{y \in Y^n | P(x)V(y|x) = P_{x,y}(x, y), \text{ for every } x \in X, y \in Y\}.$$

Lemma 63 *For every integer nwe have*

(A) $|P_n| \leq (n+1)^{|X|}$ *and* $|V_n(P)| \leq (n+1)^{|X| \cdot |Y|}$, *for every $P \in \mathcal{P}_n$.*

(B) *Let $P \in \mathcal{P}_n$, $V \in \mathcal{V}_n(P)$, $x \in T_P$, $y \in T_V(x)$. For every stochastic matrix $\hat{V} \in \mathcal{V}$*

$$\hat{V}^n(y|x) = \exp\{-n(D(V\|\hat{V}|P) + H(V, P))\}.$$

(C) *For every $P \in \mathcal{P}_n$, $V \in \mathcal{V}_n(P)$, and $x \in T_P$ er have*

$$(n + 1)^{-|X|} \exp\{nH(P)\} \leq |T_P| \leq \exp\{nH(P)\}.$$

(D) *For every $P \in \mathcal{P}_n$, $V \in \mathcal{V}_n(P)$, and $x \in T_P$ we have*

$$(n + 1^{-|Y|} \exp\{nH(V, P)\} \leq |T_V(x)| \leq \exp\{nH(P)\}.$$

Proof (A) is well-known and trivial; (B) is a consequence of the definitions. In order to prove (C), notice first that

$$1 \geq P^n(T_P) = |T_P| \cdot \exp\{-nH(P)\}$$

whence the upper bound follows. For the lower bound, it is sufficient to whoc that $P^n(T_{\hat{P}})$ is maximized by $\hat{P} = P$ whence the lower bound follows by (A). Since we have

$$\frac{P^n(T_{\hat{P}})}{P^n(T_P)} = \prod_{x \in X} \frac{(nP(x))!}{(nP(x))!} P(x)^{n(\hat{P}(x) - P(x))},$$

by applying the obvious inequality

$$\frac{k!}{l!} \leq k^{k-l}$$

we get

$$\frac{P^n(T_{\hat{P}})}{P^n(T_P)} \leq \prod_{x \in X} n^{n(P(x) - \hat{P}(x))} = 1.$$

Finally (D) follows from (C).

Lemma 64 (The direct part of the theorem) *For all $R \geq 0$*

$$\limsup_{n \to \infty} -\frac{1}{n} \log P_c(R, n) \leq \min_{P \in \mathcal{P}} \min_{V \in \mathcal{V}} \{D(V \| w | P) + (R - I(V | P))^+\}.$$

Proof The lemma is established by an improved version of Omura's idea [14].

Consider for a fixed $\delta > 0$ and large enough n a E-distribution $P \in \mathcal{P}_n$ and a "dummy channel" $V \in \mathcal{V}_n(P)$. By the maximal code lemma of Feinstein or by random selection as in Chapter ... one can get an n-length block code (f, φ) with message set M satisfying:

(a) the average probability $\bar{c}(V, f, \varphi)$ of correct decoding of (f, φ) with respect to the DMC V is at least $1 - \delta$;
(b) all the codewords, i.e., the values of f have type P;
(c) $(11/n) \log |M| \geq \min\{R, I(V | P) - \delta\}$.

Now we consider this n-length block code (f, φ) as a code for our original channel $W : X \to Y$. If $R \leq I(V | P) - \delta$, this gives

$$\overline{c}(w,f,\varphi) \geq \exp\{-n(D(V\|w|P)+\delta)\} \tag{7.51}$$

by (a)–(c) and the properties (B) and (D) of Lemma 63 if n is large enough.

If $R > I(V|P) - \delta$ we modify the code (f, φ) with message set M as follows.

We enlarge the set M to a new message set $\hat{M} \supset M$ such that $|\hat{M}|$ is the smallest number exceeding $\exp\{nR\}$. Let $\hat{f} : \hat{M} \to X^n$ be an arbitrary function satisfying $\hat{f} = f$ on M, and leave the decoder φ unchanged. Then clearly, for n large enough, the code $(\hat{f}, \varphi)$ with message set $\hat{M}$ yields

$$\overline{c}(w,\hat{f},\varphi) = \frac{1}{|\hat{M}|} \left(\sum_{m \in M} w^n\big(\varphi^{-1}(m)|f(m)\big) + \sum_{m \in \hat{M}-M} w^n\big(\varphi^{-1}(m)|\hat{f}(m)\big) \right)$$

$$\geq \frac{1}{|\hat{M}|} \sum_{m \in M} w^n\big(\varphi^{-1}(m)|f(m)\big)$$

$$\geq \exp\{-n(R+\delta)\} \cdot |M| \frac{1}{|M|} \cdot \sum_{m \in M} w^n\big(\varphi^{-1}(m)|f(m)\big)$$

$$= \exp\{-n(R+\delta)\} \cdot |M| \cdot \overline{c}(w,f,\varphi)$$

$$\geq \exp\{-n(R+\delta)\} \cdot \exp\{n(I(V|P)-\delta)\}\overline{c}(w,f,\varphi),$$

where the last inequality follows by c) and the assumption $R > I(V|P) - \delta$. Hence, by (7.51),

$$\overline{c}(w,\hat{f},\varphi) \geq \exp\{-n(D(V\|W|P) + (R - I(V|P))^+ + 3\delta)\}. \tag{7.52}$$

Equations (7.51) and (7.52) together prove the following statement.

For any $R > 0$ and $\delta > 0$ there is an integer n_0 such that for every $n \geq n_0$, every F-distribution $P \in \mathcal{P}_n$, and $V \in \mathcal{V}_n(P)$ there is an n-length block code (f, φ) with message set M satisfying

$$\frac{1}{n} \log |M| \geq R$$

and

$$\overline{c}(w,f,\varphi) \geq \exp\{-n(D(V\|w|P) + (R - I(P|V))^+\delta)\}.$$

Now observe that for large n the set $\mathcal{P}_n$ is arbitrarily close to the set $\mathcal{P}$ of all probability distributions on X, and that for any $P \in \mathcal{P}$ the set $\mathcal{V}_N(P)$ is arbitrarily close to the set V of all channels $V : X \to Y$.

Lemma 65 *For any $R > 0$ and $\delta > 0$ there is an n_0 depending on $|X|, |Y|$ and δ such that the following is true: given $n \geq n_0$, $P \in \mathcal{P}_n$, $V \in \mathcal{V}_n(P)$ and a collection of (not necessarily distinct) sequences $\{x_i | i \in M\} \subset T_P$ with $(1/n) \log |M| \geq R + \delta$ for any mapping $\varphi : Y^n \to M' \supset M$ we have*

$$\frac{1}{|M|} \sum_{i \in M} \frac{|T_V(x_i) \cap \varphi^{-1}(i)|}{|T_V(x_i)|} \leq \exp\{-n(R - I(V|P))^+\}. \tag{7.53}$$

Proof Let us denote the expression on the left side of (7.53) by F. Obviously, F cannot exceed one. Hence it suffices to prove

$$F \leq \exp\{-n(R - I(P, V))\}. \tag{7.54}$$

By (D) we have

$$F \leq (n+1)^{|Y|} \cdot \exp\{-nH(V, P)\} \cdot \frac{1}{|M|} \sum_{i \in M} |T_V(x_i) \cap \varphi^{-1}(i)|$$

$$\leq (n+1)^{|Y|} \cdot \exp\{-nH(V, P)\} \cdot \frac{1}{|M|} \cdot |T_{PV}|$$

where the last inequality follows from the fact that $x_i \in T_P$ and that the sets $T_V(x_i) \cap \varphi^{-1}(i)$ are disjoint. We further bound using (C):

$$F \leq (n+1)^{|Y|} \cdot \exp\{-nH(VP)\} \cdot \exp\{-n(R+\delta)\} \cdot \exp\{nH(P, V)\}$$

$$\leq (n+1)^{|Y|} \cdot \exp\{-n(R - I(P, V) + \delta)\}.$$

Hence if n is so large that $(n+1)^{|Y|} \cdot \exp(-n\delta) \leq 1$, we have (7.54) and then (7.53) follows.

This lemma will lead to the converse estimate. Er first establish a simple lemma concerning our exponent function.

Lemma 66 *The function*

$$G(R, w) \triangleq \min_{P \in \mathcal{P}} \min_{V \in \mathcal{V}} \{D(V \| w | P) + (R - I(P, V))^+\}$$

satisfies the inequality

$$|G(R', w) - G(R'', w)| \leq |R' - R''|$$

for every $R' \geq 0$ and $R'' \geq 0$.

Proof Suppose $R'' > R'$. Let $P' \in \mathcal{P}$ and $V' \in \mathcal{V}$ achieve $G(R', w)$. Then

$$G(R'', w) \leq D(V' \| w | P') + (R'' - I(P', V'))^+$$
$$\leq D(V' \| w | P') + (R'' - R') + (R' - I(P', V'))^+$$
$$= G(R', w) + R'' - R.$$

Since $G(R, w)$ is increasing in R, the lemma is proved.

Lemma 67 (The converse part of the theorem) *For $R \geq 0$ we have*

$$\liminf_{n \to \infty} -\frac{1}{n} \log P_c(R, n) \geq \min_{P \in \mathcal{P}} \min_{V \in \mathcal{V}} \left\{ D(V \| w | P) + (R - I(P, V))^+ \right\}.$$

Proof Fix some $\delta > 0$ and consider any n-length block code $(\hat{f}, \hat{\varphi})$ with message set M for the DMC $w : X \to Y$ such that $(1/n) \log |M| \geq R + \delta$. Clearly there is a F-distribution $P \in \mathcal{E}_n$ such that for eome new encoder $f : M \to \hat{f}(M) \cap T_P$

$$\bar{c}(w, \hat{f}, \hat{\varphi}) \leq (n + 1)^{|X|} \bar{c}(w, \hat{f}, \hat{\varphi}). \tag{7.55}$$

In fact, let P be a distribution from P_n which maximizes the expression

$$\frac{1}{|M|} \sum_{m : \hat{f}(m) \in T_P} w^n \left(\hat{\varphi}^{-1}(m) | \hat{f}(m) \right).$$

Clearly, defining

$$f(m) \triangleq \begin{cases} \hat{f}(m), & \text{if } \hat{f}(m) \in T_P, \\ \text{arbitrary element of } T_P, & \text{otherwise,} \end{cases}$$

the same decoder $\varphi \triangleq \hat{\varphi}$ as before leads to a code (f, φ) with message set M satisfying (7.55).

Fix any $V \in \mathcal{V}_n(P)$. Then the probability $w^n(y|x)$ is the same for every (x, y) with $x \in T_P$ and $y \in T_V(x)$. Using this, B) and D) of Lemma 63 yield the estimate

$$\bar{c}(w, f, \varphi) = \frac{1}{|M|} \sum_{m \in M} w^n \left(\varphi^{-1}(m) | f(m) \right)$$

$$\leq \sum_{V \in \mathcal{V}_n(P)} \frac{1}{|M|} \sum_{m \in M} w^n \left(\varphi^{-1}(m) \cap R_V \left(f(m) \right) | f(m) \right)$$

$$= \sum_{V \in \mathcal{V}_n(P)} \frac{1}{|M|} \sum_{m \in M} w^n \left(T_V \left(f(m) | f(m) \right) \right) \cdot \frac{|T_V(f(m)) \cap \varphi^{-1}(m)|}{|T_V(f(m))|}$$

$$= \sum_{V \in \mathcal{V}_n(P)} \exp\{-nD(V \| w | P)\} \cdot \frac{1}{|M|} \sum_{m \in M} \frac{|T_V(f(m)) \cap \varphi^{-1}(m)|}{|T_V(f(m))|}.$$

Hence by Lemma 65

$$\bar{c}(w, f, \varphi) \leq \sum_{V \in \mathcal{V}_n(P)} \exp\{-n\left(D(V \| w | P) + (R - I(P, V))^+\right\}.$$

By (A), we can further bound this:

$$\bar{c}(w,f,\varphi) \le (n+1)^{|X|\cdot|Y|} \cdot \exp\left\{-n\left(\min_{V\in\mathcal{V}}\left(D(V\|w|P) + (R - I(P,V))^+\right)\right)\right\}.$$

$$(7.56)$$

Since $\lim_{n\to\infty}(1/n)\log(n+1)^{|X|\cdot|Y|} = 0$, (7.55) and (7.56) complete the proof of the inequality

$$\liminf_{n\to\infty} -\frac{1}{n}\log P_c(R+\delta, n) \ge G(R, W),$$

using the notation of Lemma 66, or equivalently,

$$\liminf_{n\to\infty} -\frac{1}{n}\log P_c(R, n) \ge G(R-\delta, w).$$

However by Lemma 66 $G(R - \delta, w) \ge G(R, w) - \delta$ and thus

$$\liminf_{n\to\infty} -\frac{1}{n}\log P_c(R, n) \ge G(R, w) - \delta$$

which is the statement of the lemma.

Lemmas 64 and 67 together prove the theorem.

7.1.5.4 Discussion

It can be shown by routine analytic technique that our exponent function $G(R, w)$ coincides with the one appearing in Arimoto's upper bound [3]. However, this was not needed, for we have established asymptotically coincident bounds in the two directions without returning to Arimoto's result and have thus proved that $G(R, w)$ is the reliability function of the DMC w at rates R above capacity. Clearly for $R = C$, $G(R, w) = 0$.

Originally this theorem had been proved independently and at nearly the same time by the two authors. The present paper is a combination of their original arguments. In particular, the converse is based on the approach of the second author. The first author originally applied list codes in the manner of Ahlswede [1] and showed by a symmetrization argument that they are "essentially the best possible".

7.2 Lecture on Upper Bounds on Error Probabilities

7.2.1 The Random Coding Upper Bound on the Optimal Error Probability for the DMC

We have derived in Sect. 7.1 the sphere packing bound as a lower bound for $\lambda(n, R, W)$ and in the last lecture we saw how this name originated in the analysis of the BSC by Elias [7]. We also learned there that above a certain rate $R_{(n)}$ this bound is tight, because it coincides with an upper bound derived from the average over all possible codes. For general DMC the idea of this upper bound led to the random coding bound, originally derived by Fano [8].

We follow here the technically simple and beautiful derivation by Gallager [9], which proceeds as follows.

1. Consider any set of codewords $u_1, \ldots, u_M \in \mathcal{X}$ and use the best decoder for the average error probability, namely the ML-decoder.
2. Thus we can express λ_i, the error probability for word u_i, as

$$\lambda_i = \sum_{y \in \mathcal{Y}} W(y|u_i)\phi_i(y), \qquad (7.57)$$

where

$$\phi_i(y) = \begin{cases} 1, & \text{if } W(y|u_i) \leq W(y|u_j) \text{ for some } j \neq i \\ 0 & \text{otherwise.} \end{cases}$$

3. We upperbound now λ_i by upperbounding $\phi_i(y)$.

$$\phi_i(y) \leq \left[\frac{\sum\limits_{j \neq i} W(y|u_j)^{1/1+\rho}}{W(y|u_i)^{1/1+\rho}} \right]^{\rho}, \rho > 0. \qquad (7.58)$$

Indeed, since the right side expression is non-negative it suffices to check the case $\phi_i(y) = 1$. But here some term in the numerator is at least as large as the denominator. Raising the fraction to the ρ power keeps it greater than or equal to 1.

At the moment (7.58) is a trick falling from the sky and not at all obvious intuitively. Previous work stands behind it.

Substituting (7.58) into (7.57) gives

$$\lambda_i \leq \sum_{y \in \mathcal{Y}} W(y|u_i)^{\frac{1}{1+\rho}} \left[\sum_{j \neq i} W(y|u_i)^{\frac{1}{1+\rho}} \right]^{\rho}, \rho > 0. \qquad (7.59)$$

Clearly $\lambda_i = \lambda_i(u_1, \ldots, u_M)$ for any set of codewords and the bound is too complicated to be useful, if M is large.

4. However, by averaging over an appropriately chosen ensemble of codes the formula becomes rather smooth. Following Shannon we fix a PD P on $\mathcal{X}$ and generate an ensemble of codes by picking each codeword $U_1, \ldots, U_M$ independently, according to $(P, \mathcal{X})$. Thus $\prod_{i=1}^{M} P(u_i)$ is the probability associated with a code $\{u_1, \ldots, u_M\}$.

Clearly, at least one code in the ensemble will have a probability of error as small as the ensemble average probability of error. We now have

$$\mathbb{E}\,\lambda_i(U_1, \ldots, U_M) \leq \mathbb{E} \sum_y W(y|U_i)^{1/1+\rho} \left[\sum_{j \neq i} W(y|U_j)^{1+\rho} \right]^{\rho}$$

$$= \sum_y \mathbb{E}\, W(y|U_i)^{\frac{1}{1+\rho}} \left[\sum_{j \neq i} W(y|U_j)^{1+\rho} \right]^{\rho}$$

$$= \sum_y \mathbb{E}\, W(y|U_i)^{\frac{1}{1+\rho}} \cdot \mathbb{E} \left[\sum_{j \neq i} W(y|U_j)^{1+\rho} \right]^{\rho} . \quad (7.60)$$

(We have used that the expected value of sums of RV's equals the sum of their expected values and likewise for products), if the RV's are independent.

We require now $0 \leq \rho \leq 1$. Then by Jensen's inequality (see Fig. 7.1) $\mathbb{E}\,\xi^\rho \leq (\mathbb{E}\,\xi)^\rho$ and therefore

$$\mathbb{E} \left[\sum_{j \neq i} W(y|U_j)^{1+\rho} \right]^{\rho} \leq \left[\mathbb{E} \sum_{j \neq i} W(y|U_j)^{1+\rho} \right]^{\rho} = \left[\sum_{j \neq i} \mathbb{E}\, W(y|U_j)^{1+\rho} \right]^{\rho} .$$

$$(7.61)$$

Furthermore, since the RV's U_i have distribution P

$$\mathbb{E}\, W(y|U_i)^{1/1+\rho} = \sum_x P(x) W(y|x)^{1/1+\rho} \quad (7.62)$$

(independent of i!) and substitutions in (7.60) with (7.61) and (7.62) give for any $\rho, 0 < \rho \leq 1$,

$$\mathbb{E}\,\lambda_i \leq (M-1)^{\rho} \sum_y \left[\sum_x P(x) W(y|x)^{1/1+\rho} \right]^{1+\rho} . \quad (7.63)$$

Fig. 7.1 Convexity

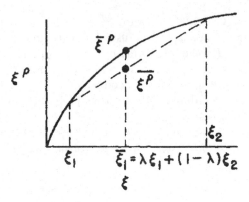

Consider now a DMC for blocklength n, and choose P^n such that $P^n(x^n) = \prod_{t=1}^n P(x_t)$, then we get from (7.63)

$$\mathbb{E}\,\lambda_i \le (M-1)^\rho \sum_{y^n} \left[\sum_{x^n} \prod_{t=1}^n p(x_t) W(y_t|x_t)^{1/1+\rho} \right]^{1+\rho} \tag{7.64}$$

by the distributive law of arithmetic

$$\mathbb{E}\,\lambda_i \le (M-1)^\rho \prod_{t=1}^n \sum_{y_t} \left[\sum_{x_t} p(x_t) W(y_t|x_t)^{1/1+\rho} \right]^{1+\rho}. \tag{7.65}$$

For stationary DMC this simplifies

$$\mathbb{E}\,\lambda_i \le (M-1)^\rho \sum_{y} \left[\sum_{x} p(x) W(y|x)^{1/1+\rho} \right]^{1+\rho}, 0 \le \rho \le 1. \tag{7.66}$$

We upperbound this by replacing $M-1$ by $M = \exp\{Rn\}$, where R is the code rate,

$$\mathbb{E}\,\lambda_i \le \exp\left\{ -n \left[-\rho R - \ell n \sum_{x} \left(\sum_{y} P(x) W(y|x)^{1/1+\rho} \right)^{1+\rho} \right] \right\}.$$

We have proved the following fundamental result.

Theorem 53 *For any DMC* $W : \mathcal{X} \to \mathcal{Y}$, *any blocklength n, any number of code-words* $M = e^{Rn}$, *and any PD P on* $\mathcal{X}$ *there exists an* $(n, M, \overline{\lambda})$-*code with*

$$\overline{\lambda} \le \exp\{-n[-\rho R + E_0(\rho, P)]\},$$

where

$$E_0(\rho, P) = -\ell n \sum_x \left(\sum_y P(x)W(y|x)^{1/1+\rho} \right)^{1+\rho} , 0 \le \rho \le 1,$$

and also an $(n, M, \overline{\lambda})$*-code with* $\overline{\lambda} \le \exp\{-n \, E_r(R)\}$*, where*

$$E_r(R) = \max_{P,\rho \in [0,1]} \left[-\rho R + E_0(\rho, P) \right]. \tag{7.67}$$

$E_r(R)$ *is called random coding exponent.*

For the maximal probability of error we have the immediate consequence.

Corollary 5   *Under the assumptions of Theorem 53 there is an* (n, M, λ)*-code with*

$$\lambda \le 4 \, e^{-n \, E_r(R)}.$$

Properties of $E_r(R)$
The random coding exponent depends on the behavior of the function $E_0(\rho, P)$*. It is
described in Theorem 54 as function of* ρ *and in Theorem 55 as function of* P*.*

Theorem 54   *For* $W : \mathcal{X} \to \mathcal{Y}$ *and* $P \in \mathcal{P}(\mathcal{X})$ *the function*

$$E_0(\rho, P) = -\ell n \sum_x \left(\sum_y P(x)W(y|x)^{1/1+\rho} \right)^{1+\rho}$$

has the properties

(i) $E_0(\rho, P) = 0$ *for* $\rho = 0$
(ii) $E_0(\rho, P) > 0$ *for* $\rho > 0$
(iii) $\left. \frac{\partial E_0(\rho,P)}{\partial \rho} \right|_{\rho=0} = I(W|P)$, $\frac{\partial E_0(\rho,P)}{\partial \rho} > 0$ *for* $\rho > 0$
(iv) $\frac{\partial^L E_0(\rho,P)}{\partial \rho^2} \le 0$ *with equality exactly if the following two conditions hold:*

  (a) $W(y|x)$ *is independent of* x *for* y, x *with* $P(x)W(y|x) \ne 0$
  (b) $\sum_{x:W(y|x) \ne 0} P(x)$ *is independent of* y.

We need an auxiliary result.

Lemma 68   *Let* $a_1, \ldots, a_L$ *be a set of non-negative reals and let* $q_1, \ldots, q_L$ *be
probabilities adding up to 1. Then*

$$f(\beta) = \ell n \left(\sum_\ell q_\ell a_\ell^{1/\beta} \right)^\beta \tag{7.68}$$

*is non-increasing in $\beta > 0$ and is strictly decreasing unless the a_ℓ for which $q_\ell \neq 0$
are all equal. Also f is convex ($\cup$) for $\beta > 0$ and is strictly convex unless all the
non-zero a_ℓ for which $q_\ell \neq 0$ are equal.*

Proof Define $r = 1/\beta$ and recall (Sect. 3.2.3) that the weighted mean $\left(\sum_\ell q_\ell q_\ell^r \right)^{1/r}$ is

nondecreasing for $r > 0$ and strictly increasing unless the a_ℓ for which $q_\ell \neq 0$ are all
equal. Another property of weighted means coming from Hölder's inequality is, that
if r and t are unequal positive numbers, θ satisfies $0 < \theta < 1$, and $s = \theta r + (1 - \theta)t$,
then

$$\sum_\ell q_\ell a_\ell^s \leq \left(\sum_\ell q_\ell a_\ell^r \right)^\theta \left(\sum_\ell q_\ell a_\ell^t \right)^{1-\theta} \tag{7.69}$$

with equality only if all the non-zero a_ℓ for which $q_\ell \neq 0$ are equal. Let α be
defined by

$$\alpha = \frac{r\theta}{r\theta + t(1-\theta)}, \quad \frac{1}{s} = \frac{\alpha}{r} + \frac{1-\alpha}{t}. \tag{7.70}$$

Substituting (7.70) into (7.69) and taking the $1/s$ power of each side, gives

$$\left(\sum_\ell q_\ell a_\ell^s \right)^{1/s} \leq \left(\sum_\ell q_\ell a_\ell^r \right)^{\alpha/r} \left(\sum_\ell q_\ell a_\ell^t \right)^{\frac{1-\alpha}{t}}. \tag{7.71}$$

Taking the logarithm of both sides and interpreting $1/r$ and $1/t$ as two different values
of z, we find that f is convex ($\cup$) with strict convexity under the stated conditions.

Proof of Theorem 54 From Lemma 68 $\left(\sum_x P(x)w(y|x)^{1/1+\rho} \right)^{1+\rho}$ is non-increasing

in ρ. Since $I(W|P) \neq 0$ by assumption, there is at least one y for which $W(y|x)$

changes with x for $P(x) \neq 0$, for that y $\left(\sum_x P(x)W(y|x)^{1/1+\rho} \right)^{1+\rho}$ is strictly decreas-

ing, and thus $E_0(\rho, P)$ is strictly increasing with ρ.

Verify directly that (i) $E_0(0, P) = 0$ holds and consequently also (ii) holds and
(iii) $\frac{\partial E_0(\rho,P)}{\partial \rho} > 0$ for $\rho > 0$ and $\frac{\partial E_0(\rho,P)}{\partial \rho} = I(W|P)$ by direct differentiation.

Finally, let ρ_1 and ρ_2 be unequal positive numbers, let α satisfy $0 < \alpha < 1$, and
let $\rho_3 = \alpha\rho_1 + (1 - \alpha)\rho_2$. From Lemma 68

$$\left(\sum_x P(x)W(y|x)^{1/1+\rho_3} \right)^{1+\rho_3}$$

$$\leq \left(\sum_x P(x)W(y|x)^{1/1+\rho_1} \right)^{\alpha(1+\rho_1)}$$

$$\left(\sum_x P(x) W(y|x)^{\frac{1}{1+\rho_2}} \right)^{(1-\lambda)(1+\rho_2)} . \tag{7.72}$$

We apply now Hölder's inequality:

$$\sum_i a_i b_i \leq \left(\sum_i a_i^{1/\alpha} \right)^{\alpha} \left(\sum_i b_i^{1/1-\alpha} \right)^{1-\alpha} , \; a_i, b_i \geq 0 \tag{7.73}$$

"=" iff a_i and b_i are proportional.

Letting a_y and b_y by the two terms at the right in (7.72), over y in (7.71) and upper bounding the new right side by Hölder we obtain

$$\sum_y \left(\sum_x P(x) W(y|x) \right)^{\frac{1}{1+\rho_3}} \leq \left[\sum_y \left(\sum_x P(x) W(y|x)^{1/1+\rho_1} \right)^{1+\rho_1} \right]^{\alpha}$$

$$\cdot \left[\sum_y \left(\sum_x P(x) W(y|x)^{1/1+\rho_2} \right)^{1+\rho_2} \right]^{1-\alpha} . \tag{7.74}$$

Taking $-\ln$ of (7.74) establishes that $E_0(\rho, P)$ is convex ($\cap$) and thus that $\frac{\partial^2 E_0}{\partial \rho^2} \leq 0$.

The convexity is strict unless both (7.72) and (7.74) are satisfied with equality. But condition (iv, a) in Theorem 54 is the condition for (7.72) to hold with equality and condition (iv, b) is the condition for a_y and b_y to be proportional when condition (iv, a) is satisfied.

Using Theorem 54 we can do the maximization of

$$E_r(R) = \max_{0 \leq \rho \leq 1, P} \left[-\rho R + E_0(\rho, P) \right]$$

over ρ for fixed P. Define

$$E(R, P) = \max_{0 \leq \rho \leq 1} \left[-\rho R + E_0(\rho, P) \right] \tag{7.75}$$

$$\frac{\partial \left[-\rho R + E_0(\rho, P) \right]}{\partial \rho} = 0 \text{ implies}$$

$$R = \frac{\partial E_0(\rho, P)}{\partial \rho}. \tag{7.76}$$

From (iv) in Theorem 54, if for some ρ in [0, 1] (7.74) holds, then that ρ must maximize (7.75).

Furthermore, from (iv) $\frac{\partial E_0(\rho, P)}{\partial \rho}$ is non-increasing with ρ, so that a solution to (7.76) exists if R lies in the range

Fig. 7.2 Geometric construction

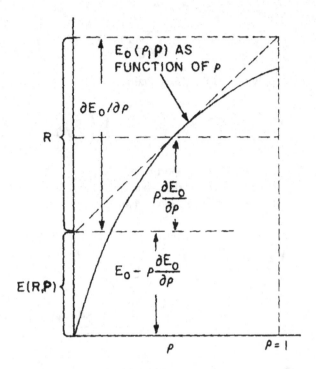

$$\left.\frac{\partial E_0(\rho, P)}{\partial \rho}\right|_{\rho=1} \leq R \leq I(W|P). \tag{7.77}$$

In this range it is most convenient to use (7.76) to relate $E(R, P)$ and R parametrically as functions of ρ. This gives us

$$E(R, \rho) = E_0(\rho, P) - \rho \frac{\partial E_0(\rho, P)}{\partial \rho} \tag{7.78}$$

$$R = \frac{\partial E_0(\rho, P)}{\partial \rho}, \ 0 \leq \rho \leq 1. \tag{7.79}$$

Figure 7.2 gives a construction for the solution of the parametric equations. For $R < \left.\frac{\partial E_0(\rho, P)}{\partial \rho}\right|_{\rho=1}$ the parametric equations (7.78) and (7.79) are not valid. In this case $-\rho R + E_0(\rho, P)$ increases with ρ in the range $0 \leq \rho \leq 1$, and the maximum occurs at $\rho = 1$. Thus

$$E(R, \rho) = E_0(1, P) - R \text{ for } R < \left.\frac{\partial E_0(\rho, P)}{\partial \rho}\right|_{\rho=1}. \tag{7.80}$$

The behavior of $E(R, P)$ as a function of R given by (7.78)–(7.80) is shown in Fig. 7.3.

Fig. 7.3 Rate curve

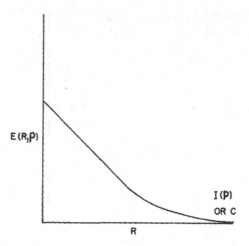

This behavior depends upon whether $\frac{\partial^2(\rho,P)}{\partial\rho^2}$ is negative or 0. If it is negative, then R as given by (7.78) is strictly decreasing with ρ. Differentiating (7.79) with respect to ρ, we get $-\rho\frac{\partial^2 E_0(\rho,P)}{\partial\rho^2}$; thus $E(R,\rho)$ is strictly increasing with ρ for $\rho \geq 0$, and is equal to 0 for $\rho = 0$. Thus if $R < I(W|P)$, then $E(R,P) > 0$, and is equal to 0 for $\rho = 0$. Thus if $R < I(W|P)$, then $E(R,P) > 0$. If P is chosen to achieve capacity C, then for $R < C$, $E(R,P) > 0$, and the error probability can be made to vanish exponentially with the blocklength.

Taking the ratio of the derivatives of (7.78) and (7.79), we see that

$$\frac{\partial E(R,P)}{\partial R} = -\rho. \tag{7.81}$$

Then the parameter ρ in (7.79) and (7.79) has significance as the negative slope of the (E,R)-curve.

From the conditions following (iv), it is clear that if $\frac{\partial^2 E_0(\rho,P)}{\partial\rho} = 0$ for one value of $\rho > 0$, it is 0 for all $\rho > 0$. Under these circumstances, R and $E(R,P)$ as given by (7.78) and (7.79), simply specify the point at which $R = I(W|P)$, $E(R,P) = 0$. The rest of the curve, as shown in Fig. 7.4 comes from (7.80).

The class of channels for which $\frac{\partial^2 E_0(\rho,P)}{\partial\rho^2} = 0$ is somewhat pathological. It includes noiseless channels, for which one can clearly achieve zero error probability for rates below the capacity. The exponential bounds here simply reflect the probability of assigning more than one message to the same codeword. The expurgated bound of Sect. 7.1.4 yields zero error probability in these cases.

As an example of a noisy channel with $\frac{\partial^2 E_0(\rho,P)}{\partial\rho^2} = 0$, see Fig. 7.4.

An alternative approach to the maximization of (19) over ρ is to regard $-\rho R + E_0(\rho,P)$ as a lower function of R with slope $-\rho$ and intercept $E_0(\rho,P)$ for fixed ρ.

Fig. 7.4 Rate curve for given channel

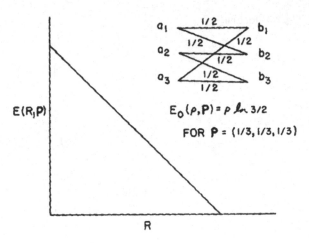

$$E_0(\rho,P) = \rho \ln 3/2$$

$$\text{FOR } P = (1/3, 1/3, 1/3)$$

Fig. 7.5 Rate curve as envelop of straight Lines

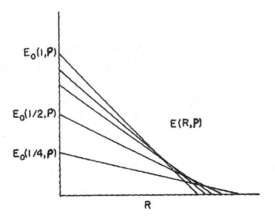

Thus $E(R, P)$ as a function of R is simply the upper envelope of this set of straight lines (see Fig. 7.5).

This picture also interprets $E_0(\rho, P)$ as the E-axis intercept of the tangent of slope $-\rho$ to the (E, R) curve.

Since $E(R, P)$ is the upper envelope of a set of straight lines, it must be a convex $(\cup)$ function of R, i.e., a function whose cords never lie below the function. This fact, of course, also follows from $\frac{\partial E(R,P)}{\partial R}$ decreasing with ρ and thus increasing with R. The function

$$E_r(R) = \max_P E(R, P) \tag{7.82}$$

is the upper envelope of all the $E(R, P)$ curves, and we have the following results.

Theorem 55 *For every $W : \mathcal{X} \to \mathcal{Y}$ $E_r(R)$ is positive, continuous, and convex $(\cup)$ for $R \in [0, C)$. Thus the error probability $\lambda \leq \exp\{-n\, E(R)\}$ is an exponentially decreasing function of the blocklength for $0 \leq R \leq C$.*

Proof If $C = 0$ nothing is to be proved. Otherwise for P with $I(W|P) = C$, we have shown that $E(R, P)$ is positive for $0 \leq R < C$, and then $E_r(R)$ is positive in the same range. Also, we have shown that for every P $E(R, P)$ is continuous in R and convex ($\cup$) with a slope between 0 and -1, and therefore the upper envelope is continuous and convex ($\cup$).

We shall show later that $E_r(R)$ does not have a continuous slope for any P, but the P that maximizes $E(R, P)$ can change with R and this can lead to discontinuities in the slope of $E(R)$.

We turn now to the actual maximization of $E(R, P)$ over P. We may write

$$E_r(R) = \max_{0 \leq \rho \leq 1} \left[-\rho R + \max_P E_0(\rho, R) \right]. \tag{7.83}$$

Now define

$$F(\rho, P) = \sum_y \left(\sum_x P(x) W(y|x)^{\frac{1}{1+\rho}} \right)^{1+\rho}. \tag{7.84}$$

So that $E_0(\rho, P) = -\ell n\, F(\rho, P)$ and minimizing $F(\rho, P)$ over P will maximize $E_0(\rho, P)$.

Theorem 56 $F(\rho, P)$ *is convex* ($\cup$) *over* $\mathcal{P}(\mathcal{X})$. *Necessary and sufficient conditions on P that minimize $F(\rho, P)$ are*

$$\sum_y W(y|x)^{1/1+\rho} \alpha_y^\rho \geq \sum_y \alpha_y^{1+\rho} \text{ for all } x \tag{7.85}$$

with equality if $P(x) \neq 0$, where

$$\alpha_y = \sum_x P(x) W(y|x)^{1/1+\rho}.$$

Remark If all the $P(x)$ are positive (7.85) is simply the result of applying a Lagrange multiplier to minimize $F(\rho, P)$ subject to $\sum P(x) = 1$. It will be shown that the same technique can be used to get necessary and sufficient conditions on P to maximize $I(W|P)$. The result is

$$\sum_y W(y|x) \ell n \frac{W(y|x)}{\sum_y P(x) W(y|x)} \leq I(W|P) \tag{7.86}$$

for all x with equality if $P(x) \neq 0$. (Shannon Lemma)

Neither (7.85) nor (7.86) is very useful in finding the maximum of $E_0(\rho, P)$ or $I(W|P)$, but both are useful theoretical tools and are useful in verifying that a hypothesized solution is indeed a solution. For channels of any complexity, $E_0(\rho, P)$ and $I(W|P)$ can usually be maximized most easily by numerical techniques.

Proof of Theorem 56 We begin by showing that $F(\rho, P)$ is a convex ($\cup$) function of P for $\rho \geq 0$. From (7.84) we can rewrite $F(\rho, P)$ as

$$F(\rho, P) = \sum_y \alpha_y^{1+\rho}, \quad \alpha_y = \sum_x P(x)W(y|x)^{1/1+\rho}. \tag{7.87}$$

For $P, Q \in \mathcal{P}(\mathcal{X})$ define in addition to α_y also $\beta_y = \sum_x Q(x)W(y|x)^{1/1+\rho}$.

For any $\gamma, 0 \leq \gamma < 1$, we have

$$F(\rho, \gamma P + (1-\gamma)Q) = \sum_y \left[\sum_x (\gamma P(x) + (1-\gamma)Q(x)W(y|x)^{1/1+\rho} \right]^{1+\rho}$$

$$= \sum_y [\gamma \alpha_y + (1-\gamma)\beta_y]^{1+\rho}. \tag{7.88}$$

Since α_y and β_y are non-negative, and since $z^{1+\rho}$ is convex ($\cup$) for $\rho \geq 0, z \geq 0$, we can upperbound the RHS of (7.88)

$$F(\rho, \gamma P + (1-\gamma)Q) \leq \sum_y \gamma \alpha_y^{1+\rho} + (1-\gamma)\beta_y^{1+\rho}$$

$$= \gamma F(\rho, P) + (1-\gamma)F(\rho, Q). \tag{7.89}$$

Thus $F(\rho, P)$ is convex ($\cup$) in P for $\rho \geq 0$.

The general problem of finding necessary and sufficient conditions for the vector that minimizes a differentiable convex ($\cup$) function over a convex region of a vector space defined by a set of inequalities has been solved by Kuhn and Tucker[13].

For the special case in which the region is constrained by $P(x) \geq 0$ for $x \in \mathcal{X}$ and $\sum_x P(x) = 1$, their solution reduces to

$$\frac{\partial F(\rho, P)}{\partial P(x)} \geq u \text{ for all } x \text{ with } P(x) \neq 0. \tag{7.90}$$

Differentiating $F(\rho, P)$ and solving for the constant u, we immediately get (7.85). Similarly, if we substitute the convex ($\cup$) function $-I(W|P)$, in (7.90), then (7.86) follows.

Finally, we obtain that $F(\rho, P)$ is continuous on $\mathcal{P}(\mathcal{X})$. It therefore has a minimum (Weierstrass), and thus (7.85) has a solution.

Given the maximization of $E_0(\rho, P)$ over P, we can find $E_r(R)$ in any of three ways.

Fig. 7.6 A special channel

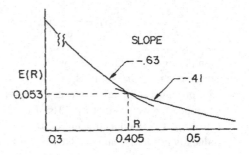

(1) From (7.83) $E_r(R)$ is the upper envelope of the family of straight lines given by

$$- \rho R + \max_P E_0(\rho, P), 0 \le \rho \le 1. \tag{7.91}$$

(2) We can also plot $\max_P E_0(\rho, P)$ as a function of ρ, and use the graphical technique of Fig. 7.2 to find $E_r(R)$.

(3) Finally, we can use the parametric equations, (7.78) and (7.79), and for each ρ use the P' that maximizes $E_0(\rho, P)$. To see that this generates the curved portion of $E_r(R)$, let ρ_0 be a fixed value of ρ, $0 < \rho_0 < 1$, and let P_0 maximize $E_0(\rho_0, P')$. We have already seen that the only point on the straight line $-\rho_0 R + E_0(\rho_0, P_0)$ that lies on the curve $E(R, P_0)$, and thus that can lie on $E(R)$, is the one given by (7.78) and (7.79). Since the straight lines generate all points on the $E_r(R)$ curve, we see that (7.78)–(7.80), with $E_0(\rho, P)$ maximized over P, generate all points on the $E(R)$ curve. These parametric equations can, under pathological conditions, also lead to some points not on the $E(R)$ curve. To see this, consider the channel of Fig. 7.6. From (7.85) we can verify that for $\rho \le 0.51$, $E_0(\rho, P)$ is maximized by $P_1 = P_2 = P_3 = P_4 = \frac{1}{4}$. For $\rho > 0.51$, $E_0(\rho, P)$ is maximized by $P_5 = P_6 = \frac{1}{2}$. The parametric equations are discontinuous at $\rho = 0.51$, where the input distribution suddenly changes.

Figure 7.6 shows the $E_r(R)$ curve for this channel and the spurious points generated by (7.78) and (7.79).

		y_1	y_2	y_3	y_4
	x_1	0.82	0.06	0.06	0.06
	x_2	0.06	0.82	0.06	0.06
W :	x_3	0.06	0.06	0.82	0.06
	x_4	0.06	0.06	0.06	0.82
	x_5	0.49	0.49	0.01	0.01
	x_6	0.01	0.01	0.49	0.49

It has been shown in Lecture that the exponent $E_{sp}(R)$ controlling the lower bound to error probability is given by

$$E_{sp}(R) = \sup_{0 < \rho < \infty} \left[-\rho R + \max_P E_0(\rho, P) \right]. \tag{7.92}$$

Comparing this with (7.83), we see that the only difference is in the range in which ρ is to be maximized.

Interpreting $E_r(R)$ and $E_{sp}(R)$ as the upper envelopes of a family of straight lines of slope $-\rho$, we see that $E_r(R) = E_{sp}(R)$ for $R_{crit} \leq R < C$, where the critical rate R_{crit} is defined as the supremum of R values for which the slope of $E_{sp}(R)$ values is not less than -1. This is a non-zero range of R unless, for the P that maximizes $E_0(\rho, P)$, we have $\frac{\partial^2 E_0(\rho, P)}{\partial \rho^2} = 0$ for $0 < \rho \leq 1$. The channels in Fig. 7.5 is such a channel.

7.2.2   Expurgation: Improvements for Low Rates

$E_r(R)$ does not yield a tight bound on error probability at low rates. The exponent is so large here that previously negligible effects such as assigning the same codeword to two messages suddenly become important.

Now we avoid this problem by expurgating those codewords for which the error probability is high. Equation (7.59) in Lecture 7.2 gives a bound on error probability for a particular code when the ith word is transmitted. With $\rho = 1$, this is

$$\lambda_i \leq \sum_y \sqrt{W(y|u_i)} \sum_{j \neq i} \sqrt{W(y|u_j)}. \tag{7.93}$$

The quantity

$$d(u_i, u_j) = \sum_y \sqrt{W(y|u_i) W(y|u_j)} \tag{7.94}$$

is very useful, especially, if applied to $(W^n, \mathcal{X}^n, \mathcal{Y}^n)$, where

$$d(u_i, u_j) = \prod_{t=1}^n \sum_{y_t \in \mathcal{Y}} \sqrt{W(y_t|u_{it}) W(y_t|u_{jt})}. \tag{7.95}$$

We call $-\ell n\, d(u_i, u_j)$ the *discrepancy* between u_i and u_j—a generalization of Hamming distance on the BSC to general DMC.

We upperbound now the RV $\lambda_i = \lambda_i(U_1, \ldots, U_M)$ by studying $\Pr(\lambda_i \geq B)$, where the constant B will be choosen later, and then expurgate codewords for which $\lambda_i \geq B$.

We obtain

$$\Pr(\lambda_i \geq B) = \mathbb{E}\phi_i, \tag{7.96}$$

where

$$\phi_i = \phi_i(U_1, \ldots, U_M) = \begin{cases} 1 & \text{if } \lambda_i \geq B \\ 0 & \text{otherwise.} \end{cases}$$

We upperbound ϕ_i using (7.93), (7.94), and (7.95)

$$\phi_i(U_1, \ldots, U_M) \leq \sum_{j \neq i} \frac{d(U_i, U_j)}{B^s}, \quad 0 < s \leq 1 \tag{7.97}$$

by the reasoning which also led to (3.2) in Lecture 3.1.
   Together with (7.96) we get

$$\Pr(\lambda_i \geq B) \leq B^{-s} \sum_{j \neq i} \mathbb{E}\, d(U_i, U_j)^s. \tag{7.98}$$

In the product space situation we have by (7.95)

$$\mathbb{E}\, d(U_i, U_j)^s = \sum_{\substack{u_{it} \in \mathcal{X} \\ u_{jt} \in \mathcal{X}}} \prod_{t=1}^{n} P(u_{it}) P(u_{jt}) \cdot \prod_{t=1}^{n} \left[\sum_{y_t} \sqrt{W(y_t | u_{it}) W(y_t | u_{jt})} \right]^s$$

$$= \prod_{t=1}^{n} \sum_{x \in \mathcal{X}} \sum_{x' \in \mathcal{X}} P(x) P(x') \left[\sum_{y \in \mathcal{Y}} \sqrt{W(y|x) W(y|x')} \right]^s.$$

This is independent of i, j and substitution in (7.98) gives for any $s, 0 < s \leq 1$,

$$\Pr(\lambda_i \geq B) \leq (M-1) B^{-s} \left[\sum_x \sum_{x'} P(x) P(x') \left(\sum_y \sqrt{Q(y|x) W(y|x')} \right)^s \right]^n. \tag{7.99}$$

Now choose B so that the right-hand side of (7.62) is equal to $\frac{1}{2}$. Thus

$$\Pr(\lambda_i \geq B) \leq \frac{1}{2}$$

$$B = [2(M-1)]^{1/s} \left[\sum_x \sum_{x'} P(x) P(x') \left(\sum_y \sqrt{W(y|x) W(y|x')} \right)^s \right]^{n/s}. \tag{7.100}$$

If we expurgate all codewords in the ensemble for which $\lambda_i \geq B$ then the expected number of codewords remaining in a code is at least $M \cdot \frac{1}{2}$, since the probability of expurgation is at most $\frac{1}{2}$ and expected values are additive.

Then there exists a code with $M' \geq M/2$ codewords with individual error probabilities bounded by

$$\lambda_i < B < (4M')^{1/s} \left[\sum_x \sum_{x'} P(x)P(x') \left(\sum_y \sqrt{W(y|x)W(y|x')} \right)^s \right]^{n/s} . \quad (7.101)$$

Of course removing words from a code cannot increase ML-error probabilities of remaining words. Write $M' = e^{nR}$ and define $\rho = 1/s$. Then (7.101) can be written

$$\lambda_i < \exp\left\{ -n \left[-\rho R + E_{ex}(\rho, P) - \rho \frac{\ell n 4}{n} \right] \right\}, \rho \geq 1 \quad (7.102)$$

$$E_{ex}(p, P) = -\rho \ell n \left[\sum_{x,x'} P(x)P(x') \left(\sum_y \sqrt{W(y|x)W(y|x')} \right)^{1/s} \right]. \quad (7.103)$$

Theorem 57 *For the DMC W, any R, and $n \in \mathbb{N}$ there exists an (n, e^{nR}, λ)-code with $\lambda = \max \lambda_i$ bounded by (7.102).*

Remark Earlier similar, but less efficient, expurgation techniques have been used by Elias [7] for the BSC and by Shannon for the additive Gaussian noise channel.

7.2.3 Properties of $E_{ex}(\rho, P)$

The interpretation of Theorem 58 is almost identical to that of Theorem 1. The exponent rate curve given by (7.102) is the upper envelope of a set of straight lines; the line corresponding to each value of $\rho \geq 1$ has slope $-\rho$ and intercept $E_{ex}(\rho, P)$ on the E-axis. We derive now properties of $E_{ex}(\rho, P)$.

Theorem 58 *Assume that $I(W|P) > 0$, then*

(i) $E_{ex}(\rho, P)$ *is strictly increasing in $\rho > 0$.*
(ii) $E_{ex}(1, P) = E_0(1, P)$, *so*

$$\left(-\ell n \sum_{x,x'} P(x)P(x') \sum_y \sqrt{W(y|x)W(y|x')} - \ell n \sum_Y \left(\sum_x P(x)W(y|x)^{1/2} \right)^2 \right).$$

(iii) $E_{ex}(\rho, P)$ *is strictly $(\cap)$-convex in ρ unless (P, W) is noiseless in the sense that for each pair of inputs, x and x', with $P(x) > 0$, $P(x') > 0$ we have either $W(y|x)W(y|x') = 0$ for all y or $W(y|x) = W(y|x')$ for all y.*

Proof We write $\mathcal{Z} = \mathcal{X} \times \mathcal{X}$, $P(x)P(x') = p(z)$ and $a(z) = \sum_y \sqrt{W(y|x)W(y|x')}$

for $z = (x, x')$. Apply Lemma 1 to $E_{ex}(\rho, P)$ with $\beta = \rho$. Since $I(W|P) > 0$ $a(z)$ cannot be independent of z, and $E_{ex}(\rho, P)$ is strictly increasing with ρ. Also, $E_{ex}(\rho, P)$ is $(\cap)$-convex in ρ, and the convexity is strict unless $a(z)$ is always 1 or 0 for $P(z) > 0$. But $a(z) = 1$ only if $W(y|x) = W(y|x')$ for all y and $a(z) = 0$ only if $W(y|x)W(y|x') = 0$ for all y.

Theorem 58 can be used in exactly the same way as Theorem 2 to obtain a parametric form of the exponent, rate curve at low rates. Let

$$E_{ex}(R, P) = \max_\rho \left[-\rho R + E_{ex}(\rho, P) - \rho \frac{\ell n 4}{n} \right]. \tag{7.104}$$

Then for R in the range

$$\lim_{\rho \to \infty} \frac{\partial E_{ex}(\rho, P)}{\partial \rho} \le R + \frac{\ell n 4}{n} \le \left. \frac{\partial E_{ex}(\rho, P)}{\partial \rho} \right|_{\rho=1} \tag{7.105}$$

we have the parametric equations in ρ

$$\begin{cases} \frac{\partial E_{ex}(\rho, P)}{\partial \rho} &= R + \frac{\ell n 4}{n} \\ E_{ex}(R, P) &= -\rho \frac{\partial E_{ex}(\rho, P)}{\partial \rho} + E_{ex}(\rho, P). \end{cases} \tag{7.106}$$

If $E_{ex}(\rho, P)$ is strictly $(\cap)$-convex in ρ, then (7.106) represents a $(\cup)$-convex curve with a continuous slope given by $-\rho$. The smallest rate for which (7.106) is applicable is

$$\lim_{\rho \to \infty} \frac{\partial E_{ex}(\rho, P)}{\partial \rho} = \lim_{\rho \to \infty} -\ell n \sum_{x,x'} P(x)P(x') \left(\sum_y \sqrt{W(y|x)W(y|x')} \right)^{1/\rho}$$

$$= -\ell n \sum_{x,x'} P(x)P(x')\varphi(x, x'), \tag{7.107}$$

where

$$\varphi(x, x') = \begin{cases} 1 & \text{if } \sum_y W(y|x)W(y|x') > 0 \\ 0 & \text{if } \sum_y W(y|x)W(y|x') = 0. \end{cases}$$

If there are two letters x, x' with $\sum_y W(y|x)W(y|x') = 0$, then the right-hand side of (7.107) is styrictly positive. It $R + \frac{\ell n 4}{n}$ is less than this quantity, then $E_{ex}(R, P)$ is ∞. This can be seen by regarding the $E_{ex}(R, P), R$ curve as the upper envelope of straight lines of slope $-\rho$; the right-hand side of (7.107) is the limit of the R intercepts of these lines as the slope approaches $-\infty$. The right-hand side of (7.15) thus gives a

Fig. 7.7 Typical exponent,
rate curves obtained by
using low-rate improvement:
a Ordinary channel, **b** Noiseless channel, **c** Channel with
zero-error capacity

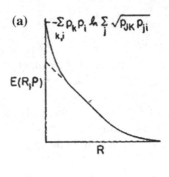

(a)

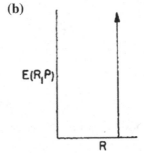

(b)

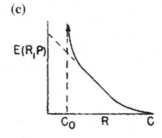

(c)

lower bound zero error capacity $C_0(W)$. Figure 7.7 shows the exponent, rate curves
given by Theorem 58 for some typical channels.

If the channel is noiseless in the sense of Theorem 58, then $E_{ex}(R, P)$ is ∞ for
$R + \ell n4/n < I(W|P)$. It is satisfying that this obvious result comes naturally out of
the general theory.

7.2.3.1 Maximization of $E_{ex}(\rho, P)$ Over P

Here little can be said. It is possible for a number of local maxima to exist, and no
general maximization techniques are known. Concerning non-stationary channels or
particularly parallel channels clearly

$$E_{ex}(\rho, P_1 \times P_2) = E_{ex}(\rho, P_1) + E_{ex}(\rho, P_2). \qquad (7.108)$$

However, in contrast to the behaviour of $E_0(\rho, P)$ now

$$\max_{P^2} E_{ex}(\rho, P^2) > \max_{P_1} E_{ex}(\rho, P_1) + \max_{P_2} E_{ex}(\rho, P_2)$$

can happen.

Example The zero error capacity bound $\lim_{\rho \to \infty} E_{ex}(\rho, P)/\rho$ for the pentagon channel

$$W = \begin{pmatrix} \frac{1}{2} & \frac{1}{2} & 0 & 0 & 0 \\ 0 & \frac{1}{2} & \frac{1}{2} & 0 & 0 \\ 0 & 0 & \frac{1}{2} & \frac{1}{2} & 0 \\ 0 & 0 & 0 & \frac{1}{2} & \frac{1}{2} \\ \frac{1}{2} & 0 & 0 & 0 & \frac{1}{2} \end{pmatrix}$$

is *ln*2, achieved by using for instance inputs 1 and 4 with equal probability in (7.107). For $W^2 = W \times W$ (7.107) yields $\ell n5$, achieved for instance by the pairs of letters (1, 1), (2, 3), (3, 5), (4, 2), (5, 4) with equal probabilities.

7.2.3.2 Consequences for the BSC

Finally, we show what Theorem 58 implies for the BSC $W = \begin{pmatrix} 1-q & q \\ q & 1-q \end{pmatrix}$, $0 \le q \le \frac{1}{2}$. Choose $P = (\frac{1}{2}, \frac{1}{2})$. Clearly

$$E_{ex}(\rho, P) = -\rho \ell n \left[2 \cdot \frac{1}{4} (2\sqrt{q(1-q)})^{1/\rho} + 2 \cdot \frac{1}{4} (\sqrt{q^2} + \sqrt{(1-q)^2})^{1/\rho} \right]$$

$$= -\rho \ell n \left[\frac{1}{2} + \frac{1}{2} [4q(1-q)^{1/2\rho}] \right]. \qquad (7.109)$$

Using this in the parametric equation (7.106) after some calculations we get

$$R + \frac{\ell n4}{n} = \ell n2 - h(\delta) \qquad (7.110)$$

$$E_{ex}(R, P) = \frac{\delta}{2} \ell n \frac{1}{4q(1-q)} \qquad (7.111)$$

where δ is related to ρ by

$$\delta/1 - \delta = [4q(1-q)]^{1/2\rho}. \qquad (7.112)$$

These equations are valid for

$$\delta \geq \sqrt{4q(1-q)}/1 + \sqrt{4q(1-q)}.$$

For large n, δ in (7.110) approaches $D_{\min}/n$ where $D_{\min}$ is the Gilbert bound [10] on minimum distance for a binary code of rate R. The exponent in (7.111) turns to be the same as the exponent for probability of confusion between two codewords at the Gilbert distance from each other. This result has also been established for parity check codes [19]. There it (Elias [7]) is relevant that the random coding method works already if the RV's are pairwise independent.

Problem For higher alphabets codes above Gilbert's bound, what is the expurgated bound for an associated symmetric channel?

7.2.4 Error Bounds for Basic Channels

7.2.4.1 The BSC

Again $W = \begin{pmatrix} 1-q & q \\ q & 1-q \end{pmatrix}$ and clearly P with $P(0) = P(1) = \frac{1}{2}$ is the input PD maximizing $E_0(\rho, P)$. Formally this follows with Theorem 56. For this P

$$E_0(\rho, P) = -\ell n \sum_y \left(\sum_x P(x) W(y|x)^{\frac{1}{1+\rho}} \right)^{1+\rho}$$

$$= \rho \ell n 2 - (1+\rho)\ell n \left[q^{1/1+\rho} + (1-q)^{1/1+\rho} \right]. \tag{7.113}$$

We evaluate now the parametric expressions for the exponent. We obtain

$$R = \ell n 2 - h(q_p) \tag{7.114}$$

$$E_r(R, P) = D\big((q_p, 1 - q_p)\|(q, 1 - q)\big) \tag{7.115}$$

where

$$q_p = \frac{q^{1/1+\rho}}{q^{11/1+\rho} + (1-q)^{1/1+q}}. \tag{7.116}$$

These equations are valid for $0 \leq \rho \leq 1$, or for

$$\ell n 2 - h\left(\frac{\sqrt{q}}{\sqrt{q} + \sqrt{1-q}} \right) \leq R \leq I(W|P) = C.$$

For rates below this we have (7.80), which becomes

$$E_r(R, P) = \ell n2 - 2\ell n\left(\sqrt{q} + \sqrt{1-q}\right) - R. \tag{7.117}$$

The random coding exponent specified in (7.94)–(7.117) is the one first derived by Elias [7].

7.2.4.2 Very Noisy Channels

Very noisy channels are considered to be those for which the probability of receiving a given output is almost independent of the input. Formally, for a $Q \in \mathcal{P}(\mathcal{Y})$

$$W(y|x) = Q(y)\left(1 + \varepsilon(y|x)\right), \tag{7.118}$$

where $\|\varepsilon(y|x)\| \ll 1$ for all x, y.

Further, we require

$$\sum_y Q(y)W(y|x) = 0 \text{ for all } x \in \mathcal{X}. \tag{7.119}$$

Then

$$E(R) \sim \left(\sqrt{C} - \sqrt{R}\right)^2, R \geq \frac{C}{4} \tag{7.120}$$

$$E(R) \sim \frac{C}{2} - R, R < \frac{C}{4}. \tag{7.121}$$

Observe that the exponent rate curve given by (7.120), (7.121) is that for orthogonal signals in white Gaussian noise [8].

These channels were first considered by Reiffen [16], who showed that the exponent corresponding to zero is $C/2$.

7.2.4.3 Parallel Channels

Consider the non-stationary DMC, periodic in W_1 and W_2. Then for $t = 1, 2$

$$E_0(\rho, P_t, W_t) \triangleq -\ell n \sum_y \left(\sum_x P_t(x)W_t(y|x)^{\frac{1}{1+\rho}}\right)^{1+\rho}.$$

Theorem 59 *For the parallel channel* $W = W_1 \times W_2$

(i) *(Fano [8]). For any* $\rho, 0 \leq \rho \leq 1$

$$\lambda(n, e^{Rn}) \leq \exp\left\{-n\left[-\rho R + E_0(\rho, P_1 \times P_2)\right]\right\} \tag{7.122}$$

where
$$E_0(\rho, P_1 \times P_2) = E_0(\rho, P_1) + E_0(\rho, P_2). \tag{7.123}$$

(ii) *If P_t maximizes $E_0(\rho, P_t, W_t)$, then*

$$E_0(\rho, P_1 \times P_2, W_1 \times W_2) = \max_{P^2} E_0(\rho, P^2, W_1 \times W_2). \tag{7.124}$$

Proof (i) is the random coding bound if the U_i are choosen as $\left((U_{i1}^{(1)}, U_{i2}^{(2)}), \ldots, (U_{in}^{(1)}, U_{in}^{(2)})\right)$ with pairwise independent RV's of joint distributions $P_1 \times P_2$.

(ii) Regarding the parallel channels as a single channel with input probability distribution P^2, we get

$$E_0(\rho, P^2) = -\ell n \sum_{y_1, y_2} \left[\sum_{x_2, x_2} P^2(x_1, x_2) \left(W_1(y_1|x_1) W(y_2|x_2) \right)^{\frac{1}{1+\rho}} \right]^{1/\rho}. \tag{7.125}$$

For $P^2 = P_1 \times P_2$ we get by seperation of sums (7.87).

The main task is to show that $P^2 = P_1 \times P_2$ maximizes $E_0(\rho, P^2)$, if P_t maximizes $E_0(\rho, P_t)$ for $t = 1, 2$.

We know from (7.85) in Theorem 56 that these P_1, P_2 must satisfy

$$\sum_y W_t(y|x)^{\frac{1}{1+\rho}} \alpha_{ty}^\rho \geq \sum_y \alpha_{ty}^{1+\rho} \text{ for all } x \tag{7.126}$$

with equality if $P_t(x) > 0$ where $\alpha_{ty} \triangleq \sum_x P_t(x) W_t(y|x)^{\frac{1}{1+\rho}}$ $(t = 1, 2)$.

Multiplying the two left-hand expressions of the two inequalities in (7.126), respectively, we get

$$\sum_{y_1, y_2} \left(W_1(y_1|x_1) W_2(y_2|x_2) \right)^{\frac{1}{1+\rho}} (\alpha_{1y_1} \cdot \alpha_{2y_2})^\rho \geq \sum_{y_1, y_2} (\alpha_{1y_1} \cdot \alpha_{2y_2})^{1+\rho} \tag{7.127}$$

with equality if $P^2(x_1, x_2) = P_1(y_1), P_2(x_2) > 0$.

Remark Theorem 57 has an interesting geometrical interpretation.

Let $E(\rho, W_1 \times W_2)$ and $R(\rho, W_1 \times W_2)$ be the exponent and rate for the parallel combination, as parametrically related by (7.102) and (7.103) with the optimum choice of P^2 for each ρ. Let $E(\rho, W_t)$ and $R(\rho, W_t)$ $(t = 1, 2)$ be the equivalent quantities for the individual channel. From (7.123)

$$E(\rho, W_1 \times W_2) = \sum_{t=1}^2 E(\rho, W_t)$$
$$R(\rho, W_1 \times W_2) = \sum_{t=1}^2 R(\rho, W_t). \tag{7.128}$$

Thus the parallel combination is formed by vector addition of points of the same slope from the individual $E(\rho)$, $R(\rho)$ curves. Theorem 57 applies to any number of channels in parallel. If we consider a block of length n as a single use of n identical parallel channels, then Theorem 57 justifies our choice of independent identically distributed symbols in the ensemble of codes.

7.2.5 Another Derivation of the Random Coding Exponent: A Combinatorial Approach

The form of the sphere packing exponent

$$E_{sp}(R) = E_{sp}(R, W) = \max_{P \in \mathcal{P}(\mathcal{X})} E_{sp}(R, P, W), \tag{7.129}$$

where

$$E_{sp}(R, P, W) = \min_{V:I(V|P) \leq R} D(V\|W|P) \tag{7.130}$$

due to Haroutunian [12] has an appealing structure in terms of information measures relating to channel probabilities of sequences with frequency structure described by another channel. Intuitively it is far more transparent than the original analytical form involving variations over auxiliary parameters.

It suggests to look for an analogous form of the random coding exponent and a related approach to establish its validity as an upper bound on the optimal error probability $\lambda(n, e^{Rn})$ for n, e^{Rn})-codes of the DMC W.

Clearly it suffices to consider only codes with words of the same composition or ED, because there are only polynomially many EDs. This was earlier done in the feedback scheme of [1], where T_P^n, the sequences of ED P were used. We can estimate probabilities by counting where they are equal.

Concerning decoding a nice idea is due to Goppa [11], who used the rule g that a received word y^n should be decoded into the codeword $u_i \in \mathcal{U} \subset T_P^n$ with *maximal non probabilistic mutual information* with y^n, that is,

$$g(y^n) = u_i, \text{ if } I(u_i \wedge y^n) > \max_{j \neq i} I(u_j \wedge y^n), \tag{7.131}$$

where for $x^n \in \mathcal{X}^n$ and $y^n \in \mathcal{Y}^n$

$$I(x^n \wedge y^n) = I(P_{y^n|x^n}|P_{x^n}) \tag{7.132}$$

and P_{x^n} is the ED of x^n and $P_{y^n|x^n}$ is the conditional ED of y^n given x^n.

Then Goppa's rule takes the form

$$g(y^n) = u_i \text{ iff } I(P_{y^n|u_i}|P_{x^n}) > I(P_{y^n|u_j}|P_{u_j}) \text{ for all } j \neq i. \tag{7.133}$$

Hence

$$\lambda_i = W^n(\{y^n : g(y^n) \neq u_i\} | u_i) \leq \sum_{\substack{V, \hat{V} \in \mathcal{V} \\ I(\hat{V}|P) \geq I(V|P)}} W^n(T_V(u_i) \cap \bigcup_{j \neq i} T_{\hat{V}}(u_j) | u_i), \quad (7.134)$$

where as in (7.130) $\mathcal{V}$ is the set of channels with alphabets $\mathcal{X}, \mathcal{Y}$. Of course $T_V(x^n)$ is empty, if V is not a conditional ED. Now by (7.130) for $x^n \in T_P^n$ and $y^n \in T_V^n(x^n)$

$$W^n(y^n | x^n) = \exp\{-n[D(V\|W|P) + H(V|P)]\}. \quad (7.135)$$

Therefore

$$W^n\left(T_V(u_i) \cap \bigcup_{j \neq i} T_{\hat{V}}(u_j) | u_i\right)$$

$$= |T_V(u_i) \cap \bigcup_{j \neq i} T_{\hat{V}}(u_j))| \exp\{-n[D(V\|W|P) + H(V|P)]\}. \quad (7.136)$$

In order to make $|T_V(u_i) \cap \bigcup_{j \neq i} T_{\hat{V}}(u_j)|$ small we choose the codewords at random uniformly from T_P^n. We get as consequence of the hypergraph coloring lemma ([2]) the

Lemma 69 (Packing Lemma 2) *For every $R > 0, \delta > 0, n \geq n_o(|\mathcal{X}|, |\mathcal{Y}|, \delta)$ and every ED $P \in \mathcal{P}_o(n)$ with $H(P) > R$ there exists a set of codewords $\mathcal{U} \subset T_P^n$, $|\mathcal{U}| \geq \exp\{u(R - \delta)\}$, such that for every pair of stochastic matrices $V : \mathcal{X} \to \mathcal{Y}$, $\hat{V} : \mathcal{X} \to \mathcal{Y}$ and every i, j*

$$|T_V^n(u_i) \cap T_{\hat{V}}^n(u_j)| \leq |T_V^n(u_i)| \exp\{-nI(\hat{V}|P)\} \quad (7.137)$$

provided that $n \geq n_o(|\mathcal{X}|, |\mathcal{Y}|, \delta)$. Consequently

$$|T_V^n(u_i) \cap \bigcup_{j \neq i} T_{\hat{V}}^n(u_j)| \leq |T_V^n(u_i)| \exp\{-n(I(\hat{V}|P) - R)^+\}, \text{ if } |\mathcal{U}| = e^{Rn}. \quad (7.138)$$

From (7.134), (7.136), (7.137) and $|T_V^n(u_i)| \leq \exp\{nH(V|P)\}$ we get

$$\lambda_i \leq \sum_{\substack{V, \hat{V} \in \mathcal{V} \\ I(\hat{V}|P) \geq I(V|P)}} \exp\{-n[D(V\|W|P) + (I(\hat{V}|P) - R)^+]$$

and by ED counting and since $I(\hat{V}|P) \geq I(V|P)$

$$\lambda_i \leq (n+1)^{2|\mathcal{X}||\mathcal{Y}|} \exp\left\{-n\left[\min_{V \in \mathcal{V}} D(V\|W|P) + (I(V|P) - R)^+\right]\right\}. \quad (7.139)$$

Now

$$E_r(R, P, W) \triangleq \min_V \left(D(V \| W | P) + \left(I(V|P) - R \right)^+ \right) \qquad (7.140)$$

was recognized by Haratounian as the random coding exponent rate function for channel W with input distribution P. We summarize our results.

Theorem 60 (Random Coding bound for constant compositions) *For every ED $P \in \mathcal{P}_0(\mathcal{X}, n)$, every $R < H(P)$ and every $\delta > 0$ there is a set of codewords $\mathcal{U} \subset T_p^n$ of*

$$rate(\mathcal{U}) \geq R$$

such that for the MMI decoder and every DMC $W : \mathcal{X} \to \mathcal{Y}$

$$\lambda(\mathcal{U}, W) \leq \exp\{-n[Er(R, P, W) - \delta]\}$$

for $n \geq n_0(|\mathcal{X}|, |\mathcal{Y}|, \delta)$.

Remark The result is universal in the sense that encoding and decoding don't depend on the channel.

7.2.6 The Equivalence of the Different Definition of Error Exponents

Recall the definition of sphere packing and random coding exponent functions in terms of relative entropy

$$E_{sp}(R, P, W) = \min_{V: I(V|P) \leq R} D(V \| W | P) \qquad (7.141)$$

$$E_r(R, P, W) = \min_V \left(D(V \| W | P) + \left(I(V|P) - R \right)^+ \right). \qquad (7.142)$$

We first connect these two functions.

Lemma 70 *For fixed P and W, $E_{sp}(R, P, W)$ is a $(\cup)$-convex function in $R \geq 0$, positive for $R < I(W|P)$ and 0 else. It is strictly decreasing where it is finite and positive. Moreover,*

(i) $E_r(R, P, W) = \min_{R' \geq R} \left(E_{sp}(R', P, W) + R' - R \right)$

(ii) $E_r(R, P, W) = \begin{cases} E_{sp}(R, P, W) & \text{if } R \geq R_{crit}(P, W) \\ E_{sp}(R, P, W) + R_{crit}(P, W) - R, & \text{if } 0 \leq R \leq R_{crit}(P, W) \end{cases}$

where $R_{crit}(P) = R_{crit}(R, P, W)$ is the smallest R at which the convex $E_{sp}(R, P, W)$ meets its supporting line of slope -1.

Proof For two rates $R_1 \neq R_2$ and $0 < \alpha_1, \alpha_2 < 1$, $\sum_{i=1}^2 \alpha_i = 1$ let V_i achieve the minimum in the definition of $E_{sp}(R_i, P)$ $(i = 1, 2)$. Consider

$$V = \sum_{i=1}^{2} \alpha_i V_i.$$

By the ($\cup$)-convexity of divergence and therefore also of conditional divergence

$$D(V \| W | P) \leq \sum_{i=1}^{2} \alpha_i D(V_i \| W | P) = \sum_{i=1}^{2} \alpha_i \cdot E_{sp}(R_i, P)$$

and since

$$I(V | P) \leq \sum_{i=1}^{2} \alpha_i I(V_i | P) \leq \sum_{i=1}^{2} \alpha_i R_i$$

we get

$$E_{sp}\left(\sum_{i=1}^{2} \alpha_i R_i, P\right) \leq \sum_{i=1}^{2} \alpha_i E_{sp}(R_i, P).$$

Obviously, if $R \geq I(W | P)$, then the choice $V = W$ gives $E_{sp}(R, P) = 0$. Furthermore, if $I(V | P) \leq R < I(W | P)$, then necessarily for some x we have $P(x) > o$ and $V(\cdot | x) \neq W(\cdot | x)$, implying $D(V \| W | P) > o$ and thus $E_{sp}(R, P) > 0$ if $R < I(W | P)$.

The convexity and positivity properties imply the claimed monotonicity property. Hence

$$E_{sp}(R, P) = \min_{V : I(V | P) = R} D(V \| W | P), \tag{7.143}$$

$R_\infty(P, W) \leq R \leq I(W | P)$, where $R_\infty(P, W)$ is the smallest rate such that

$$E_{sp}(R, W) > o \text{ for all } R \geq R_\infty(P, W).$$

As a V achieving the minimum in the definition of $E_r(R, P)$ satisfies

$$R_\infty(P, W) \leq I(V | P) \leq I(W | P),$$

$$E_r(R, P) = \min_{R' : R_\infty(P, W) \leq R' \leq I(W | P)} \min_{I(V | P) = R'} \left(D(V \| W | P) + (R' - R)^+\right)$$

and by (7.143)

$$E_r(R, P) = \min_{R' : R_\infty(P, W) \leq R' \leq I(W | P)} \left(E_{sp}(R', P) + (R' - R)^+\right)$$

$$= \min_{R'} \left(E_{sp}(R', P) + (R' - R)^+\right)$$

where the last step follows as $E_{sp}(R', P) = 0$ for $R' \geq I(W|P)$. Finally, since $E_{sp}(R', P)$ decreases in R' the values $R' < R$ with $(R' - R)^+ = 0$ don't contribute in the minimization and (i) follows.

The established properties yield (ii).

Remark Notice that for fixed P $E_r(R, P, W)$ is a straight line at the left of $R_{crit}(P, W)$.

Recall the Forms of the Exponent Used in (7.75), (7.78), (7.80).

They involve

$$E_o(\rho, P) = -\ell n \sum_x \left(\sum_y P(x) W(y|x)^{1/1+\rho} \right)^{1+\rho}. \tag{7.144}$$

Recall from Lecture 7.2

$$E(R, P) = \max_{0 \leq \rho \leq 1} \left[-\rho R + E_o(\rho, P) \right] \tag{7.145}$$

and the random coding exponent

$$E_r(R) = \max_P E(R, P). \tag{7.146}$$

Further

$$E_{sp}(R) = \sup_{\ll \rho < \infty} \left[-\rho R + \max_P E_o(\rho, P) \right]. \tag{7.147}$$

We show now that corresponding *main* concepts are indeed equal. For this purpose we assign stars to these "old" definitions in (7.146) and (7.147): $E_r^*(R)$, $E_{sp}^*(R)$.

Lemma 71 (Equivalence of exponent definitions)

(i) $E_{sp}(R, P) \geq E(R, P)$ *(new worse)*
(ii) $E_{sp}(R) = E_{sp}^*(R)$
(iii) $E_r(R, P) \geq E_r^*(R, P)$ *(new better)*
(iv) $E_r(R) = E_r^*(R)$.

References

1. R. Ahlswede, A constructive proof of the coding theorem of discrete memoryless channels in case of complete feedback. in *TSixth Prague Conference on Information Theory, Statistical Decision Function's and Random Process*, 1971, pp. 1–22
2. R. Ahlswede, Coloring hypergraphs: a new approach to multi-user source coding I. J. Combin. Inform. Syst. Sci. **4**(1), 76–115 (1979)
3. S. Arimoto, On the converse to the coding theorem for discrete memoryless channels. IEEE Trans. Inform. Theory, **IT-19**, 357–359 (1973)

4. R.E. Blahut, A hypothesis-testing approach to information theory. Ph.D. Dissertation, Cornell University, Ithaca, NY (1972)
5. R.E. Blahut, Hypothesis testing and information theory. IEEE Trans. Inform. Theor. **20**, 405–417 (1974)
6. G. Dueck, J. Körner, Reliability function of a discrete memoryless channel at rates above capacity. IEEE Trans. Inform. Theor. **25**, 82–85 (1979)
7. P. Elias, Coding for noisy channels. IRE Conv. Rec. **Part 4**, 37–46 (1955)
8. R.M. Fano, *Transmission of Information: A Statistical Theory of Communications* (The M.I.T. Press, Wiley, Cambridge, New York-London, 1961)
9. R.G. Gallager, A simple derivation of the coding theorem and some applications. IEEE Trans. Inform. Theor. **IT-11**, 3–18 (1965)
10. E.N. Gilbert, A comparison of signalling alphabets. Bell Syst. Tech. J. **31**, 504–522 (1952)
11. V.D. Goppa, Nonprobabilistic mutual information without memory. Probl. Contr. Inform. Theory/Problemy Upravlenija i Teorii Informacii **4**(2), 97–102 (1975)
12. E.A. Haroutunian, Bounds to the error probability exponent for semicontinuous memoryless channels. Prob. Peredachi Inform. **4**, 37–48 (1968)
13. H.W. Kuhn, A.W. Tucker, Nonlinear programming. in *Proceedings of the 2nd Berkeley Symposium on Mathematical Statistics and Probability, 1950* (University of California Press, Berkeley and Los Angeles, 1951), pp. 481–492
14. J.K. Omura, A lower bounding method for channel and source coding probabilities. Inform. Control **27**, 148–177 (1975)
15. M. Plotkin, Binary codes with specified minimum distance. IRE Trans. Inform. Theor. **6**, 445–450 (1960)
16. B. Reiffen, A note on "very noisy" channels. Inform. Control **6**, 126–130 (1963)
17. C.E. Shannon, R.G. Gallager, E.R. Berlekamp, Lower bounds to error probability for coding on discrete memoryless channels I. Inform. Control **10**, 65–103 (1967)
18. C.E. Shannon, R.G. Gallager, E.R. Berlekamp, Lower bounds to error probability for coding on discrete memoryless channels II. Inform. Control **10**, 522–552 (1967)
19. R.R. Varshamov, Estimate of the number of signals in error correcting codes. Dokl. Acad. Nauk **117**, 739–741 (1957)
20. J. Wolfowitz, The coding of messages subject to chance errors. Ill. J. Math. **1**(4), 591–606 (1957)

Part III
Appendix

Part II.
Appendix

Chapter 8
Inequalities

8.1 Lecture on Analytic Inequalities

8.1.1 Jensen's Inequality (Convexity)

Definition 36 A function $f : A \to \mathbb{R}$, $A \subset \mathbb{R}^n$, is convex on A, if for all $x, y \in A$ and for all $\alpha \in [0, 1]$

$$f(\alpha \cdot x + (1 - \alpha) \cdot y) \leq \alpha \cdot f(x) + (1 - \alpha) \cdot f(y).$$

Accordingly, f is said to be *concave*, if $f(\alpha \cdot x + (1 - \alpha) \cdot y) \geq \alpha \cdot f(x) + (1 - \alpha) \cdot f(y)$, and strictly convex (strictly concave), if strict inequality holds.

For a function $f : \mathbb{R} \to \mathbb{R}$ convexity (concavity) just means, that the line connecting $f(x)$ and $f(y)$ is always above (below) the graph of f.

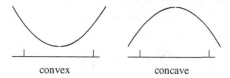

convex   concave

The following lemmas, which we will state without proof, are well known from Analysis.

Lemma 72 *Let $f : \mathbb{R} \to \mathbb{R}$ be differentiable on the interval (a, b). Then f is convex on (a, b), exactly if $f'(x) \leq f'(y)$ for all $a < x \leq y < b$.*

Lemma 73 *Let $f : \mathbb{R} \to \mathbb{R}$ be twice differentiable on the interval (a, b). Then f is convex on (a, b), exactly if $f''(x) \geq 0$ for all $x \in (a, b)$.*

A. Ahlswede et al. (eds.), *Storing and Transmitting Data*,
Foundations in Signal Processing, Communications and Networking 10,
DOI: 10.1007/978-3-319-05479-7_8, © Springer International Publishing Switzerland 2014

Immediately from the definition of convexity follows

Lemma 74 (Jensen's inequality) *Let f be a convex function on $A \subset \mathbb{R}^n$, $(x_1, \ldots, x_n) \in A$ and $\alpha_1, \ldots, \alpha_n \in [0, 1]$ such that $\sum_{i=1}^{n} \alpha_i = 1$. Then*

$$f\left(\sum_{i=1}^{n} \alpha_i x_i\right) \leq \sum_{i=1}^{n} \alpha_i f(x_i).$$

Accordingly, for a concave function f $\left(\sum_{i=1}^{n} \alpha_i x_i\right) \geq \sum_{i=1}^{n} \alpha_i f(x_i)$.

8.1.2 Arithmetic-Geometric Mean Inequality

Since the logarithm obviously is concave on $\mathbb{R}^+$, Jensen's inequality can be applied to obtain for $x_1, \ldots, x_n > 0$, $\alpha_1 + \cdots + \alpha_n = 1$

$$\log\left(\sum_{i=1}^{n} \alpha_i x_i\right) \geq \sum_{i=1}^{n} \alpha_i \cdot \log x_i.$$

The exponential function is monotone and hence

$$\sum_{i=1}^{n} \alpha_i x_i = \exp\left\{\log \sum_{i=1}^{n} \alpha_i x_i\right\} \geq \exp\left\{\sum_{i=1}^{n} \alpha_i \cdot \log x_i\right\} = \prod_{i=1}^{n} x_i^{\alpha_i}.$$

Moreover, since the logarithm is strictly concave and since the exponential function is strictly monotone, equality only holds, if $\alpha_1 = \cdots = \alpha_n$. So we have derived from Jensen's inequality.

Lemma 75 (Arithmetic-geometric mean inequality) *For all $x_1, \ldots, x_n > 0$ and $\alpha_1, \ldots, \alpha_n \in [0, 1]$ with $\sum_{i=1}^{n} \alpha_i = 1$ holds*

$$\sum_{i=1}^{n} \alpha_i x_i \geq \prod_{i=1}^{n} x_i^{\alpha_i} \tag{8.1}$$

with equality, exactly if $\alpha_1 = \cdots = \alpha_n = \frac{1}{n}$.

As an application we shall compare the entropy $H(P) = -\sum_{x \in \mathcal{X}} P(x) \cdot \log P(x)$ and the Rényi-entropy $H_\alpha(P) = \frac{1}{1-\alpha} \cdot \log \sum_{x \in \mathcal{X}} P(x)^\alpha$ for some $\alpha > 0, \alpha \neq 1$ (see also Chap. 4 on measures of uncertainty).

First observe that

$$H(P) = -\sum_{x \in \mathcal{X}} P(x) \cdot \log P(x) = \sum_{x \in \mathcal{X}} P(x) \cdot \log \frac{1}{P(x)}$$

$$= \sum_{x \in \mathcal{X}} \log\left(\frac{1}{P(x)}\right)^{P(x)} = \log \prod_{x \in \mathcal{X}} \left(\frac{1}{P(x)}\right)^{P(x)}$$

and that

$$H_\alpha(P) = \frac{1}{1-\alpha} \cdot \log \sum_{x \in \mathcal{X}} P(x)^\alpha = \frac{1}{1-\alpha} \cdot \log \sum_{x \in \mathcal{X}} P(x) \cdot P(x)^{\alpha-1}.$$

Since of course $\sum_{x \in \mathcal{X}} P(x) = 1$, the entropy can hence be regarded as the logarithm of a geometric mean, whereas the Rényi-entropy is the logarithm of an arithmetic mean. Application of the arithmetic-geometric mean inequality yields

$$\sum_{x \in \mathcal{X}} P(x) \cdot P(x)^{\alpha-1} \geq \prod_{x \in \mathcal{X}} \left(P(x)^{\alpha-1}\right)^{P(x)}, \text{ hence}$$

$$H_\alpha(P) \geq \frac{1}{1-\alpha} \cdot \log \prod_{x \in \mathcal{X}} \left(P(x)^{\alpha-1}\right)^{P(x)} = \log \prod_{x \in \mathcal{X}} \left(P(x)^{\alpha-1}\right)^{\frac{P(x)}{1-\alpha}}$$

$$= \log \prod_{x \in \mathcal{X}} \left(\frac{1}{P(x)}\right)^{P(x)} = H(P).$$

Let us further mention that several important inequalities such as Hölder's inequality, Minkowski's inequality and the Cauchy-Schwarz inequality can be derived from the arithmetic-geometric mean inequality by an appropriate choice of the parameters x_i and α_i.

8.1.3 Hölder's, Cauchy-Schwarz, and Minkowski's Inequalities

A very systematic discussion of those inequalities, their proofs, extensions, and also their relationships can be found in [5]. Our exposition closely resembles the one given in [4].

For various different proofs we refer again to [5]. But we shall show now that from (8.1) one easily obtains Hölder's inequality, Minkowski's inequality, various extensions of those, and all the other inequalities discussed in this lecture.

Lemma 76 (Hölder's inequality) *For $i = 1, \ldots, n$ let now a_i, b_i, p_i, q_i be non-negative numbers and $\sum p_i = \sum q_i = 1$.*

$$\sum_i a_i b_i \leq \left(\sum_i a_i^{1/\lambda}\right)^\lambda \left(\sum_i b_i^{1/(1-\lambda)}\right)^{1-\lambda} \tag{8.2}$$

with equality iff for some c $a_i^{1-\lambda} = b_i^\lambda c$ for all i.

Proof As a special case of (8.1) one obtains

$$\prod_{i=1}^{n} \left(\frac{p_i}{q_i}\right)^{q_i} \leq 1. \tag{8.3}$$

Another useful relationship is

$$\sum_i p_i^{\lambda} q_i^{1-\lambda} \leq 1, \ 0 < \lambda < 1, \text{with equality iff } p_i = q_i \text{ for all } i, \tag{8.4}$$

which is again readily obtained, because (8.1) implies $p_i^{\lambda} q_i^{1-\lambda} \leq \lambda p_i + (1-\lambda)q_i$
and $\sum_i \lambda p_i + (1-\lambda)q_1 = 1$.

Actually, $\sum_i p_i^{\lambda} q_i^{1-\lambda}$ can be shown to be convex ($\cup$) in λ and since the sum takes
the value 1 for $\lambda = 0$ and $\lambda = 1$, this directly gives the inequality.

Choosing now

$$p_i = a_i^{1/\lambda} \bigg/ \left(\sum_i a_i^{1/\lambda}\right)$$

$$q_i = b_i^{1/(1-\lambda)} \bigg/ \sum_i b_i^{1/(1-\lambda)}$$

we obtain the lemma from (8.4).   □

Remark For $\lambda = \frac{1}{2}$, Hölder's inequality yields the Cauchy-Schwarz inequality.

Corollary 6 (Variants of Hölder's inequality)

$$\sum_i p_i a_i b_i \leq \left(\sum_i p_i a_i^{1/\lambda}\right)^{\lambda} \left(\sum_i p_i b_i^{1/1-\lambda}\right)^{1-\lambda} \tag{8.5}$$

with equality iff for some c, $p_i a_i^{1/\lambda} = p_i b_i^{1/1-\lambda} c$ for all i.
  The following special cases are sometimes of interest:

(i)

$$\left(\sum_i p_i a_i\right)^r \leq \sum_i p_i a_i^r \quad r > 1$$

$$\geq \sum_i p_i a_i^r \quad r < 1$$

with equality iff the a_i are equal for those i's with $p_i > 0$.

(ii)

$$\left(\sum_i p_i a_i^r\right)^{1/r} \leq \left(\sum_i p_i a_i^s\right)^{1/s} \; ; \; 0 < r < s$$

with equality iff a_i's are equal for i with $p_i > 0$.

(iii)

$$\left(\sum_i a_i\right)^r \leq \sum_i a_i^r, \; r \leq 1,$$

$$\geq \sum_i a_i^r, \; r \geq 1;$$

with equality iff $r = 1$ or if only one a_i is non-zero.

Proof To prove (8.5), choose $p_i^\lambda a_i$ for a_i and $p_i^{1-\lambda}$ for b_i in Hölder's inequality.

(i) For $r > 1$, use (8.5) with $b_i = 1$, for $r < 1$ use a_i^r for a_i.
(ii) Use (8.5) with $b_i = 1$, $\alpha = r/s$.
(iii) Choose $p_i = \frac{a_i}{\sum a_i}$ in (8.5).

□

Lemma 77 (Minkowski's inequality) *Let* $(a_{jk})_{\substack{j=1,\ldots,n \\ k=1,\ldots,m}}$ *be a matrix with non-negative entries, then*

$$\left[\sum_j \left(\sum_k a_{jk}\right)^{1/r}\right]^r \leq \sum_k \left(\sum_j a_{jk}^{1/r}\right)^r, \; r < 1,$$

$$\geq \sum_k \left(\sum_j a_{jk}^{1/r}\right)^r, \; r > 1$$

with equality iff a_{jk}'s are independent of k.

Proof For $r < 1$, use the factorization

$$\left(\sum_k a_{jk}\right)^{1/r} = \left(\sum_k a_{jk}\right)\left(\sum_i a_{ji}\right)^{(1-r)/r}$$

to obtain

$$\sum_j \left(\sum_k a_{jk}\right)^{1/r} = \sum_k \left[\sum_j a_{jk}\left(\sum_i a_{ji}\right)^{(1-r)/r}\right].$$

Apply Hölder's inequality to the term in brackets for each k, getting

$$\sum_j \left(\sum_k a_{jk} \right)^{1/r} \leq \sum_k \left(\sum_j a_{jk}^{1/r} \right)^r \left[\sum_j \left(\sum_i a_{ji} \right)^{1/r} \right]^{1-r}.$$

Divide both sides by the term in brackets to obtain the result.

For $r > 1$, apply the first inequality with $r' < 1$ and a_{jk} replaced by $a_{jk}^{1/r}$.

By replacing a_{jk} by $q_j^r a_{jk}$ in (8.6) one gets

$$\left[\sum_j a_i \left(\sum_k a_{jk} \right)^{1/r} \right]^r \leq \sum_k \left(\sum_j q_j a_{jk}^{1/r} \right)^r \quad \text{for} \quad r < 1 \qquad (8.6)$$

and the reverse inequality for $r > 1$.   □

8.2 Lecture on Inequalities Involving Entropy and Relative Entropy

Remember that the entropy $H(P)$ and the relative entropy $D(P\|Q)$ are defined for probability distributions P and Q on $\mathcal{X} = \{1, \dots, a\}$ as

$$H(P) \triangleq - \sum_{x \in \mathcal{X}} P(x) \cdot \log P(x),$$

$$D(P\|Q) \triangleq \sum_{x \in \mathcal{X}} P(x) \cdot \log \frac{P(x)}{Q(x)}.$$

Central in this lecture is the following log-sum-inequality, from which several important inequalities can be derived.

Lemma 78 (Log-sum inequality) *For non-negative numbers a_i and b_i, $i = 1, \dots, n$ holds*

$$\sum_{i=1}^{n} a_i \log \frac{a_i}{b_i} \geq \left(\sum_{i=1}^{n} a_i \right) \cdot \log \frac{\sum_{i=1}^{n} a_i}{\sum_{i=1}^{n} b_i},$$

with equality exactly if $a_i \sum_{j=1}^{n} b_j = b_i \sum_{j=1}^{n} a_j$ for $i = 1, \dots, n$.

Proof Let $a \triangleq \sum_{i=1}^{n} a_i$ and $b \triangleq \sum_{i=1}^{n} b_i$. Then

$$\sum_{i=1}^{n} a_i \log \frac{a_i}{b_i} = -a \cdot \sum_{i=1}^{n} \frac{a_i}{a} \cdot \log \frac{b_i}{a_i} \geq -a \cdot \log \left(\sum_{i=1}^{n} \frac{a_i}{a} \cdot \frac{b_i}{a_i} \right) = -a \cdot \log \frac{b}{a},$$

since $\sum_{i=1}^{n} \frac{a_i}{a} = 1$ and since the logarithm is concave. □

From the log-sum inequality follows immediately.

Lemma 79 $D(P||Q) \geq 0$ *with equality, exactly if* $P = Q$.

Proof $D(P||Q) = \sum_{x \in \mathcal{X}} P(x) \cdot \log \frac{P(x)}{Q(x)} \geq \left(\sum_{x \in \mathcal{X}} P(x) \right) \cdot \log \frac{\sum_{x \in \mathcal{X}} P(x)}{\sum_{x \in \mathcal{X}} Q(x)} = 0$,

since $\sum_{x \in \mathcal{X}} P(x) = \sum_{x \in \mathcal{X}} Q(x) = 1$. □

A second proof of $D(P||Q) \geq 0$ is obtained by application of the arithmetic-geometric mean inequality, from which follows that

$$\left(\frac{Q(1)}{P(1)} \right)^{P(1)} \cdot \cdots \cdot \left(\frac{Q(a)}{P(a)} \right)^{P(a)} \leq \sum_{x \in \mathcal{X}} P(x) \cdot \frac{Q(x)}{P(x)} = \sum_{x \in \mathcal{X}} Q(x) = 1.$$

Taking the logarithm on both sides yields

$$\sum_{x \in \mathcal{X}} P(x) \cdot \log \frac{Q(x)}{P(x)} \leq 0 \text{ and hence}$$

$$D(P||Q) = \sum_{x \in \mathcal{X}} P(x) \cdot \log \frac{P(x)}{Q(x)} \geq 0.$$

Although the relative entropy is not a metric, it possesses several useful properties which justify to introduce $D(P||Q)$ as a distance function for the probability distributions P and Q. The most important of these properties is stated in Lemma 6, of course. Another such property is the statement in the following lemma.

Lemma 80 (Data Processing Lemma) *Let* $\mathcal{X} = \{1, \ldots, a\}$, P *and* Q *be two probability distributions on* $\mathcal{X}$ *and let* $A_1, \ldots, A_t$ *be a partition of* $\mathcal{X}$ *(hence* $\mathcal{X} = \bigcup_{i=1}^{t} A_i$ *and* $A_i \cap A_j = \emptyset$ *for* $i \neq j$). *Then*

$$D(P||Q) = \sum_{x \in \mathcal{X}} P(x) \cdot \log \frac{P(x)}{Q(x)} \geq \sum_{i=1}^{t} P(A_i) \cdot \log \frac{P(A_i)}{Q(A_i)}.$$

Proof By definition $D(P||Q) = \sum_{x \in \mathcal{X}} P(x) \cdot \log \frac{P(x)}{Q(x)} = \sum_{i=1}^{t} \sum_{x \in A_i} P(x) \cdot \log \frac{P(x)}{Q(x)}$.

Now the log-sum inequality can be applied to each $i = 1, \ldots, t$ to obtain

$\sum_{x \in A_i} P(x) \cdot \log \frac{P(x)}{Q(x)} \geq \left(\sum_{x \in A_i} P(x) \right) \cdot \log \frac{\sum_{x \in A_i} P(x)}{\sum_{x \in A_i} Q(x)} = P(A_i) \cdot \log \frac{P(A_i)}{Q(A_i)}$ and

the lemma is proved. □

Another distance for probability distributions is the L_1-*norm* (or *total variation* in Statistics), defined by

$$||P - Q||_1 = \sum_{x \in \mathcal{X}} | P(x) - Q(x) | .$$

Remember from Analysis that for a continuous function f (under appropriate constraints) the L_p-norm is defined as $||f||_p = \left(\int_{-\infty}^{\infty} |f(y) - g(y)|^p dy\right)^{\frac{1}{p}}$, $p > 0$. The L_p-norm gives rise to a Banach space and especially for $p = 2$ to a Hilbert space. So here we have the discrete version of the L_1-norm.

The following inequality relates the relative entropy and the metric induced by the L_1-norm. It was discovered by Pinsker [9]. Further proofs are due to Kemperman [6] and Csiszár [1], who found the best constant $\frac{1}{2 \cdot \ell n 2}$.

Lemma 81 (Pinsker's inequality, [9]) *For two probability distributions P and Q on $\mathcal{X} = \{1, \ldots, a\}$*

$$D(P||Q) \geq \frac{1}{2 \cdot \ell n 2} ||P - Q||_1^2.$$

Proof First we define $A \triangleq \{x : P(x) \geq Q(x)\}$ in order to obtain two probability distributions $\hat{P} \triangleq (P(A), P(A^c))$ and $\hat{Q} \triangleq (Q(A), Q(A^c))$, where, of course, $P(A) = \sum_{x \in A} P(x)$, etc.

From the Data Processing Lemma follows that

$$D(P||Q) \geq D(\hat{P}||\hat{Q}).$$

On the other hand we have

$$||P - Q||_1 = ||\hat{P} - \hat{Q}||_1,$$

since

$$||P - Q||_1 = \sum_{x \in \mathcal{X}} |P(x) - Q(x)|$$

$$= \sum_{x : P(x) \geq Q(x)} (P(x) - Q(x)) - \sum_{x : P(x) < Q(x)} (P(x) - Q(x))$$

$$= |P(A) - Q(A)| + |P(A^c) - Q(A^c)| = ||\hat{P} - \hat{Q}||_1.$$

So we only have to prove Pinsker's inequality for $\hat{P}$ and $\hat{Q}$ and can hence assume w.l.o.g. that $\mathcal{X} = \{0, 1\}$.

Therefore we define $p \triangleq P(0) (= P(A))$ and $q \triangleq Q(0) (= Q(A))$. Since now $D(P||Q) = p \cdot \log \frac{p}{q} + (1 - p) \log \frac{1-p}{1-q}$ and $||P - Q||_1^2 = [p - q + (p - q)]^2 = [2 \cdot (p - q)]^2$ it suffices to find some constant c, such that for all $q \in [0, 1]$

$$f(q) \triangleq p \cdot \log \frac{p}{q} + (1 - p) \log \frac{1-p}{1-q} - c \cdot [2 \cdot (p - q)]^2 \geq 0.$$

W.l.o.g. we can assume that $0 \leq q \leq p \leq 1$.

Observe that for $p = q = \frac{1}{2}$ we have $f(q) = 0$ for arbitrary c. Our intention now is to make f monotonically decreasing on the interval $(0, \frac{1}{2}]$, since in this case $f(q) \geq f(\frac{1}{2}) = 0$ obviously holds.

Of course, f is differentiable. Hence we are done, if we find some constant c that guarantees

$$\frac{d\, f(q)}{d\, q} = \left(-\frac{p}{q} + \frac{1-p}{1-q}\right) \cdot \frac{1}{\ell n 2} + 8c(p - q) \leq 0.$$

An easy calculation yields that in this case

$$\frac{q \cdot (1 - p) - p(1 - q)}{q \cdot (1 - q)} \cdot \frac{1}{\ell n 2} \leq 8c \cdot (q - p),$$

hence $\frac{q-p}{q(1-q)} \cdot \frac{1}{\ell n 2} \leq 8c \cdot (q - p)$, which is equivalent to $\frac{1}{q \cdot (1-q) \cdot \ell n 2} \leq 8c$ and $c \leq \frac{1}{8q(1-q)\ell n 2}$.

Since the denominator is maximal for $q = \frac{1}{2}$, $f(q)$ is monotonically decreasing for $c \leq \frac{1}{2 \cdot \ell n 2}$ and hence Pinsker's inequality holds. □

Pinsker's inequality relates the L_1-norm and the relative entropy. In the following lemma a relation between the entropy difference and the L_1-norm is presented, from which we can conclude that the entropy function H is L_1-norm-continuous. (It is understood here that the L_1-norm induces a topology on $P(\mathcal{X})$, the set of all probability distributions on $\mathcal{X}$, and that the real numbers are equipped with the usual interval topology.)

Lemma 82 *For two probability distributions P and Q on $\mathcal{X} = \{1, \ldots, a\}$ with $\|P - Q\|_1 \leq \alpha \leq \frac{1}{2}$ it holds $\mid H(P) - H(Q) \mid \leq -\alpha \cdot \log \frac{\alpha}{|\mathcal{X}|}$.*

Proof Let us first take a closer look at the function $f(t) = -t \cdot \log t$ on the interval $(0, 1]$. It is $f'(t) = -\log t - \frac{1}{\ell n 2}$ and $f''(t) = -\frac{1}{t \cdot \ell n 2} < 0$ for all $t \in (0, 1]$. So f is concave on $(0, 1]$ with a maximum in $t = \frac{1}{e}$ and can hence be represented by the graph

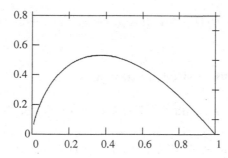

With the convention $f(0) = 0$ we now know

(i) $f(0) = f(1) = 0$

(ii) for $0 \le t \le 1 - s, 0 \le s \le \frac{1}{2}$ it is $\mid f(t) - f(t+s) \mid \le \max(f(s), f(1-s)) = -s \cdot \log s$.

The second assertion is immediate since the interval $[t, t + s]$ with maximal $\mid f(t) - f(t+s) \mid$ cannot be in the interior of $[0, 1]$, hence $t = 0$ to $t = 1 - s$.

Now

$$\mid H(P) - H(Q) \mid = \mid \sum_{x \in \mathcal{X}} f(P(x)) - f(Q(x)) \mid$$

$$\le \sum_{x \in \mathcal{X}} \mid f(P(x)) - f(Q(x)) \mid .$$

With $t = P(x), s = Q(x) - P(x)$, if $Q(x) \ge P(x)$, or $t = Q(x), s = P(x) - Q(x)$, if $P(x) > Q(x)$, we obtain

$$\mid H(P) - H(Q) \mid \le \sum_{x \in \mathcal{X}} f(\mid P(x) - Q(x) \mid)$$

$$= \sum_{x \in \mathcal{X}} - \mid P(x) - Q(x) \mid \cdot \log \mid P(x) - Q(x) \mid$$

and with $g(x) \triangleq \mid P(x) - Q(x) \mid$

$$\mid H(P) - H(Q) \mid \le - \sum_{x \in \mathcal{X}} g(x) \cdot \log g(x)$$

$$= -\alpha \cdot \left(\sum_{x \in \mathcal{X}} \frac{g(x)}{\alpha} \cdot \log \left(\frac{g(x)}{\alpha} \right) + \log \alpha \right).$$

Since $\sum_{x \in \mathcal{X}} g(x) = \|P - Q\|_1 \le \alpha$, of course $\sum_{x \in \mathcal{X}} \frac{g(x)}{\alpha} \le 1$.

Hence $- \sum_{x \in \mathcal{X}} \frac{g(x)}{\alpha} \cdot \log \frac{g(x)}{\alpha} \le \log \mid \mathcal{X} \mid$ and

$$\mid H(P) - H(Q) \mid \le \alpha \cdot (\log \mid \mathcal{X} \mid - \log \alpha) = -\alpha \cdot \log \frac{\alpha}{\mid \mathcal{X} \mid}. \qquad \square$$

8.3 Lecture on Stochastic Inequalities

8.3.1 Chebyshev's Inequality, Weak Law of Large Numbers

For a discrete RV $X : \Omega \to \mathcal{X}$ ($\mathcal{X} \subset \mathbb{Z}$, say) the expected value $\mathbb{E} X$ and the variance Var X are defined as follows

$$\mathbb{E}\,X = \sum_{x \in \mathcal{X}} x\, P(x)$$

$$\text{Var}\,X = \mathbb{E}(X - \mathbb{E}\,X)^2.$$

As usual, $\sigma_X \triangleq \sqrt{\text{Var}\,X}$ denotes the standard deviation.

Lemma 83 (Markov's inequality) *Let X be a RV on (Ω, P) with $X(w) \geq 0$ for all $w \in \Omega$.*
Then for all $a \in \mathbb{R}^+$:

$$P(X \geq a) \leq \frac{\mathbb{E}\,X}{a}.$$

Proof

$$P(X \geq a) = \sum_{x \geq a} P(X = x) \leq \sum_{x \geq a} P(X = x)\frac{x}{a} \leq \sum_{x \in X} P(X = x)\frac{x}{a}$$

$$= \frac{1}{a}\mathbb{E}\,X. \qquad \square$$

Theorem 61 (Chebyshev's inequality) *Let X be a RV on (Ω, P). Then for all $a > 0$*

$$P(|X - \mathbb{E}\,X| > a) \leq \frac{\text{Var}\,X}{a^2}.$$

Proof We consider the RV $(X - \mathbb{E}\,X)^2$ and apply Markov's inequality to obtain for every $a > 0$

$$P\big((X - \mathbb{E}X)^2 \geq a^2\big) \leq \frac{\mathbb{E}(X - \mathbb{E}\,X)^2}{a^2} = \frac{\text{Var}\,X}{a^2}.$$

This is equivalent to

$$P(|X - \mathbb{E}X| \geq a) \leq \frac{\text{Var}X}{a^2}. \qquad \square$$

From Chebyshev's inequality the following "weak law of large numbers" follows immediately, since $\sum_{i=1}^{n} \text{Var}\,X_i = \text{Var}\sum_{i=1}^{n} X_i$ for independent RV's $X_1, \dots, X_n$.

Theorem 62 (Weak law of large numbers) *Let $X_1, \dots, X_n$ be independent RV's on (Ω, P). Then for all $a > 0$*

$$P\Big(|\sum_{i=1}^{n} X_i - \mathbb{E}(\sum_{i=1}^{n} X_i)| \geq a\Big) \leq \frac{\sum_{i=1}^{n} \text{Var}\,X_i}{a^2}.$$

Most interesting for us is the special case where the X_i's are independent and identically distributed ($P_{X_i} = P_X$, say, for $i = 1, \dots, n$). Then since $\mathbb{E}(\frac{1}{n}X) = \frac{1}{n}\mathbb{E}\,X$ and since $\text{Var}(\frac{1}{n}X) = \frac{1}{n^2}\text{Var}\,X$

$$P\left(\left|\frac{1}{n}\sum_{i=1}^{n}X_i - \mathbb{E}\,X\right| \geq a\right) \leq \frac{\operatorname{Var} X}{n\,a^2}.$$

8.3.2 Bernstein's Trick, Chernoff Bound

Lemma 84 (Bernstein, Chernoff) *Let again X be a RV. Then by the monotonicity of the exponential function*

$$exp\{\alpha X\} \geq exp\{\alpha a\}$$

is for $\alpha > 0$ equivalent to $X \geq a$ and for $\alpha < 0$ equivalent to $X \leq a$. Therefore

(i) $P(X \geq a) = P\left(exp\{\alpha X\} \geq exp\{\alpha a\}\right) \leq \frac{\mathbb{E}exp\{\alpha \cdot X\}}{exp\{\alpha a\}}$ *for $\alpha > 0$;*

(ii) $P(X \leq a) = P\left(exp\{\alpha X\} \geq exp\{\alpha a\}\right) \leq \frac{\mathbb{E}exp\{\alpha X\}}{exp\{\alpha a\}}$ *for $\alpha < 0$.*

If $X = \sum_{i=1}^{n} X_i$ is the sum of independent RV's the product rule for the exponential function yields

$$\mathbb{E}\left(exp\{\alpha \sum_{i=1}^{n} X_i\}\right) = \mathbb{E}\left(exp\{\alpha X_1\}\right) \cdots \cdots \mathbb{E}\left(exp\{\alpha X_n\}\right). \tag{8.7}$$

Thus we obtain from Lemma 84,

Theorem 63 *Let $X_1, \ldots, X_n$ be independent RV's, and let $a \in \mathbb{R}$. Then*

(i) $P\left(\sum_{i=1}^{n} X_i \geq a\right) \leq exp\{-\alpha a\} \cdot \prod_{i=1}^{n} \mathbb{E}(exp\{\alpha X_i\})$ *for all $\alpha > 0$*

(ii) $P\left(\sum_{i=1}^{n} X_i \leq a\right) \leq exp\{-\alpha a\} \cdot \prod_{i=1}^{n} \mathbb{E}(exp\{\alpha X_i\})$ *for all $\alpha < 0$.*

Observe that we are free to choose the parameter α. As an application we now present a consequence that turns out to be a very useful tool in the proofs of coding theorems.

Corollary 7 (Large deviation bound for the binomial distribution) *Let $X_1, \ldots, X_n$ be i.i.d. RV's with values in $\{0, 1\}$.*
Suppose that the expected value $\mathbb{E}\,X$ satisfies $\mathbb{E}\,X \leq \mu < \lambda < 1$; then

$$P\left(\sum_{i=1}^{n} X_i > n \cdot \lambda\right) \leq exp\{-n \cdot D(\lambda||\mu)\},$$

where $D(\lambda||\mu)$ denotes the relative entropy for the probability distributions $(\lambda, 1-\lambda)$ and $(\mu, 1 - \mu)$. Here the exp-function is to the base 2.

Proof Since the conditions of Theorem 63 are fulfilled we have with $a \triangleq \lambda \cdot n$ for every $\alpha > 0$

$$P\left(\sum_{i=1}^{n} X_i \geq n \cdot \lambda\right) \leq \exp\{-\alpha \cdot \lambda n\} \cdot \left(\mathbb{E}(\exp\{\alpha X\})\right)^n.$$

Now $\mathbb{E}(\exp\{\alpha X\}) = P(X = 0) \cdot 2^0 + P(X = 1) \cdot 2^\alpha = 1 - \mathbb{E}\, X + 2^\alpha \mathbb{E}\, X$.
Since $\alpha > 0$, we have $2^\alpha > 1$. Hence $\mathbb{E}(\exp\{\alpha X\}) \leq 1 - \mu + 2^\alpha \cdot \mu$, because $\mathbb{E}\, X \leq \mu$ by assumption. We choose

$$\alpha' \triangleq -\log\left(\frac{1-\lambda}{\lambda} \cdot \frac{\mu}{1-\mu}\right).$$

Since $\mu < \lambda$, also $\frac{1-\lambda}{\lambda} < \frac{1-\mu}{\mu}$; therefore, $s' > 0$. We get easily

$$\mathbb{E}(\exp\{\alpha' X\})^n = \left(1 - \mu + \mu \cdot \frac{\lambda}{1-\lambda} \cdot \frac{1-\mu}{\mu}\right)^n$$

$$= \exp\{-n(\log(1-\lambda) - \log(1-\mu))\}$$

and also

$$\exp\{-\alpha' \lambda n\} = \exp\left\{-n\left(-\lambda \log\left(\frac{1-\lambda}{\lambda} \cdot \frac{\mu}{1-\mu}\right)\right)\right\}.$$

Note finally that the product of the last two right sides expressions equals $\exp\{-n\, D(\lambda\|\mu)\}$. $\qquad\square$

This can be generalized under a suitable regularity to the case where the RV's X_i take any values in $[0, 1]$.

Corollary 8 *Let $X_1, X_2, \ldots, X_n$ be i.i.d. RV's with values in $[0, 1]$ and let*

$$\frac{d\mathbb{E}2^{\alpha x}}{d\alpha} = \int_0^1 \frac{d2^{\alpha x}}{d\alpha} dP_X(x).$$

(i) *If $\mathbb{E}X \leq \mu < \lambda \leq 1$, then*

$$Pr\left(\sum_{i=1}^{n} X_i > n\lambda\right) \leq 2^{-nD(\lambda\|\mu)}.$$

(ii) *If $\mathbb{E}X_i = \mu$ and $0 < a < 1$, then*

$$Pr\left\{\frac{1}{n}\sum_{i=1}^{n}X_i > (1+a)\mu\right\} \le exp(-nD((1+a)\mu||\mu)),$$

$$Pr\left\{\frac{1}{n}\sum_{i=1}^{n}X_i < (1-a)\mu\right\} \le exp(-nD((1-a)\mu||\mu)).$$

Since

$$D((1+a)\mu||\mu) \ge \frac{a^2\mu}{2\ln 2} \quad for \ |a| \le \frac{1}{2},$$

also

$$Pr\left\{\frac{1}{n}\sum_{i=1}^{n}X_i \notin [(1-a)\mu, (1+a)\mu]\right\} \le 2exp\left(-n\cdot\frac{a^2\mu}{2\ln 2}\right).$$

Proof To verify (i) notice that

$$\mathbb{E}2^{\alpha x} = \int_0^1 2^{\alpha x}dP_X(x) \le 1 - \mu + \mu 2^{\alpha}, \tag{8.8}$$

because the function $f(\alpha) = 1 - \mu + \mu 2^{\alpha} - \int_0^1 2^{\alpha x}dP_X(x)$ has the properties $f(0) = 0$ and

$$f'(\alpha) = \mu \ln 2\left(2^{\alpha} - \frac{1}{\mu}\int_0^1 x2^{\alpha x}dP_X(x)\right)$$

$$\ge \mu \ln 2\left(2^{\alpha} - \frac{1}{\mu}\int_0^1 x2^{\alpha}dP_X(x)\right)$$

$$= \mu \ln 2\left(2^{\alpha} - \frac{1}{\mu}2^{\alpha}\cdot\mu\right) = 0$$

and (8.8) holds for every $\alpha \ge 0$ and the proof can be completed as for the previous corollary.

The first two inequalities in (ii) follow by the same reasoning (also a consequence of Sanov's Theorem [10]) and the lower bound is obtained by elementary calculus.   □

Now we consider RV's which need not be independent and also not identically distributed.

Lemma 85 *Let $X_1, \ldots, X_n$ be a sequence of discrete RV's, then*

(i) $\Pr\left\{\dfrac{1}{n}\sum_{i=1}^{n} X_i \geq a\right\} \leq e^{-\alpha a n} \prod_{i=1}^{n} \max_{x^{i-1}} \mathbb{E}(\exp_e\{\alpha X_i\}|X^{i-1} = x^{i-1})$ *for*
$\alpha > 0$.

(ii) $\Pr\left\{\dfrac{1}{n}\sum_{i=1}^{n} X_i \geq a\right\} \leq e^{-\frac{an}{2b}} \prod_{i=1}^{n} \max_{x^{i-1}} \mathbb{E}(1+b^{-1}X_i|X^{i-1} = x^{i-1})$ *if* $X_1, \ldots, X_n$
take values in $[0, b]$.

(iii) $\Pr\left\{\dfrac{1}{n}\sum_{i=1}^{n}(X_i - \mathbb{E}X_i) \geq a\right\} \leq \exp\left[(-\alpha a + \alpha^2 b^2)n\right]$ *for* $0 < \alpha \leq \min$
$\left(1, \dfrac{b^2}{2}e^{-2b}\right)$ *if* $X_1, \ldots, X_n$ *are independent and take values in* $[-b, b]$.
Further, the exponent is negative if $\alpha < ab^{-2}$.

(iv) $\Pr\left\{\dfrac{1}{n}\sum_{i=1}^{n}(X_i - \mathbb{E}T_i) \leq a\right\} \leq \exp\left[(-\alpha a + \alpha^2 b^2)n\right]$ *for* $\alpha < 0$, $|\alpha| \leq$
$\min\left(1, \dfrac{b^2}{2}e^{-2b}\right)$, *if* $X_1, \ldots, X_n$ *are independent and take values in* $[-b, b]$.
The exponent is negative if $a < 0$ *and* $|\alpha| \leq |a|b^{-2}$.

Proof

$$\Pr\left\{\frac{1}{n}\sum_{i=1}^{n} X_i \geq a\right\} \leq e^{-\alpha a n}\mathbb{E}\left(\exp\left\{\alpha\sum_{i=1}^{n} X_i\right\}\right)$$

$$\leq e^{-\alpha a n}\prod_{i=1}^{n}\max_{x^{i-1}}\mathbb{E}(\exp\{\alpha X_i\}|x^{i-1}).$$

(i) is now readily established by using the expansion

$$\mathbb{E}(\exp\{\alpha X_i\}|x^{i-1}) = \sum_{s=0}^{\infty}\frac{\alpha^s}{s!}\mathbb{E}(X_i^s|x^{i-1}).$$

Since $X_i^s \leq b^{s-1}X_i$ for $s \geq 1$ we conclude that for $\alpha = (2b)^{-1}$

$$\sum_{s=0}^{\infty}\frac{\alpha^s}{s!}\mathbb{E}(X_i^s|x^{i-1}) \leq 1 + (2b)^{-1}\sum_{s=1}^{\infty}\frac{1}{s!}\mathbb{E}(X_i|x^{i-1})$$

$$\leq 1 + b^{-1}\mathbb{E}(X_i|x^{i-1}) = \mathbb{E}(1 + b^{-1}X_i|x^{i-1}).$$

We show now (iii). Starting with

$$\Pr\left\{\sum_{i=1}^{n}(X_i - \mathbb{E}X_i) \geq an\right\} \leq e^{-\alpha a n}\mathbb{E}\exp\left\{\alpha\sum_{i=1}^{n}(X_i - \mathbb{E}X_i)\right\}$$

we upper bound next

$$\mathbb{E}\exp\{\alpha(X_i - \mathbb{E}X_i)\} = \sum_{s=0}^{\infty} \frac{\alpha^s}{s!} \mathbb{E}(X_i - \mathbb{E}X_i)^s \leq 1 + 0 + \frac{\alpha^2}{2}b^2 + \sum_{s=3}^{\infty} \frac{\alpha^s}{s!}(2b)^s$$

$$\leq 1 + \frac{\alpha^2}{2}b^2 + \alpha^3 e^{2b} \quad \text{for } \alpha \leq 1$$

$$\leq 1 + \alpha^2 b^2 \quad \text{for } \alpha \leq \min\left(1, \frac{b^2}{2}e^{-2b}\right).$$

Since the X_i's are independent we get

$$\Pr\left\{\sum_{i=1}^{n}(X_i - \mathbb{E}X_i) \geq an\right\} \leq e^{-\alpha an}(1 + \alpha^2 b^2)^n \leq \exp[(-\alpha a + \alpha^2 b^2)n].$$

The proof for (iv) is almost identical. Just observe that for $\alpha > 0$

$$\Pr\left\{\sum_{i=1}^{n}(X_i - \mathbb{E}X_i) \leq an\right\} \leq e^{-\alpha an}\mathbb{E}\exp\left\{\alpha \sum_{i=1}^{n}(X_i - \mathbb{E}X_i)\right\}$$

and replace in the argument above α^3 by $|\alpha^3|$.   □

Lemma 86 *Let* $Z_1, \ldots, Z_n$ *be RV's and let* $f_i(Z_1, \ldots, Z_i)$ *be arbitrary with* $0 \leq f_i \leq 1$, $i = 1, \ldots, n$. *Then the condition*

$$\mathbb{E}\big(f_i(Z_1, \ldots, Z_i | Z_1, \ldots, Z_{i-1})\big) \leq a \ (a.s.) \tag{8.9}$$

implies that

$$\Pr\left\{\frac{1}{n}\sum_{i=1}^{n} f_i(Z_1, \ldots, Z_i) > t\right\} \leq \exp\left\{-n\left(\frac{\log e}{2}t - a\log e\right)\right\}.$$

Proof Let $X_i = f_i(Z_1, \ldots, Z_i)$ in Lemma 85(ii). Then $\mathbb{E}(1 + X_i) = \mathbb{E} = \big(\mathbb{E}(1 + X_i | Z^{i-1})X^{i-1}\big)$ since T^{i-1} is a function of Z^{i-1}. Therefore

$$\max_{x^{i-1}} \mathbb{E}(1 + X_i | X_1 = x_1, \ldots, X_{i-1} = x_{i-1}) \leq \max_{z^{i-1}} \mathbb{E}(1 + X_i | Z^{i-1} = z^{i-1}).$$

Thus it follows from Lemma 85(ii)

$$\Pr\left\{\frac{1}{n}\sum_{i=1}^{n} f_i(Z_1, \ldots, Z_i) > t\right\} \leq \exp_e\left\{-\frac{nt}{2} + na\right\} = \exp\left\{-n\left(\frac{\log e}{2}t - a\log e\right)\right\}.$$

□

Lemma 87 *For RV's $X_1, X_2, \ldots, X_n$ with values in $\{0, 1\}$ and conditional probabilities lower bounded by*

$$Pr(X_i = 1 | X_{i-1} = x_{i-1}, \ldots, X_1 = x_1) \geq 1 - q$$

for all $i \in [n]$ and all values of $X_1, X_2, \ldots, X_{i-1}$ we have for weights $p_1, p_2, \ldots, p_n > 0$ with $p_i \leq b \sum_{i-1}^n p_i$ for $i \in [n]$, if $\alpha < 0$ and $q < \lambda$

$$Pr\left(\sum_{i=1}^n p_i X_i < (1 - \lambda) \sum_{i=1}^n p_i\right) \leq \exp\left\{\alpha(\lambda - q) \sum_{i=1}^n p_i + b\frac{\alpha^2}{2}\left(\sum_{i=1}^n p_i\right)^2\right\}.$$

Proof From Lemma 84

$$Pr\left(\sum_{i=1}^n p_i X_i < (1 - \lambda) \sum_{i=1}^n p_i\right) \leq \exp\left\{\alpha(1 - \lambda) \sum_{i=1}^n p_i\right\} \cdot \mathbb{E}\exp\left\{\alpha \sum_{i=1}^n p_i X_i\right\}$$

$$(8.10)$$

and by the bound on the conditional probabilities we get for the second factor the upper bound

$$\prod_{i=1}^n (q + (1 - q)e^{\alpha p_i}) \leq \prod_{i=1}^n\left(1 + (1 - q)\left(\alpha p_i + \frac{(\alpha p_i)^2}{2}\right)\right)$$

and since $\ln(1 + x) \leq x$

$$\leq \exp\left\{(1 - q)\left(\alpha \sum_{i=1}^n p_i + \sum_{i=1}^n \frac{(\alpha p_i)^2}{2}\right)\right\}$$

and since $\sum_{i=1}^n p_i^2 \leq b\left(\sum_{i=1}^n p_i\right)^2$

$$\leq \exp\left\{(1 - q)\left(\alpha \sum_{i=1}^n p_i + b\frac{\alpha^2}{2}\left(\sum_{i=1}^n p_i\right)^2\right)\right\}.$$

This together with the first factor in (8.10) gives the upper bound

$$\exp\left\{\alpha(\lambda - q) \sum_{i=1}^n p_i + b\frac{\alpha^2}{2}\left(\sum_{i=1}^n p_i\right)^2\right\},$$

that is (i). □

Remark Alternatively, we can bound

$$\exp\left\{-\alpha(1-\lambda)\sum_{i=1}^{n}p_i\right\}\prod_{i=1}^{n}(q+(1-q)e^{\alpha p_i})$$

with the choice $\alpha = \log\frac{q}{1-q}\frac{1-\lambda}{\lambda}$, if $\frac{q}{1-q}\frac{1-\lambda}{\lambda} \leq 1$, that is, $q \leq \lambda$.

8.3.3 Lovasz Local Lemma

Finally, we present a very useful tool.

Lemma 88 (Lovasz local lemma [3]) *Let $A_1, A_2, \ldots, A_n$ be events in an arbitrary probability space such that $Pr(A_i) \leq p$ for all $1 \leq i \leq n$ and each event A_i is mutually independent of a set of all but at most d of the other events. If*

$$e \cdot p \cdot (d+1) \leq 1$$

then $Pr(\bigcup_{i=1}^{n} \bar{A}_i) > 0$.

Here e denotes Euler's number $\exp(1)$. We shall sometimes replace the above condition by

$$4 \cdot p \cdot d \leq 1. \tag{8.11}$$

8.4 Lecture on Schur-Convexity

Definition 37 For $x^n = (x_1, \ldots, x_n) \in \mathbb{R}^n$ let $x_{[1]} \geq x_{[2]} \geq \cdots \geq x_{[n]}$ be a permutation such that the $x_i's$ are ordered nonincreasingly.

On $\mathbb{R}^n$ a relation $\prec$ is defined by

$$x^n \prec y^n \text{ exactly if } \sum_{i=1}^{k} x_{[i]} \leq \sum_{i=1}^{k} y_{[i]} \quad \text{for all } k = 1, \ldots, n-1$$

$$\text{and } \sum_{i=1}^{n} x_{[i]} = \sum_{i=1}^{n} y_{[i]}.$$

If $x^n \prec y^n$, we say that x^n is majorized by y^n.

Definition 38 A Dalton transfer is a transfer from $x^n = (x_1, \ldots, x_n)$ to $x'^n = (x'_1, \ldots, x'_n)$ with

(i) $x'_i = x_i + a, x'_j = x_j - a$ for two indices $i, j \in \{1, \ldots, n\}$ and some $a \in \mathbb{R}$, $0 \le a \le x_j - x_i$

(ii) $x'_k = x_k$ for all $k \in \{1, \ldots, n\} \setminus \{i, j\}$.

Lemma 89 (Muirhead, 1903, [8]) *If $x^n \prec y^n$ ($x^n, y^n \in \mathbb{R}^n$), then x^n can be obtained from y^n by a finite number of Dalton transfers.*

Lemma 90 *Let W be a doubly stochastic matrix of order n and $x^n \in \mathbb{R}^n$. Then $x^n W \prec x^n$.*

Definition 39 A function $f : A \to \mathbb{R}, A \subset \mathbb{R}^n$, is called Schur-convex on A [11] if

$$x^n \prec y^n \text{ on } A \Rightarrow f(x) \le f(y).$$

Accordingly, a function $f : A \to \mathbb{R}, A \subset \mathbb{R}^n$ is Schur-concave on A if

$$x^n \prec y^n \text{ on } A \Rightarrow f(x) \ge f(y)$$

Lemma 91 *The entropy function H is Schur-concave.*

Lemma 92 *Let W be a doubly stochastic matrix of order a and $P = (P(1), \ldots, P(a))$ be a probability distribution on $\mathcal{X} \triangleq \{1, \ldots, a\}$. Then $H(PW) \ge H(P)$.*

Proof By Lemma 90 $PW \prec P$. From the Schur-concavity of the entropy function the statement follows immediately. $\qquad\square$

8.5 Lecture on Properties of Mutual Information

8.5.1 A Characterisation of Optimal Sources, Shannon's Lemma

As usual we denote by $W = (w(y|x))_{x \in \mathcal{X}, y \in \mathcal{Y}}$ a discrete memoryless channel and let P be the distribution of an input variable X and $Q = P \cdot W$ the distribution of an output variable Y. For the mutual information we shall use the notations $I(X \wedge Y) = I(P, W) = \sum_x \sum_y P(x)w(y|x) \log \frac{w(y|x)}{Q(y)}$.

Lemma 93 *For any distribution Q^* on $\mathcal{Y}$*

$$I(P, W) + D(Q||Q^*) = \sum_{x \in \mathcal{X}} P(x)D(w(\cdot|x)||Q^*)$$

Proof $I(P, W) + D(Q||Q^*)$

$$= \sum_{y \in \mathcal{Y}} \sum_{x \in \mathcal{X}} P(x)w(y|x) \log \frac{w(y|x)}{Q(y)} + \sum_{y \in \mathcal{Y}} \sum_{x \in \mathcal{X}} P(x)w(y|x) \log \frac{Q(Y)}{Q^*(Y)}$$

$$= \sum_{y \in \mathcal{Y}} \sum_{x \in \mathcal{X}} P(x)w(y|x)\big[\log w(y|x) - \log Q(y) + \log Q(y) - \log Q^*(y)\big]$$

$$= \sum_{y \in \mathcal{Y}} \sum_{x \in \mathcal{X}} P(x)w(y|x) \log \frac{w(y|x)}{Q^*(y)} = \sum_{x \in \mathcal{X}} P(x)D\big(w(\cdot|x)\|Q^*\big)$$

$\square$

Theorem 64 *Let Q^* be an arbitrary distribution on $\mathcal{Y}$ and $\lambda_1, \ldots, \lambda_r \in \mathbb{R}$ with $\sum_{i=1}^{r} \lambda_i = 1$. For distributions P_i and $Q_i = P_i W$, $i = 1, \ldots, r$ then*

$$\sum_{i=1}^{r} \lambda_i I(P_i, W) + \sum_{i=1}^{r} \lambda_i D(Q_i \| Q^*) = I\Big(\sum_{i=1}^{r} \lambda_i P_i, W\Big) + D\Big(\sum_{i=1}^{r} \lambda_i Q_i \| Q^*\Big)$$

Proof By the preceding lemma for every $i = 1, \ldots, r$

$$I(P_i, W) + D(Q_i \| Q^*) = \sum_{x \in \mathcal{X}} P_i(x)D\big(w(\cdot|x)\|Q^*\big)$$

So we can conclude

$$\sum_{i=1}^{r} \lambda_i I(P_i, W) + \sum_{i=1}^{r} \lambda_i D(Q_i \| Q^*) = \sum_{x \in \mathcal{X}} \sum_{i=1}^{r} \lambda_i P_i(x)D\big(w(\cdot|x)\|Q^*\big)$$

$$= I\Big(\sum_{i=1}^{r} \lambda_i P_i, W\Big) + D\Big(\sum_{i=1}^{r} \lambda_i Q_i \| Q^*\Big)$$

by Lemma 93. $\square$

From Theorem 64 the following corollaries follow immediately.

Corollary 9 *The mutual information is concave in P, i.e.,*

$$I\Big(\sum_{i=1}^{r} \lambda_i P_i, W\Big) \geq \sum_{i=1}^{r} \lambda_i I(P_i, W).$$

Furthermore equality holds, exactly if $Q_i = Q^*$ for all i with $\lambda_i > 0$.

Corollary 10 *An optimal distribution on $\mathcal{Y}$ is uniquely determined, i.e., for distributions P, P^* with*

$$I(P, W) = I(P^*, W) = \max_P I(P, W) = C \text{ it is } PW = P^*W.$$

Proof Let $P_1, \ldots, P_r$ be distributions on $\mathcal{X}$ with $I(P_i, W) = C$. Then also $\sum_{i=1}^{r} \lambda_i I(P_i, W) = C$ and because of the concavity $I\big(\sum_{i=1}^{r} \lambda_i P_i, W\big) = C$.
Hence we have in Theorem 64

$$\sum_{i=1}^{r} \lambda_i D(Q_i \| Q^*) = D\left(\sum_{i=1}^{r} \lambda_i Q_i \| Q^*\right),$$

and since the relative entropy is strictly convex in this case, $Q_1 = Q_2 = \cdots = Q_r$ must hold.   □

Theorem 65 (Shannon's lemma [12]) *A necessary and sufficient condition for* P^* *to be optimal, i.e.,* $I(P^*, W) = \max_P I(P, W) = C$, *is:*

For some constant K *we have for* $Q^* = P^*W$

$$D\big(w(\cdot|x)\|Q^*\big) = K, \text{ if } P^*(x) > 0$$
$$D\big(w(\cdot|x)\|Q^*\big) \le K, \text{ if } P^*(x) = 0.$$

In this case $K = C$, *the capacity of the channel* W.

Proof From Lemma 93 we know that $I(P, W) + D(Q\|Q^*) \le K$. Since $D(Q\|Q^*) \ge 0$, obviously $I(P, W) \le K$. The sufficiency part now follows immediately, since for $P = P^*$

$$I(P^*, W) = \sum_{x \in \mathcal{X}} P^*(x) D\big(w(\cdot|x)\|Q^*\big) = K.$$

In order to prove the necessity, we assume that P^* is an optimal distribution and that $Q^* = P^* \cdot W$. Now we define for every $x \in \mathcal{X}$

$$C_x \triangleq D\big(w(\cdot|x)\|Q^*\big).$$

Obviously $\sum_{x \in \mathcal{X}} P^*(x) C_x = C$, the capacity.
Suppose now that $C_{x_0} = \max_x C_x > C$.

Then we define the probability distribution δ_{x_0} by $\delta_{x_0}(x) = \begin{cases} 1, x = x_0 \\ 0, x \ne x_0 \end{cases}$ and the distributions $P_\Theta (0 \le \Theta \le 1)$ as convex combinations $P_\Theta \triangleq \Theta \delta_{x_0} + (1 - \Theta) P^*$. Let $Q_\Theta \triangleq P_\Theta W$, hence $Q_\Theta(y) = (1 - \Theta) Q^*(y) + \Theta w(y|x_0)$.
Now we can apply Theorem 64 with Q_Θ taking the role of Q^* and obtain

$$\Theta \cdot I(\delta_{x_0}, W) + (1 - \Theta) \cdot \big(I(P^*, W) + D(Q^*\|Q_\Theta)\big) + \Theta \cdot D\big(w(\cdot|x_0)\|Q_\Theta\big)$$
$$= I(P_\Theta, W) + D(Q_\Theta\|Q_\Theta) = I(P_\Theta, W)$$

Thus

$$(1 - \Theta)C + \Theta D\big(w(\cdot|x_0)\|Q_\Theta\big) + (1 - \Theta)D(Q^*\|Q_\Theta) = I(P_\Theta, W)$$

and hence

$$I(P_\Theta, W) \ge (1 - \Theta)C + \Theta \, D\big(w(\cdot|x_0)\|Q_\Theta\big)$$

Since by assumption $D\big(w(\cdot|x_0)\|Q^*\big) > C$, there exists some $\Theta' > 0$ with $D\big(w(\cdot|x_0)\|Q_{\Theta'}\big) > C$ (by reasons of continuity). Hence $I(P_{\Theta'}, W) > C$, which is not possible.

So we can conclude that for every $x \in \mathcal{X}$ $C_x \leq C$. Since $C = \sum_{x \in \mathcal{X}} P^*(x)C_x$, obviously $C_x = C$ for all x with $P^*(x) > 0$. $\qquad\square$

Remark The above proof of Shannon's Lemma is due to F. Topsøe [13]. It makes only use of properties of the information measures. Usually Theorem 65 is proven with the help of the Kuhn-Tucker theorem ([7], see also Gallager [4], p. 91).

8.5.2 Convexity

Lemma 94 $D(P\|Q)$ *is convex in the pair* (P, Q), *so for two pairs* (P_1, Q_1) *and* (P_2, Q_2) *it holds*

$$D\big(\lambda P_1 + (1 - \lambda) P_2 \| \lambda Q_1 + (1 - \lambda) Q_2\big) \leq \lambda D(P_1\|Q_1) + (1 - \lambda)D(P_2\|Q_2)$$

for all $0 \leq \lambda \leq 1$, *with equality, exactly if* $P_1(x) \cdot Q_2(x) = P_2(x) \cdot Q_1(x)$ *for all* $x \in \mathcal{X}$.

Proof Application of the log-sum inequality yields

$$\big(\lambda P_1(x) + (1 - \lambda) P_2(x)\big) \log \frac{\lambda P_1(x) + (1 - \lambda) P_2(x)}{\lambda Q_1(x) + (1 - \lambda) Q_2(x)}$$

$$\leq \lambda P_1(x) \log \frac{\lambda P_1(x)}{\lambda Q_1(x)} + (1 - \lambda) P_2(x) \log \frac{(1 - \lambda) P_2(x)}{(1 - \lambda) Q_2(x)}$$

for every $x \in \mathcal{X}$, with equality, if $P_1(x) \cdot Q_2(x) = P_2(x) \cdot Q_1(x)$. Summing up over all $x \in \mathcal{X}$ we have the desired result. $\qquad\square$

Remark When $P \triangleq P_1 = P_2$ or $Q \triangleq Q_1 = Q_2$ are fixed, then $D(P\|Q)$ is strictly convex.

Lemma 95 $I(P, W)$ *is concave in* P *and convex in* W.

Proof (a) We already saw in Corollary 1 of Theorem 1 that $I(P, W)$ is concave in P. A second proof makes use of the linearity (in P) of the expressions $P \cdot W$ and $\sum_{x \in \mathcal{X}} P(x)H\big(w(\cdot|x)\big)$ and of the concavity of the entropy function H. Therefore let $\lambda_1, \ldots, \lambda_r \in [0, 1]$ with $\sum_{i=1}^r \lambda_i = 1$.

$$I\Big(\sum_{i=1}^r \lambda_i P_i, W\Big) = H\Big(\big(\sum_{i=1}^r \lambda_i P_i\big) \cdot W\Big) - \sum_{x \in \mathcal{X}} \big(\sum_{i=1}^r \lambda_i P_i(x)\big) \cdot H\big(w(\cdot|x)\big)$$

$$= H\Big(\sum_{i=1}^r \lambda_i (P_i \cdot W)\Big) - \sum_{i=1}^r \lambda_i \sum_{x \in \mathcal{X}} P_i(x) H\big(w(\cdot|x)\big)$$

$$\geq \sum_{i=1}^{r} \lambda_i H(P_i \cdot W) - \sum_{i=1}^{r} \lambda_i \sum_{x \in \mathcal{X}} P_i(x) H\big(w(\cdot|x)\big)$$

$$= \sum_{i=1}^{r} \lambda_i \Big[H(P_i \cdot W) - \sum_{x \in \mathcal{X}} P_i(x) H\big(w(\cdot|x)\big) \Big] = \sum_{i=1}^{r} \lambda_i I(P_i, W).$$

(b) In order to prove the convexity in W, remember that

$$I(P, W) = \sum_{x \in \mathcal{X}} \sum_{y \in \mathcal{Y}} P(x) w(y|x) \log \frac{w(y|x)}{Q(y)}$$

$$= \sum_{x \in \mathcal{X}} \sum_{y \in \mathcal{Y}} P_{XY}(x, y) \log \frac{P_{XY}(x, y)}{P(x) \cdot Q(y)}$$

$$= D(P_{XY} \| P \cdot Q)$$

is the relative entropy between the joint distribution P_{XY} and the product of the distributions P (on $\mathcal{X}$) and Q (on $\mathcal{Y}$). Now let $W = \sum_{i=1}^{r} \lambda_i W_i$, $\sum_{i=1}^{r} \lambda_i = 1$, $\lambda_1, \ldots, \lambda_r \in [0, 1]$, be the convex combination of the channels $W_i = (w_i(y|x))_{x \in \mathcal{X}, y \in \mathcal{Y}}$, $i = 1, \ldots, r$. Then $P_{XY} \triangleq \sum_{i=1}^{r} \lambda_i P_{XY,i}$ defined by $P_{XY,i}(x, y) = P(x) w_i(y|x)$ and $Q \triangleq \sum_{i=1}^{r} \lambda_i Q_i$ defined by $Q_i(y) = \sum_{x \in \mathcal{X}} P(x) w_i(y|x)$ are convex combinations. The statement now follows from the convexity of the relative entropy. $\square$

8.5.3 Conditional Mutual Information

The conditional information is defined by

$$I(X \wedge Y|Z) \triangleq H(X|Z) - H(X|Y, Z)$$

Recall that $H(X|Z) \geq H(X|YZ)$ with equality, exactly if X and Y are conditionally (given Z) independent or, equivalently, $X \ominus Z \ominus Y$ is a Markov chain. So we immediately have that

$$I(X \wedge Y|Z) \geq 0$$

with equality exactly if $X \ominus Z \ominus Y$ is a Markov chain.

In general it is not possible to compare $I(X \wedge Y)$ and $I(X \wedge Y|Z)$, but

Lemma 96 $I(X \wedge Y|Z) \geq I(X \wedge Y)$, if Z and X or Z and Y are independent.

Proof We shall only consider the case where Z, Y are independent. If Z and X are independent, the proof is analogous.

$$I(X \wedge Y|Z) = H(Y|Z) - H(Y|X, Z)$$
$$\geq H(Y) - H(Y|X) = I(X \wedge Y). \qquad \square$$

The following lemma states that data processing reduces the information.

Lemma 97 (Data processing lemma for the information) *For a Markov chain* $X \ominus$ $Y \ominus Z$ *we have*

(i) $I(X \wedge Y) \geq I(X \wedge Z)$
(ii) $I(Y \wedge Z) \geq I(X \wedge Z)$.

Proof

(i)

$$I(X \wedge Y) = H(X) - H(X|Y)$$
$$= H(X) - H(X|YZ) \text{ (because of the Markov property)}$$
$$= I(X \wedge YZ)$$
$$= I(X \wedge Z) + I(X \wedge Y|Z) \text{ (Kolmogorov's identity)}$$
$$\geq I(X \wedge Z)$$

(ii) can be shown analogously.

$\square$

Corollary 11 $I(X \wedge Y) \geq I\big(f(X) \wedge g(Y)\big)$

Proof Choosing $U = f(X)$ and $V = g(Y)$ we obtain the Markov chains $U \ominus X \ominus Y$ and $U \ominus Y \ominus V$.

$$I(X \wedge Y) \geq I(U \wedge Y) \geq I(U \wedge V) = I\big(f(X) \wedge g(Y)\big). \qquad \square$$

Lemma 98 (Dobrushin's inequality [2]) *Let* $X^n \triangleq (X_1, \ldots, X_n)$ *and* $Y^n \triangleq$ $(Y_1, \ldots, Y_n)$ *be sequences of (finite-valued) RV's (not necessarily connected by a DMC). We further require that the X_t's are independent. Then*

$$I(X^n \wedge Y^n) \geq \sum_{t=1}^{n} I(X_t \wedge Y_t)$$

Proof (Induction on n) Obviously, it suffices to prove Dobrushin's inequality for $n = 2$, since $X^n = (X^{n-1}, X_n)$, $Y^n = (Y^{n-1}, Y_n)$ and hence the statement would follow from $I(X^n \wedge Y^n) \geq I(X^{n-1} \wedge Y^{n-1}) + I(X_n \wedge Y_n)$.

By Kolmogorov's identity we have for discrete RV's X, Y, Z

$$I(XY \wedge Z) = I(Y \wedge Z) + I(X \wedge Z|Y) \quad \text{and}$$
$$I(X \wedge YZ) = I(X \wedge Y) + I(X \wedge Z|Y),$$

and it is easily seen that

$$I(XY \wedge Z) + I(X \wedge Y) = I(X \wedge YZ) + I(Y \wedge Z)$$

Now we choose $X \triangleq X_1$, $Y \triangleq X_2$, and $Z \triangleq Y_1 Y_2$ and obtain

$$I(X_1 X_2 \wedge Y_1 Y_2) + I(X_1 \wedge X_2) = I(X_1 \wedge X_2 Y_1 Y_2) + I(X_2 \wedge Y_1 Y_2)$$

From this immediately follows

$$I(X_1 X_2 \wedge Y_1 Y_2) \geq I(X_1 \wedge Y_1) + I(X_2 \wedge Y_2),$$

since

(1) $I(X_1 \wedge X_2) = 0$ because of the independence,
(2) $I(X_1 \wedge X_2 Y_1 Y_2) = I(X_1 \wedge Y_1) + I(X_1 \wedge X_2 Y_2 | Y_1) \geq I(X_1 \wedge Y_1)$
(3) $I(X_2 \wedge Y_1 Y_2) = I(X_2 \wedge Y_2) + I(X_2 \wedge Y_1 | Y_2) \geq I(X_2 \wedge Y_2)$
  by Kolmogorov's identity.

$\square$

References

1. I. Csiszár, Information-type measures of difference of probability distributions and indirect observations. Stud. Sci. Math. Hung. **2**, 299–318 (1967)
2. R.L. Dobrushin, General formulation of Shannon's basic theorem in information theory. Usp. Mat. Nauk **14**(6), 3–104 (1959). (in Russian)
3. P. Erdös, L. Lovász, *Problems and Results on 3-chromatic Hypergraphs and Some Related Questions*, ed. by A. Hajnal, R. Rado, V. T. Sós. Infinite and Finite Sets (to Paul Erdős on his 60th birthday), vol. 2 (North Holland, Amsterdam, 1975), pp. 609–627
4. R.G. Gallager, *Information Theory and Reliable Communication* (Wiley, New York, 1968)
5. G.H. Hardy, J.E. Littlewood, G. Pólya, *Inequalities*, 1st edn. (2nd edn. 1952) (University Press, Cambridge, 1934)
6. J.H.B. Kemperman, On the optimal rate of transmitting information. Ann. Math. Stat. **40**, 2156–2177 (1969)
7. H.W. Kuhn, A.W. Tucker, Nonlinear programming, in *Proceedings of the Second Berkeley Symposium on Mathematical Statistics and Probability, 1950*, University of California Press, Berkeley and Los Angeles, pp. 481–492, 1951
8. R. Muirhead, Some methods applicable to identities and inequalities of symmetric algebraic functions of n letters. Proc. Edinb. Math. Soc. **21**, 144–157 (1903)
9. M.S. Pinsker, *Information and Information Stability of Random Variables and Processes*, trans. and ed. by A. Feinstein (San Francisco, Holden-Day, 1964)
10. I.N. Sanov, On the probability of large deviations of random variables. Mat. Sb. **42**, 11–44 (1957)
11. I. Schur, Über eine Klasse von Mittelbildungen mit Anwendungen auf die Determinantentheorie. Sitzunsber. Berlin. Math. Ges. **22**, 9–20 (1923)
12. C.E. Shannon, Geometrische Deutung einiger Ergebnisse bei der Berechnung der Kanalkapazität. Nachr. Tech. Z. **10**, 1–4 (1957)
13. F. Topsøe, A new proof of a result concerning computation of the capacity for a discrete channel. Z. Wahrscheinlichkeitstheorie Verw. Gebiete **22**, 166–168 (1972)

Rudolf Ahlswede 1938–2010

We, his friends and colleagues at the Department of Mathematics at the University of Bielefeld are terribly saddened to share the news that Professor Rudolf Ahlswede passed away in the early hours of Saturday morning 18th December, 2010.

Rudolf Ahlswede had after an excellent education in Mathematics, Physics, and Philosophy almost entirely at the University of Göttingen and a few years as an Assistant in Göttingen and Erlangen received a strong push towards research, when he moved to the US, taught there at the Ohio State University in Columbus and greatly profited from joint work in *Information Theory* with the distinguished statistician Jacob Wolfowitz at Cornell and the University of Illinois during the years 1967–1971 (see the obituary [A82]).

The promotion to full professor in Mathematics followed in 1972, but only after Rudolf Ahlswede convinced his faculty by his work in Classical Mathematics. Information Theory was not yet considered to be a part of it. A problem in p-adic analysis by K. Mahler found its solution in [AB75] and makes now a paragraph in Mahler's book [M81].[1]

For a short time concentrating on Pure Mathematics and quitting Information Theory was considered. But then came strong responses to multi-way channels [A71] and it became clear that Information Theory would always remain a favorite subject—it looked more interesting to Rudolf Ahlswede than many areas of Classical Mathematics. An account of this period is given in the books [W78], [CK81] and [CT06]. However, several hard problems in Multi-user Information Theory led Rudolf Ahlswede to *Combinatorics*, which became the main subject in his second research stage starting in 1977. Writing joint papers, highly emphasized in the US, helped Rudolf Ahlswede to establish a worldwide network of collaborators. Finally, an additional

---

[1] Ingo Althöfer heard this story from Rudolf Ahlswede in a version with more personal flavour: Rudolf Ahlswede was in the Math department of Ohio State, but his Mathematics (= Information Theory) was not fully accepted by some traditionalists in the department. Rudi decided to ask them: "Who is the strongest mathematician in the department?" Answer: "Kurt Mahler." So, he stepped in Mahler's office and asked him: "Please, give me an interesting open problem of yours." So did Mahler (1903–1988), and Ahlswede solved it within a few weeks. After that demonstration there were no problems for him to get full Professorship.

A. Ahlswede et al. (eds.), *Storing and Transmitting Data*,
Foundations in Signal Processing, Communications and Networking 10,
DOI: 10.1007/978-3-319-05479-7, © Springer International Publishing Switzerland 2014

fortunate development was an offer from the Universität Bielefeld in 1975, which for many years was the only research university in Germany with low teaching obligations, implying the possibility to teach only every second year. In a tour de force within half a year Rudolf Ahlswede shaped a main part of the Applied Mathematics Division with Professorships in Combinatorics, Complexity Theory (first position in Computer Science at the university), and Statistical Mechanics.

Among his students in those years were Ingo Althöfer (Habilitationspreis der Westfälisch-Lippischen Universitätsgesellschaft 1992), Ning Cai (IEEE Best Paper Award 2005), Gunter Dueck (IEEE Best Paper Award 1990; Wirtschaftsbuchpreis der Financial Times Deutschland 2006), Ingo Wegener (Konrad-Zuse-Medaille 2006), Andreas Winter (Philip Leverhulme Prize 2008) and Zhen Zhang. In the second stage 1977–1987 the AD-inequality was discovered, made it into many text books like [B86], [A87], [AS92], [E97], and found many generalizations and number theoretical implications [AB08].

We cite from Bollobas' book [B86], §19 The Four Function Theorem:

At the first glance the FFT looks too general to be true and, if true, it seems too vague to be of much use. In fact, exactly the opposite is true: the Four Function Theorem (FFT) of Ahlswede and Daykin is a theorem from "the book". It is beautifully simple and goes to the heart of the matter. Having proved it, we can sit back and enjoy its power enabling us to deduce a wealth of interesting results.

Combinatorics became central in the whole faculty, when the DFG-Sonderforschungsbereich 343 "Diskrete Strukturen in der Mathematik" was established in 1989 and lasted till 2000. Rudolf Ahlswede started a strong and successful cooperation with the Armenien scientists Harout Aydinian and Levon Khachatrian. The highlight of that third stage is among solutions of several number theoretical and combinatorial problems of P. Erdős [A01]. The most famous is the solution of the $4m$-Conjecture from 1938 of Erdős/Ko/Rado (see [E97], [CG98]), one of the oldest problems in combinatorial extremal theory and an answer to a question of Erdős (1962) in combinatorial number theory "What is the maximal cardinality of a set of numbers smaller than n with no $k + 1$ of its members pairwise relatively prime?".

As a model most innovative seems to be in that stage Creating Order [AYZ90], which together with the Complete Intersection Theorem demonstrates two essential abilities, namely to shape new models relevant in science and/or technology and solving difficult problems in Mathematics.

In 1988 (with Imre Csiszár) and in 1990 (with Gunter Dueck) Rudolf Ahlswede received the Best Paper Award of the IEEE Information Theory Society. He received the Claude Elwood Shannon Award 2006 of the IEEE information Theory Society for outstanding achievements in the area of the information theory (see his Shannon Lecture [A06]).

A certain fertility caused by the tension between these two activities runs like a red thread through Rudolf Ahlswede's work, documented in 235 published papers in roughly four stages from 1967–2010.

The last stage 1997–2010 was outshined by Network Information Flow [ACLY00] (see also [FS07a], [FS07b], [K]) and GTIT-updated [A08], which together with

Creating Order [AYZ90] was linked with the goal to go from Search Problems to a Theory of Search.

The seminal paper [ACLY00] founded a new research direction in the year 2000, with many applications especially for the internet. It has been identified by Essential Science Indicators^{SM} as one of the most cited papers in the research area of "NET-WORK INFORMATION FLOW". Research into network coding is growing fast, and Microsoft, IBM and other companies have research teams who are researching this new field. The most known application is the Avalanche program of Microsoft.

Rudolf Ahlswede had just started a new research project about quantum repeaters to bring his knowledge about physics and information theory together. Unfortunately he cannot work for the project anymore.

We lost a great scientist and a good friend. He will be missed by his colleagues and friends.

Bibliography

[A71] R. Ahlswede, Multi-way communication channels, Proceedings of 2nd International Symposium on Inf. Theory, Thakadsor, Armenian SSR, 1971, Akademiai Kiado, Budapest, 23–52, 1973.

[A82] R. Ahlswede, Jacob Wolfowitz (1910–1981), IEEE Trans. Inf. Theory 28, No. 5, 687–690, 1982.

[A01] R. Ahlswede, Advances on extremal problems in Number Theory and Combinatorics, European Congress of Mathematics, Barcelona 2000, Vol. I, 147–175, Carles Casacuberta, Rosa Maria Miró-Roig, Joan Verdera, Sebastiá Xambó-Descamps (Eds.), Progress in Mathematics 201, Birkhäuser, Basel-Boston-Berlin, 2001.

[A06] R. Ahlswede, Towards a General Theory of Information Transfer, Shannon Lecture at ISIT in Seattle 13th July 2006. IEEE Inform. Theory Society Newsletter, Vol. 57, No. 3, 6–28, 2007.

[A08] R. Ahlswede, General Theory of Information Transfer: updated. General Theory of Information Transfer and Combinatorics, Special Issue of Discrete Applied Mathematics, Vol. 156, No. 9, 1348–1388, 2008.

[AB08] R. Ahlswede, V. Blinovsky, Lectures on Advances in Combinatorics, Universitext, Springer, 2008.

[AB75] R. Ahlswede, R. Bojanic, Approximation of continuous functions in p-adic analysis, J. Approximation Theory, Vol. 15, No. 3, 190–205, 1975.

[ACLY00] R. Ahlswede, Ning Cai, S.Y. Robert Li, and Raymond W. Yeung, Network information flow, IEEE Trans. Inf. Theory 46, No. 4, 1204–1216, 2000.

[AYZ90] R. Ahlswede, J.P. Ye, Z. Zhang, Creating order in sequence spaces with simple machines, Information and Computation, Vol. 89, No. 1, 47–94, 1990.

[AZ89] R. Ahlswede, Z. Zhang, Contributions to a theory of ordering for sequence spaces, Problems of Control and Information Theory, Vol. 18, No. 4, 197–221, 1989.

[AS92] N. Alon, J. Spencer, The Probabilistic Method, Wiley-Interscience Series in Discrete Mathematics and Optimization, Wiley & Sons Inc., New York, 1992

[A87]   I. Anderson, Combinatorics of Finite Sets, Oxford Science Publications, The Clarendon Press, Oxford University Press, New York , 1987.

[B86]   B. Bollobás, Combinatorics, Set Systems, Hypergraphs, Families of Vectors and Combinatorial Probability, Cambridge University Press, Cambridge, 1986.

[CG98] F. Chung and R. Graham, Erdős on Graphs: His Legacy of Unsolved Problems, AK Peters, 1998.

[CT06] T. Cover, J. Thomas, Elements of Information Theory, 2nd ed., Wiley, New York, 2006.

[CK81] I. Csiszár, J. Körner, Information Theory Coding Theorems for Discrete Memoryless Systems, Probability and Mathematical Statistics, Academic Press, New York-London, 1981.

[E97]   K. Engel, Sperner Theory, Encyclopedia of Mathematics and its Applications, 65, Cambridge University Press, Cambridge, 1997.

[FS07a] C. Fragouli, E. Soljanin, Network Coding Fundamentals, Foundations and Trends in Networking, Publisher Inc., 2007.

[FS07b] C. Fragouli, E. Soljanin, Network Coding Applications, Foundations and Trends in Networking, Publisher Inc., 2007.

[K]   R. Kötter, The Network Coding Home Page, http://www.ifp.illinois.edu/~koetter/NWC/.

[M81]   K. Mahler, P-adic Numbers and their Functions, 2nd ed., Cambridge University Press, 1981.

[W78]   J. Wolfowitz, Coding Theorems of Information Theory, 3rd ed., Springer, New York, 1978.

Comments by Holger Boche

In 2008, Rudolf Ahlswede began holding a series of lectures at the Technische Universität Berlin on Information Theory. These lectures were primarily addressed to PhD students at the Technische Universität Berlin and the Heinrich Hertz Institute. The book presented here is put together from these fundamental lectures. In his lectures he also responded to concrete questions he had been discussing with researchers from the Heinrich Hertz Institute. In 2009 he held an additional lecture on selected topics in discrete mathematics. Rudolf Ahlswede's lectures were immensely inspiring and stimulating. Through the many applications that he presented, and through his boundless energy, Rudolf Ahlswede was able to convey an indescribable feeling for how powerful, and at the same time, how beautiful information theory can be.

Parallel to the lectures, we had started a joint research seminar on quantum information theory. The focus of this seminar formed a theory of quantum channels. Different coding concepts for arbitrarily varying quantum channels were developed and their corresponding capacities calculated. One key aspect of the seminar presentations was to develop concepts of symmetrization of quantum channels. Additionally, questions on resource framework for arbitrarily varying quantum channels, list decoding concepts for quantum channels, and characterization of quantum channels with positive capacity for the transmission of quantum information were investigated. Rudolf Ahlswede's lectures on arbitrarily varying classical channels as well as the work of Imre Csiszár and Prakash Narayan on arbitrarily varying classic channels formed the basis of this research seminar. Additionally, new communication tasks were being investigated, for example, identification over quantum channels that was built on Rudolf Ahlswede's own work with Gunter Dueck, as well as Te Sun Han and Sergio Verdu's work on identification over classical channels. This research seminar was the foundation for the project, "Information Theory of the Quantum Repeater," that starting in 2010 was promoted and is now being sponsored by the German Federal Ministry of Research and Technology within the priority program on Quantum Technologies. In this program, a very close relationship is developing among researchers in experimental physics.

A. Ahlswede et al. (eds.), *Storing and Transmitting Data*,
Foundations in Signal Processing, Communications and Networking 10,
DOI: 10.1007/978-3-319-05479-7, © Springer International Publishing Switzerland 2014

In 2011, it was planned that Rudolf Ahlswede continue as a Fellow at the Institute of Advanced Studies at the Technische Universität München—a teaching and research program that normally runs for a span of least three years. This plan was not able to be fulfilled due to his unexpected and untimely death. He was, still full of life, plans, research and ideas, simply taken from us.

Index

A

Absolute empirical, 25

B

Bayes formula, 121, 122
Bernstein's trick, 121, 276
Boltzmann's **H**-function, 70
Bound
    Chernoff, 276
    error, 254
    expurgated, 226
    Gilbert, 254
    large deviation, 276
    Omura, 228
    Plotkin, 224–226
    random coding, 236, 256, 259
    sphere packing, 219, 226, 227
    sphere packing for BSC, 216
    stirling, 225
    straight line lower bound, 222

C

Capacity
    α, 190
    λ, 190
    E, 190
    optimistic, 176, 177, 188, 200
    pessimistic, 176, 177, 179, 183, 200
    zero-error, 189, 190, 203
Capacity function, 180, 200, 204
    pessimistic, 185
Cesaro mean, 181
Channel, 169
    AB, 202, 206
    arbitrarily varying (AVC), 163, 190, 200

average discrete memoryless (ADMC),
    170, 174, 175, 184
binary symmetric (BSC), 215, 224, 248,
    253, 254
broadcast, 177
compound, 204
discrete memoryless (DMC), 116, 118,
    170, 174, 191, 192, 228
discrete memoryless with feedback
    (DMCF), 117, 122, 142
Gaussian, 74
multiple-access (MAC), 189
nonstationary discrete memoryless
    (NDMC), 170, 174, 175, 202, 221
parallel, 202, 255
Code
    Huffman, 18, 19
    instantaneous, 12
    prefix, 11–13
    uniquely decipherable, 11, 15, 17, 18

D

Dalton transfer, 282, 283

E

Empirical distribution (ED), 25
Entropy, 10, 51, 69, 266
    conditional, 51
    function, 39, 63
    Rényi, 266
    relative, 17
Error exponent, 259
Euler number, 282

F
Faddeev, D. K., 42
Fano source, 198
Fano, R. M., 19
Feedback, 117, 164
Feinstein, A., 131
Function
    rate distortion, 191
    rate sequence-error, 191
    rate-error, 189, 190
    reliability, 228

G
Generated sequence, 118, 128

H
Hamming boundary, 146
Hamming distance, 33
Hamming metric, 146, 156
Hypergraph, 117, 128

I
Identification, 183
Identity
    Kolmogorov, 288, 289
Inequality
    arithmetic-geometric mean, 266, 267,
        271
    Bernoulli, 126
    Cauchy-Schwarz, 267, 268
    Chebyshev, 28, 118, 124, 129, 137, 139,
        141, 143, 194, 209, 274, 275
    Dobrushin, 34, 55, 288
    entropy power (EPI), 66, 68, 75
    Fano, 173
    Hölder, 241, 267–270
    Jensen's, 265, 266
    Kraft, 13, 15, 17, 18
    log-sum, 270, 271, 286
    Markov, 275
    Minkowski, 75, 267, 269
    Muirhead, 283
    Pinsker, 272, 273
Isoperimetric problem, 146

K
Kemperman, J. H. B., 136

L
Lemma

blowing up, 146
coloring, 258
covering, 36–38
data processing, 271, 272, 288
duality, 121, 129
Fano, 114, 135, 139, 156, 158, 160, 161,
    164, 192
list reduction, 122, 127
Lovasz local, 282
packing, 135, 139, 141, 142, 258
Shannon, 138, 143, 283, 285, 286
Stein, 218
List code, 117
List reduction, 117

M
Markov chain, 50, 287, 288
Mutual information, 52, 54, 283
    conditional, 52, 287

R
Random coding exponent, 257
Rate distortion, 10, 33
    function, 34

S
Schur-concave, 283
Schur-convex, 282, 283
Shadow, 14
Shannon, C. E., 134
Shannon-Fano procedure, 19
Source
    discrete memoryless (DMS), 9, 22, 33
    Fano, 140
Source coding
    noiseless, 10
    noisy, 10
Stirling's formula, 217

T
Taylor's formula, 150
Theorem
    coding theorem for DMC, 113, 114, 127,
        128, 131
    coding theorem for DMCF, 122–124
    coding theorem for DMS, 22, 23, 26
    Hadamard, 75
    HSW, 202
    Lagrange, 152
    noiseless coding, 13, 16–18
    rate distortion, 34, 113–115

Shannon, 186
Shannon's coding theorem, 115, 116
source coding theorem for DMS, 29, 30,
    113
Theorem, Kuhn-Tucker, 286
Time structure, 205
Time-sharing, 189
Transmission matrix, 118
Tree

binary, 13
Typical sequence, 25, 118, 128

W
Weak capacity, 161
Weak law of large numbers, 118, 274, 275
Wolfowitz, J., 131

Author Index

A

Ahlswede, R., 31, 128, 134, 135, 141, 156, 206, 235
Alon, N., 186
Amari, S. I., 192
Arimoto, S., 134, 143, 228, 235
Augustin, U., 134, 135, 139–141, 175, 195, 198, 202, 220

B

Balatoni, J., 69
Beck, J., 218
Berger, T., 191
Berlekamp, E. R., 222, 224
Blackwell, D., 131
Blahut, R. E., 228
Bohr, H., 181
Boltzmann, L., 18, 33, 69
Breiman, L., 131
Burnashev, M., 192

C

Clausius, R., 18, 69
Cover, T. M., 177
Csiszár, I., 161, 175, 188, 192, 200

D

De Bruijn, N. G., 66
Dobrushin, R. L., 175
Dodunekov, S., 188
Doob, J. L., 188
Dueck, G., 134, 135, 141, 228
Dyachkov, A. G., 192

E

Elias, P., 117, 236, 250, 254, 255
Erdös, P., 44, 192

F

Faddeev, D. K., 39, 42, 44, 50
Fan, K., 75
Fano, R. M., 116, 158, 177, 236
Feinstein, A., 114, 116, 131
Feller, W., 148

G

Gács. P., 156
Gallager, R. G., 175, 222, 224, 236, 286
Goethe, J. W., 51
Gray, R. M., 206

H

Haemers, W. H., 186, 189, 202
Han, T. S., 183, 185, 186, 192, 206, 207
Haroutunian, E. A., 190, 219, 228, 257
Harsany, J. C., 33
Hastings, M. B., 202
Horodecki, M., 189

K

Körner, J., 156, 161, 175, 188, 192, 228
Katona, G. O. H., 192
Kautz, W., 192
Kemperman, J. H. B., 134, 135, 138–140, 142, 175, 193, 218, 219
Khinchin, A. Y., 39, 42
Kieffer, J. C., 206, 207

A. Ahlswede et al. (eds.), *Storing and Transmitting Data,*
Foundations in Signal Processing, Communications and Networking 10,
DOI: 10.1007/978-3-319-05479-7, © Springer International Publishing Switzerland 2014

Kolmogorov, A. N., 53, 165
Kuhn, H. W., 246

L
Lóvasz, L., 202
Lindström, D., 192
Lovász, L., 189

O
Omura, J. K., 228, 231
Ornstein, D. S., 206

P
Parthasarathy, K. R., 206
Pinsker, M. S., 175
Plotkin, M., 224

R
Rényi, A., 44, 69

S
Shannon, C. E., 10, 19, 33, 34, 39, 42, 50, 66,
          69, 113, 115–117, 123, 169, 172,

          173, 186, 188, 189, 191, 202, 222,
          224, 237
Singleton, R., 192
Stam, A., 66

T
Thomasian, A. J., 131
Tichler, K., 192
Topsøe, F., 56, 286
Tucker, A. W., 246

V
Verdú, S., 183, 186, 206, 207

W
Wolfowitz, J., 116, 128, 133–135, 175, 188,
          193, 202, 206, 207, 219, 228
Wyner, A. D., 178, 184, 191, 195–197, 202

Z
Ziv, J., 188

Printed in the United States
By Bookmasters